McGRAW-HILL SERIES IN PROBABILITY AND STATISTICS
DAVID BLACKWELL and HERBERT SOLOMON, Consulting Editors

BHARUCHA-REID   Elements of the Theory of Markov Processes and Their Applications
DE GROOT   Optimal Statistical Decisions
DRAKE   Fundamentals of Applied Probability Theory
EHRENFELD and LITTAUER   Introduction to Statistical Methods
GIBBONS   Nonparametric Statistical Inference
GRAYBILL   Introduction to Linear Statistical Models, Volume I
HODGES, KRECH, and CRUTCHFIELD   STATLAB: An Empirical Introduction to Statistics
JEFFREY   The Logic of Decision
LI   Introduction to Experimental Statistics
MOOD, GRAYBILL, and BOES   Introduction to the Theory of Statistics
MORRISON   Multivariate Statistical Methods
PEARCE   Biological Statistics: An Introduction
PFEIFFER   Concepts of Probability Theory
RAJ   The Design of Sample Surveys
RAJ   Sampling Theory
THOMASIAN   The Structure of Probability Theory with Applications
WADSWORTH and BRYAN   Applications of Probability and Random Variables
WASAN   Parametric Estimation
WOLF   Elements of Probability and Statistics

AN EMPIRICAL
INTRODUCTION TO
STATISTICS STATLAB

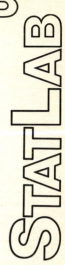

J. L. HODGES, JR.
PROFESSOR OF STATISTICS
UNIVERSITY OF CALIFORNIA, BERKELEY

DAVID KRECH
PROFESSOR EMERITUS OF PSYCHOLOGY
UNIVERSITY OF CALIFORNIA, BERKELEY

RICHARD S. CRUTCHFIELD
PROFESSOR OF PSYCHOLOGY
UNIVERSITY OF CALIFORNIA, BERKELEY

McGRAW-HILL BOOK COMPANY
NEW YORK   ST. LOUIS   SAN FRANCISCO
AUCKLAND   DÜSSELDORF   JOHANNESBURG
KUALA LUMPUR   LONDON   MEXICO   MONTREAL
NEW DELHI   PANAMA   PARIS   SÃO PAULO
SINGAPORE   SYDNEY   TOKYO   TORONTO

to the memory of
JACOB YERUSHALMY
1904 – 1973

Library of Congress Cataloging in Publication Data

Hodges, Joseph Lawson, date
StatLab: an empirical introduction to statistics.

(McGraw-Hill series in probability and statistics)
1. Statistics. I. Krech, David, joint author.
II. Crutchfield, Richard S., joint author. III. Title.
HA29.H657   519.5   74–18116
ISBN 0–07–029134–9

STATLAB
AN EMPIRICAL
INTRODUCTION TO
STATISTICS

1234567890VHVH798765

This book was set in Primer by Black Dot, Inc. The editors were Robert H. Summersgill, Laura Warner, and Carol First; the designer was Janet Durey Bollow; the production supervisor was Sam Ratkewitch. The drawings were done by Katherine C. Eardley.
Von Hoffmann Press, Inc., was printer and binder.

CONTENTS

A preface, by definition and custom, serves to introduce a book, explain its intentions, justify its publication, and acknowledge those who have helped in its making. Because some of this is done elsewhere in this book, and some cannot be done at all, this preface can fulfill its remaining tasks quickly.

STATLAB is a textbook in a field already in good supply of textbooks. Its justification is that it undertakes to teach statistics in an altogether novel manner, and one that—we believe and hope—will encourage the student gladly to learn and the instructor gladly to teach. Because so grandiose a claim needs more justification than its mere assertion, we have devoted the opening unit of this book to that task. There the reader will find a detailed discussion of the STATLAB concept.

Unique or not, STATLAB is a textbook and therefore has borrowed heavily from those who have contributed to the art and science of statistics. By the very nature of the case we cannot thank these many unwitting (and even unknown) collaborators of ours, except through this inadequate broadside.

We can, however, express our thanks here to four of our more immediate collaborators. It is pleasant to acknowledge our debt to Katherine C. Eardley, scientific illustrator at the University of California, who rendered the charts and graphs of STATLAB. She is a friend and colleague of long standing, and we have drawn liberally on her patience and talents. Throughout our work we have profited greatly from the generous help provided by Lucille Milkovich, supervising statistician of the Child Health and Development Studies of the School of Public Health, University of California. Her understanding interest in what we were attempting, and her initial preparation of the study data, made STATLAB a feasible enterprise. Additional statistical analysis was done by Susan M. Hopkin, senior statistician of the Institute of Personality Assessment and Research, University of California. To her we owe not only our appreciation for a job well done, but also our gratitude for her readiness to come to our rescue in a manner and spirit above and beyond any reasonable call of duty. Claire E. Almada, secretary in the Psychology Department of the University of California, was responsible for the typing and retyping of the many versions of the STATLAB manuscript. We are grateful for her high standards of performance, her dedication to the task we all shared, and her high intelligence in divining (we have no better word for it) what we wanted (or *should* want) done with the foolscap we turned over to her. We salute her!

J. L. HODGES, JR.
DAVID KRECH
RICHARD S. CRUTCHFIELD

STATLAB: AN OVERVIEW

The word *statistics* is often used to refer to simple enumeration: the number of births per year, the number of people killed in wars annually, etc. But the major part of the field of statistics consists of:

1 Arithmetic and graphic ways for deriving manageable, relatively simple, and useful descriptive summaries from large and heterogeneous collections of data

2 Methods for inferring from a few observed events (called a *sample*) the characteristics of a large set of as yet unobserved or even of never-to-be-observed similar events (called a *population*)

3 Procedures for testing a wide variety of hypotheses

4 Techniques for discovering whether and to what degree one set of events is contingent upon—or related to—another set of events

Statistical methods provide powerful analytic tools for medical research, economics, sociology, business, physics, psychology, agriculture, genetics, and almost every other human enterprise that can state its observations in numbers. A critical understanding of statistics—its limitations as well as its potentials—is almost as essential for modern man as is the ability to read and write.

STATLAB is designed to help you achieve such an understanding. The guiding principle behind the STATLAB design is that an active learner achieves understanding more quickly and more surely than even the most absorbent but passive receiver of information. Whatever you learn about methods or theory, you will learn in the process of making your own statistical investigation of some of the medical, physical, intellectual, and socioeconomic characteristics of the STATLAB population—a genuine population of children, women, and men. This will involve you in a continuing investigation in which the cumulative results you obtain from your own data as you work through unit after unit will be used in a progressively more sophisticated analysis and reanalysis.

The purpose of this first unit is threefold: (1) to familiarize you with the STATLAB population, with which you will be spending much time; (2) to present a concise preview of your next several months of work with STATLAB, so that knowing the plot you may more easily recognize how each succeeding unit carries you forward to a working understanding of statistics; and (3) to introduce you to some of the "hardware," or devices, which will permit you to collect, store, retrieve, and analyze your accumulating data.

THE STATLAB POPULATION

The first step in statistical analysis is to define the population being analyzed. When it is reported, for example, that a child's school achievement increases as the income of the family increases (a statistical statement), does this refer to children the world over, or only to those of a given country? Does it refer only to children of the middle class, or only to those living in urban areas? Does it refer only to children of certain age groups? Obviously the importance, meaning, and usefulness of any such statement depends upon the population of children covered by the statement. We therefore start our STATLAB investigation with a definition of our STATLAB population.

The STATLAB population consists of 1296 member families of the Kaiser Foundation Health Plan (a prepaid medical care program) living in the San Francisco Bay Area during the years 1961–1972. These families were, during that period, participating members of the Child Health and Development Studies conducted under the supervision of Professor Jacob Yerushalmy of the School of Public Health, University of California, Berkeley. The study sought to investigate how certain biological and socioeconomic attributes of parents affect the development of their offspring. The procedures which are relevant to our purposes can be briefly described. On her first visit to the Oakland hospital of the Health Plan, each pregnant woman was interviewed intensively on a wide range of medical and socioeconomic matters relating both to herself and to her husband. In addition, various physical and physiological measures were made. When the child was born, further information about the mother was recorded, as well as information about the newborn baby. Nine or ten years later, within two weeks of the child's birthday, the child and mother were called in for additional testing, interviewing, and measurement.

Approximately 1400 families completed the study in most of the details and met certain other requirements which we set down for our STATLAB population. For reasons which will become clear in Unit 2, the STATLAB scheme requires exactly 1296 families, divided into two equal subpopulations. Therefore, from the 1400 thoroughly interviewed, measured, and tested California families we drew 648 families to whom a baby girl had been born and an equal number of families to whom a baby boy had been born. Here, in summary form, is a description of the *people* of STATLAB:

1 There are 648 girls and 648 boys who were born in the Kaiser Foundation Hospital, Oakland, California, between 1 April 1961 and 15 April 1963.
2 There are 2592 parents of these infants.
3 These 2592 parents remained together during the 10-year period —no divorced or otherwise separated families were admitted to the STATLAB population.
4 These 3888 individuals (parents and offspring) survived all the vicissitudes of life for at least 9 years and were thus available for the second set of measurements made during the period 5 April 1971 through 4 April 1972.

The STATLAB population, then, consists of 1296 families, each composed of one child and two parents, who completed the Child Health and Development Studies in most details and who met the requirements listed above. Note that the STATLAB families, as defined for our purposes, do not include any other children who may also have existed in these families.

Thus far we have defined our population in terms of the families who make up our study group. Statistical usage, however, does not restrict the word *population* to people. The term can refer to nonhuman or inanimate "individuals" or even to disembodied numbers. For statistical analysis, it is proper to speak of a collection of stones retrieved from the moon as a "population of moon stones," and this is equally true of a set of measures. For example: For each of the 1296 STATLAB families 32 separate measures were taken, and each of these sets of measures makes up a separate population. Thus we can speak of the "STATLAB population of infant blood types" or the "STATLAB population of family incomes." We may if we wish (and we shall wish) refer to subpopulations and speak of the "STATLAB population of girl birthweights" and the "STATLAB population of boy birthweights." On pages 319 to 321 are given the definitions of each of the 32 *variables* (or measures) and therefore of the many measurement populations with which you will be working. These should be studied carefully so that you will be familiar with the many populations of STATLAB you will examine, measure, and speculate about during the next several months.

A PREVIEW

Very few statistical analyses—whether directed toward the solution of scientific or business or governmental problems—are carried out on the entire populations concerned. Almost all statistical analyses depend upon samples; the statistician determines the nature of the whole from a careful examination of a carefully chosen part. You will therefore start your statistical analysis of the STATLAB population by addressing yourself to the sampling problem: What is a good sample ("good" in that it will do the job you want done), and how do you obtain a good sample from the STATLAB population? Once having solved these problems you will find that even a good sample presents you with a somewhat confusing and jumbled collection of numbers. The next several units are therefore devoted to the problem of organizing and ordering this jumble and extracting from it the most useful summary data. Sometimes a single number can do an adequate summarizing job and sometimes several values will be needed.

Once you have learned how to tame your sample and extract from it various useful summary values, you will be ready for a great leap forward: making creditable inferences from these sample summaries to the statistical nature of the entire STATLAB population. This process involves more than merely saying "I would infer that what is true of my sample is true of the entire population." By a "creditable" inference is meant one to which is attached some information about its *accuracy*; an estimate or inference without such information has limited value. Your next task will therefore be to reanalyze the sample data upon which you based your inferences, and you will learn how to determine the degree of certainty or confidence you can place in your estimates. In learning how this can be done you will work through and rediscover some of the greatest achievements of statistical theory.

Knowing how to estimate useful summary values for any STATLAB measurement population you wish, you will be ready for the final tasks. First you will learn to use statistical methods to check on the validity of various hypotheses one can make about the STATLAB population (e.g., are the mental test scores of the STATLAB girls higher than those of the STATLAB boys?). Second you will learn to discover relations which may exist among various STATLAB measures (e.g., what relations, if any, exist between a family's income and a child's score on a mental test?). The ability to make such analyses is, for many people, the ultimate reason for their concern with statistics. This is where the payoff comes—the use of statistics for testing hypotheses about the wide, wide world and for detecting still unknown relations among the multitude of events in nature.

AN INVITATION TO RESEARCH   One final note about your work with STATLAB which we cannot preview for you because we have not "programmed" it: It is work which will be done (if it is done at all) only through self-initiation and under self-direction. As you become increasingly familiar with the analytic power of the various statistical tools at your disposal, you will have an increasing opportunity to test, on your own, various private hunches, guesses, and hypotheses you may have gradually formulated about your STATLAB people and their behavior, their socioeconomic attributes, their physical characteristics, their scores on mental tests, their earning power, etc. The data you have available for such investigations describe a number of significant aspects of a large group of stable families during one of their most active and vital decades of life and during a period marked by social, political, and economic change. You are invited, in other words, to enlarge the scope of your investigation of the STATLAB population beyond the topics listed in the following units.

STATLAB HARDWARE

STATLAB is a textbook made up of six different parts or devices— all designed to help you achieve the objectives we have been discussing. These six devices are (1) STATLAB Census, (2) a pair of dice, (3) Units, (4) Worksheets, (5) Inquiries, and (6) a Databank. As you work with STATLAB, all these devices will become thoroughly familiar to you. For the moment we merely wish to give you a running start by describing each one briefly.

STATLAB CENSUS   The STATLAB population of 1296 families and their 32 variables are all listed in the 36 pages of the STATLAB Census (following page 321). The listing of the families is neither haphazard nor random, but carefully ordered. The first 18 pages consist of 648 families with a baby girl; the second 18 pages consist of 648 families with a baby boy. Within each of these two major divisions the families are listed in order of the mother's age, with the youngest mothers appearing first and the oldest mothers last. In these 36 pages, then, you have all the data you will need for your STATLAB work. To know how to read the data, however, you will have to make use of the material on pages 319 to 321, where the definitions of all 32 variables are given. This is especially important for those variables whose values are presented in a coded manner (e.g., see the variable labeled "laterality").

One final note: The 36 pages of the STATLAB Census are numbered in consecutive *dice numbers* rather than in the usual sequence. Thus the first eight pages, for example, run 11, 12, 13, 14, 15, 16, 21, 22, ... Similarly, the 36 families on each page are designated by consecutive dice numbers. The reasons for this numbering will become amply clear in Unit 2.

AN HONEST PAIR OF COLORED DICE   Each of you has been provided with a pair of what the manufacturer has assured us are honest dice: one red, the other green. These dice will be used for drawing your various samples—as will be explained in detail in Unit 2. For reasons which will become apparent much later in the course, it is highly desirable that you use these same dice throughout your STAT-LAB work. Consider this as a plea to take special precautions not to mislay or damage your set of dice.

UNITS   The work in STATLAB is divided into 25 units. Although the sequence of units is more or less fixed, some units involve more work (or more thought) than others and therefore the number of units undertaken for one assignment may be a matter of negotiation between your instructor and you. Although your instructor may wish to enrich the content of the units with his own material, the text of each unit is written to be self-sufficient. At the end of each unit there is a step-by-step specification of the work to be done. Thus if for any reason you miss one or more classes, you should be able (for most of the units) to carry on with your work on your own.

WORKSHEETS   For each unit, starting with Unit 2, you will find prepared worksheets on which you can do all your calculations and graphic work. These worksheets are carefully designed to maximize the efficiency of your work and to introduce you to a standard style of preparing statistical material.

INQUIRIES   In addition to the worksheets, each unit is provided with an inquiry. This consists of a number of queries addressed to you. Some queries are there to enable you (and your instructor) to check on your understanding of the work you have completed in your worksheet; some are there to encourage disciplined speculation on your part about the matters discussed in the text of the unit; and some are there to initiate class discussion. We trust you will find many of these queries helpful and challenging. Both the worksheets and inquiries are punched so that you may file them in your note-book.

DATABANK We have said that with STATLAB you are embarking on an enterprise that will involve you in a continuing investigation in which the *cumulative* results you obtain from your own data will be analyzed and reanalyzed in progressively more sophisticated ways. To help you file the data you accumulate in each unit in an efficient and easily retrievable manner, Databank files are provided. You will find these files on pages 289 to 313. You should note three things about your Databank:

1 The data filed there are primarily from your own unique sam-ples; your fellow students' Databanks will contain information different from yours. It is therefore essential that when you enter the data in your Databank you make certain the data are cor-rectly entered—if they are incorrectly entered, they may foul up the work of several succeeding units.

2 Cherish your Databank—if you should lose it, you will have lost a great deal. It may even be wise to have a duplicate copy of your Databank in some safe place; alternatively, if you save your worksheets and inquiries, a lost Databank can be reconstructed from them.

3 Most of the data filed in your Databank will come from your own worksheets, but some of the data will come from class distribu-tions (i.e., the data reported by your fellow students) and some from your inquiries. Your instructor will provide you with the class data, as required. You will also find that some of your answers in your inquiry are to be filed in the appropriate Data-bank file. This will be clearly indicated at the appropriate places. A final note: Not *all* the data you collect need be filed in your Databank. Each unit will clearly specify which data will be used in succeeding units and will therefore require filing. We have also provided some extra Databank sheets that can be used for filing any additional information that you and your instructor may wish (Databank, File X, page 314).

☐ You are now ready to begin your survey of the STATLAB popula-tion by familiarizing yourself, in an impressionistic manner, with the data available in the STATLAB Census. In a sense you are now going to make a quick, preliminary scouting expedition.

WORK TO BE DONE

STEP 1 Study pages 319 to 321 until you are certain that you can read—with understanding—all the values in the STATLAB Census.

STEP 2 Complete the inquiry.

UNIT 2 DRAWING A RANDOM SAMPLE

The families of STATLAB differ among themselves in many attributes—mother's or father's education and occupation, family income, the child's mental test scores, etc.—and we shall examine and analyze many of these differences in the course of our statistical investigation of the world of STATLAB. We begin this investigation with a simple physical attribute of the mothers of STATLAB: height. As we shall see later, to obtain a good description of these heights we need to measure only a very small fraction of the 1296 STATLAB mothers. But this small fraction, which can describe so many from so few, must be a very special kind of fraction: It must be a *random sample* of the population.

To know what a random sample is, and how to draw one, is the beginning of all wisdom in statistics. Almost every principle of statistical inquiry is based on the assumption (or at least the pretense) that the data being analyzed were drawn randomly. Throughout our statistical analysis of the STATLAB people we shall repeatedly be faced with the problem of collecting a random sample. The reason why it is a problem is that random sampling is not merely haphazard selection as the word in its common usage might imply.

Random sampling, as used in statistics, is a demanding and precisely defined procedure.

In this unit we shall examine the cardinal principle of random sampling and describe the procedures for drawing a random sample from our STATLAB population. Then you will put these procedures into effect by selecting your own first random sample—and with that you will be launched on your STATLAB investigation.

TO SELECT ONE AT RANDOM

When a statistician says that a person is chosen at random from a population, he means that the procedure used in making that choice guarantees to each person in the population the same chance of being chosen, no one being favored over anyone else. Consider the problem of choosing one person at random from among a group of six people. If we securely blindfold the six, and each in turn draws from among a collection of straws differing in length, and we then select that person who drew the shortest straw, the statistician would object. He would argue that he simply does not have enough evidence to know that this procedure (the particular way of arranging the straws, holding them, and presenting them to the blindfolded person, etc.) actually guarantees to each of the six persons (bright or stupid, sensitive in touch or dulled) the very same chance of getting the shortest straw.

Let us try another approach: You number the persons from 1 through 6; you then pick one die from an honest pair of dice, roll it properly, and select that person whose number comes up on the die. Now the statistician would be quite ready to certify that you have indeed made a random choice. He could do so because the human race has an incredible amount of experience to assure the statistician that an honest well-made die, properly thrown, will yield its six faces with (almost) exactly equal frequencies. That is, any one of the six numbers is just as likely to come up as any other. Therefore this method, limited only by the unavoidable small imperfections of the die, *guarantees* to each of the six persons the same opportunity to be chosen.

We now turn to populations larger than 6. Rolling two dice (let us say a red one and a green one) and reading off the resulting numbers in a specified sequence (first the red number, then the green) will yield $6 \times 6 = 36$ different two-digit dice numbers, the lowest of which will be 11 and the highest 66. (See Table 1.) Any one of these 36 dice numbers is just as likely to come up at one throw of the two dice as any other number. Therefore, with one honest throw of two well-made dice we could select at random one item from a population of 36 items.

But suppose we wanted to work with a population much larger

TABLE 1 STATLAB dice numbers

Reading on green die

		1	2	3	4	5	6
	1	11	12	13	14	15	16
	2	21	22	23	24	25	26
Reading on red die	3	31	32	33	34	35	36
	4	41	42	43	44	45	46
	5	51	52	53	54	55	56
	6	61	62	63	64	65	66

Thirty-six two-digit "dice numbers" can be generated by a throw of two dice. Thus if the red die shows a 4 and the green die a 3, then number 43 has been generated. Any one of these 36 numbers, from 11 through 66, is equally likely to come up with one honest throw of a pair of honest dice.

than 36, say something over a thousand. As we shall see in a moment, by building on the fact that one throw of two dice can select at random one item from a population of 36, we can devise a system whereby two throws of two dice will draw a random item from a population of 1296. Our STATLAB population consists of exactly 1296 families for two reasons. First it permits us to draw, by a pair of throws of two dice, one of these 1296 families in a random fashion; and second this population is large enough to permit us to examine many statistical principles.

Here is the system we have devised which permits one to deal with the population of 1296. First we distributed our total STATLAB population by listing 36 different families on each of 36 different pages (that is how we obtain our total of 1296: 36 × 36 = 1296). These pages are numbered in consecutive dice numbers, with the first page bearing the number 11 and the last 66. Then each family on each page was given its dice number (again from 11 through 66). All that is now required is to select one of these pages at random, and then within that page one family at random, since that will give *every family* on *every page* an equal chance of being drawn. The random selection is made by means of a pair of throws of the dice, with the first throw selecting a page and the second throw selecting a family on that page. Thus if your first throw gives 1 on the red die and 4 on the green die, while your second throw gives 5 on the red and 3 on the green, you will have selected family 53 on page 14 at random from all 1296 families in STATLAB. For subsequent analyses this family would carry the ID number (identifying number) 14–53.

TO SELECT MANY AT RANDOM

So far we have discussed the random choice of a single family out of our population of 1296 families. Phrased in statistical terms, we have discussed the drawing of a random sample of $n = 1$. (In statistical nomenclature, the letter n stands for sample size.) To draw a random sample of $n = 2$, two pairs of throws are needed. If the first pair of throws selects family 14–53 and the second pair selects family 51–26, then these two families will constitute a random sample of $n = 2$. It is of course conceivable that the second pair of throws may give results identical with the first pair; there is only one chance out of 1296 that this would happen, but it could happen. If it does, it will only be necessary to ignore the duplicate result, throw the dice again to get one of the 1295 families not chosen on the first pair of throws, and hence get two distinct families.

Since there are 1296 possible choices for the first family, and for each of these there are 1295 choices for the second family, there are $1296 \times 1295 = 1,678,320$ possible outcomes when drawing a random sample of $n = 2$. When honest dice are properly thrown, the resulting sample of $n = 2$ is equally likely to be any one of these 1,678,320 outcomes. This being so, we say that the sample of $n = 2$ families has been chosen at random.

From what we have said about a random sample of $n = 1$ and a random sample of $n = 2$, we can now generalize a definition which will hold for samples of any size: *A sample of size n is said to be random if it is chosen in such a way that every possible sample of that size n is guaranteed the same chance of being chosen.* Thus if you want a random sample of $n = 20$, you must guarantee that every possible sample of 20 in the total population has an equal chance of being chosen. Our STATLAB population has been so arranged that a random sample of size n from that population can be obtained simply by n pairs of throws of the dice, not counting duplicates. As you will discover if you go on with your study of statistics, a set of honest dice (or its equivalent) is the statistician's best and most useful friend. For a glimpse at the lineage of STATLAB's dice as randomizing agents, see Box 1.

THE VIRTUES OF RANDOMNESS

There are two reasons why statisticians want their samples to be randomly drawn:

1   Samples selected by the exercise of judgment, or taken haphazardly by means of some unanalyzed process, often turn out to be distorted one way or another and thus do not represent the population fairly—whereas we may rely on the *impartiality* of honest dice.

HUCKLE-BONES, GALILEO, AND StatLab DICE NUMBERS

StatLab's randomizing agent—a pair of honest dice, one red and one green—is descended from the astragalus, the heel-bone of the mammal, or the "huckle-bone" as it is frequently called. This bone has four flat and two rounded sides so that when it is thrown it can come to rest on one of four flat sides. Collections of astragali have been uncovered at many archaeological digs and it has been conjectured that they were somehow involved in prehistoric man's play. These conjectures probably arose from the fact that the Egyptians of 3500 B.C. used astragali in their board games. That our "modern" six-sided die evolved from the four-sided astragalus is indicated by the existence of many astragali whose rounded sides had been rubbed almost flat and thus converted into six-sided bones. The earliest real dice so far found date from about 3000 B.C. and are made from hard, well-fired pottery. By 300 B.C., dice and astragali were used in gaming as well as in games, for gambling had now become a significant activity of the civilized world. Thus according to F. N. David, who has written on the history of probability, in the Roman period four astragali were used in a simple gambling game called "rolling of the bones"—a name which carries immediate meaning to the crap-shooters of today.

Although it is certain that the human race has had an incredible amount of experience with dice, the understanding that "an honest well-made die, properly thrown, will yield its six faces with (almost) exactly equal frequencies" is a relatively recent discovery that dates from about 1560. And this poses a problem for the historian. We must remember, David points out, that the Roman dice were carefully made and threw absolutely true. Dice were used in divination rites in the Greek and Roman temples. The die would have been cast so often that the priests (who, on the whole, were the educated class) would have inevitably accumulated considerable experience about, and insight into, the regularity of the fall of dice. It seems curious, therefore, that the realization that a die would yield its six faces with equal frequencies did not occur until well over a thousand years after the Greek and Roman priests had stopped throwing dice. David suggests several explanations:

> It is possible that . . . the priest was taught to manipulate the fall of the dice . . . to achieve a desired result. This being so, there would be no need to calculate probabilities. . . . Another possibility is that the calculation of

Reproduced by permission of the publishers, Charles Griffin and Company, Ltd., of London, and High Wycombe, from F. N. David, "Games, Gods and Gambling," 1962.

2   The fact that every possible sample of a given size is equally likely to be drawn provides the mathematical basis for the statistical methods we shall be studying.

Let us see how random sampling applies to two fairly common investigatory fields: public opinion surveys and clinical experi-

these probabilities was in fact carried out but that it was part of the mystery of the craft and not divulged. This is not so likely as the first, for with the changeover to Christianity someone would have talked. A third and more plausible possibility is that speculation on such a subject might bring with it a charge of impiety in that it could be represented as an attempt to penetrate the mysteries of the deity.

Whatever the reason, it remained for the Italian mathematicians of the sixteenth century to note the regularity of the fall of the dice and to derive a theoretical concept therefrom. The major credit for this discovery probably belongs to Girolamo Cardano (1501–1576), physician, professor of geometry, and almost-pathological gambler (in 1534 he and his wife and child had to take refuge in Milan's poorhouse). In Cardano's *Liber de Ludo Alaea* (Book on Games of Chance) appears this passage in his discussion of ''On the Cast of One Die'': ''One-half the total number of faces represents equality. . . . For example I can as easily throw one, three, or five as two, four, or six. The wagers therefore are laid in accordance with this equality if the die is honest.''

Cardano has drawn, out of his ample experience, a mathematical concept of equally likely sides of the die. With this concept he has calculated that the theoretical probability of the occurrence of *any one* face, where *all six* faces are equally likely, is $\frac{1}{6}$. Mathematician and gambler that he is, he is fully aware of the difference between a theoretical probability and the actual behavior of a die when he warns us that the die will yield its six faces with equality only ''if the die is honest.''

No one knows how this discovery of Cardano's spread so quickly through the mathematical world, since his book was not published until some 85 years after his death. But spread it did, for we find that when Galileo prepared his brief *Thoughts about Dice Games* in 1620—about 40 years before Cardano's book was published— he wrote as though the general method of calculating a probability from the mathematical concept of equality of the sides of a die was well known. In this work Galileo considers the problem of two dice and anticipates STATLAB's invention of ''dice-numbers'':

> Since a die has six faces, and when thrown it can equally well fall on any one of these, only 6 throws can be made with it, each different from all the others. But if together with the first die we throw a second, which also has six faces, we can make 36 throws each different from all the others, since each face of the first die can be combined with each of the second, and in consequence can make 6 different throws, whence it is clear that such combinations are 6 times 6, i.e., 36.

Think on this whenever you roll the bones to draw a STATLAB sample: In your act is the culmination of thousands of years of development—from prehistoric man's idle play with the huckle-bone of the deer to the mathematical thoughts of Galileo Galilei the Magnificent.

ments. A street-corner pollster who selects "at randon" (in the everyday sense of haphazard choice) those people he will quiz may be unintentionally biased toward picking those persons who look cooperative. He may thus not get opinions which are representative of the entire population—the curmudgeon *and* the friendly looking. Persons in jails or hospitals, or stay-at-homes, will certainly be un-

Table 2 125 random digits

09806	70502	11899	94694	71101
96779	86208	20640	18257	95384
53381	78237	89301	47170	88823
42232	68071	39145	85458	87721
91450	26931	03819	60155	17679

Starting with any digit and proceeding in any direction, the 10 digits will occur with equal frequency.

derrepresented in his sample. A medical researcher who tests a new drug for high blood pressure may give it to some patients (who constitute the experimental group) and withhold it from others (the control group) and then compare the subsequent blood pressures in the two groups. If the developer of the drug decides "at random" to which group a given patient is assigned, he may tend unconsciously to assign the sicker patients to the control group so that his results will appear favorable. The only safe procedure is to make the selection (of interviewees or of treated patients) by means of some impartial mechanism rather than by human choice so that the samples are truly random in the statistician's sense of that term.

We have already seen how we can achieve the ideal of randomness by the use of dice. The method we have devised for STATLAB can be used for sampling from any population once the individuals of the population have been assigned dice numbers. But in most situations where sampling is done the literal throwing of dice may be unwieldy and, further, people are used to working with ordinary digits rather than with our dice numbers. For these reasons an alternative method, one which both guarantees randomness and uses ordinary digits, is customarily employed. This method involves the use of *tables of random numbers*. Suppose that we had a fair die with 10 sides labeled 0, 1, 2, . . . , 8, 9. A throw with such a die would yield each of the 10 common digits with equal frequency and could be used to generate a sequence of randomly drawn numbers. This is, in essence, the way in which one of the published random number tables, containing a million random digits, has been made up. Table 2 is an excerpt from such a table. These random digits are then used to direct the drawing of a random sample.

It is of course true that statistical methods are used all the time on data that were not drawn randomly or were not even taken from any real population. Statistical methods are applied by the economist to a study of wheat prices, by the geologist to a distribution of earthquakes, by the astronomer to sunspot cycles. Yet whenever these methods are used, it is tacitly assumed that the data behave like randomly drawn samples—that "nature has thrown the dice." When that assumption is wrong, the theoretical justification for

TABLE 3 Random data from the STATLAB Census

ID no.	Mother's height (to tenth inch)
14–53	70.0
51–26	63.9
~~14–53~~	
46–16	65.5
46–32	61.7

Note that the third throw originally duplicated the first throw. This duplicate throw was crossed out; the dice were thrown again and yielded family 46–16. The recorded height (65.5 in) is the mother's height of family 46–16 (the replacement).

applying the usual statistical methods is dubious indeed. The kind of sampling with which you are becoming familiar in STATLAB underlies, explicitly or implicitly, *all statistical analysis*.

☐   You are now ready to begin a statistical analysis of the heights of the STATLAB mothers by drawing a random sample of $n = 40$ families and recording the heights of the 40 mothers of those families.

WORK TO BE DONE

STEP 1   Make 40 pairs of throws of your dice, and record on your worksheet the family ID numbers thus selected.

STEP 2   Examine your 40 numbers carefully to see whether you have any duplicates. If you do, cross out the duplicate (*do not erase*) and throw the dice again to obtain an unduplicated entry as a replacement family for each duplicate (see Table 3).

STEP 3   Look up each of the 40 families in the STATLAB Census, and record on the worksheet, opposite the ID number, the height of the mother of that family.

STEP 4   Copy in Databank, File B-1, the ID numbers of the families in your sample and the corresponding heights of the mothers. This is your own first random sample, of $n = 40$, with which you will be working for several succeeding units.

STEP 5   Complete the inquiry.

UNIT 3

MEASUREMENT AND ROUNDING

You now have in Databank, File B-1, the heights of a random sample of **StatLab** mothers. To extract usable information from this jumble of 40 numbers, you must simplify them and then organize them into a coherent pattern, using the techniques of statistical analysis. The next several units will make it clear how all this can be done.

But no information which can be extracted by statistical operations is going to be more dependable than the original data themselves. Unfortunately there are many sources of error in the collection of data; you have already seen examples in Unit 1 of obviously erroneous reports of smoking habits. Although some errors can be avoided, others cannot. The statistician, however, can use various principles of probability and statistics to detect some of these errors and sometimes even to estimate their probable size and cause. Therefore the first task of the statistician is to examine his *raw data* (data he has not yet simplified or otherwise processed) to get some feel for their reliability, for he believes that data of little dependability merit little statistical attention. Or, at the very least, knowing something about the dependability of the raw data, the

18

prudent and experienced statistician can be properly cautious in drawing general conclusions from his statistical analysis.

The way to become a "prudent and experienced statistician" is to behave like one from the very beginning of your training. Therefore in this lesson you will have a threefold mission: (1) to begin the process of sensitizing yourself to errors of measurement; (2) to make some assessment of the dependability of the height measures which have been reported for the STATLAB mothers; and (3) to simplify the numbers in your random sample so that your data will be in shape for further processing and efficient analysis.

TWO TYPES OF DATA

Statistical data may be either *discrete* or *continuous*, and in STAT-LAB we have both kinds. Each is susceptible to error, both in measurement and in recording, and they must be examined somewhat differently.

A datum is called *discrete* if its possible values are distinct and isolated, not grading imperceptibly into each other. To take an example from one kind of discrete data (called *categorical*), if you were asked which politician you would prefer as the next governor, your answer would be the name of a particular and unique person, and this is a discrete datum. (We shall take up the analysis of such categorical or qualitative data in later units.) A second kind of discrete data (called *numerical*) arises by counting naturally separate and distinct entities to answer the question "How many?" Its possible values are the distinct integers 0, 1, 2, ... Examples are the numbers of points showing when a die is thrown, the number of books on the top library shelf, the number of people in the United States at this moment. Although in practice many errors can—and do—occur in counting and recording, in principle, at least, a discrete value can be determined and recorded without error. In contrast, for a continuous variable *we cannot avoid error*.

In measuring a continuous variable we try to answer the question "How much?" To that question there usually is no possible *exact* answer. This follows because, by definition, such a variable is not made up of naturally separate or distinct units. This is simply another way of saying that a continuous variable is continuous. Examples are time, length, weight. How much time has elapsed between the moment you started to read this book and this very instant? Shall we measure the time in days? Or in hours? Or minutes? Or seconds? Or hundredths of seconds—or what? There is no natural unit into which you can divide time and then proceed to count the units. No matter how fine a measuring instrument you use, and no matter how fine a scale, there can be a finer one still. Time is, in theory, infinitely divisible.

Since we have to record *some* value when we measure the passage of time, what we do is first to agree (on the basis of considerations we shall discuss in a moment) on some specific unit on some scale. Once having agreed we measure the passage of time in terms of units, count them, and record the passage of time *in terms of a whole number of those units* (for example, 15 days, or 325 milliseconds, etc.) as though time could indeed be divided into the units we have chosen for our convenience. The necessity of measuring and recording a continuous variable as though it were a discrete variable (i.e., counting units) makes error unavoidable, and anyone who deals with continuous variables—physicist, psychologist, or businessman—is doomed to work with fallible measurements. Let us look at some of the problems this predicament creates for the statistician and some of his ways of dealing with them. For the statistician, accepting the fact that he lives in a fallible world, strives to make the best of it.

THE DISCRETE RECORD OF A CONTINUOUS VARIABLE

Let us illustrate the problem with the height of the mother of STATLAB family 35–15, which is recorded, as you can check, as 68.3 in. How was this value obtained? It is customary in measuring height to have the person stand next to a scale marked in the chosen units, say in tenths of an inch. Since the person's height will not ordinarily correspond exactly to a marked point on the scale but rather fall between two such points, and since all heights must be recorded in discrete units of tenths of an inch, the technician is forced to make a *judgment* and decide which discrete scale value to record. It is at this point that the statistician begins to deal with and manage the inevitable errors deriving from measuring continuous variables in discrete terms. The statistician has developed a simple set of rules which are designed to do two things: to guide the technician's judgment so as to arrive at the best approximation (in discrete units) to the true value of the measure; and to avoid consistent and repeated distortions when making these approximations.

RECORDING RULES

The first rule recommended by the statistician is that we record whichever discrete scale value appears *closest* to the perceived height (using that as our illustrative example). Thus, to return to the mother of STATLAB family 35–15, if the procedure recommended above were followed in her case, the record 68.3 would mean that the technician perceived the woman's height to be closer

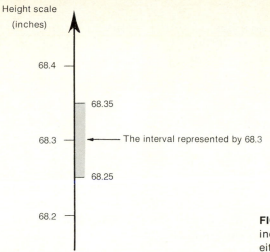

Height scale
(inches)

68.4

68.35

68.3 — The interval represented by 68.3

68.25

68.2

FIGURE 1 Height measured to the nearest tenth of an inch. Any mother perceived to be closer to 68.3 than to either 68.2 or 68.4 should be recorded as 68.3 in tall.

to the mark indicating 68.3 in than to either 68.2 or 68.4. That is, any height perceived to be above 68.25 and below 68.35 would be recorded as 68.3. As can be seen from Fig. 1, the value of 68.3 would be recorded for any height falling within an entire interval of heights of which 68.3 is the *midpoint*. It follows that any height recorded to the nearest tenth of an inch may be in error by as much as 0.05 in, even if no mistakes were made by the technician in his perception of the woman's height or in the recording. The statistician must learn to live with this kind of error. However, when the "nearest point" rule is followed not only are these inevitable errors minimized but they are also as likely to be in the upward as in the downward direction—and this is a very important desideratum because we thus avoid making a *systematic error:* a regular and repeated distortion caused by having errors occur more frequently in one direction than the other. Thus we shall not consistently—or systematically—overestimate or underestimate, and the average of a set of measures is likely to be about right.

The problem of avoiding systematic error is also involved in dealing with the troublesome borderline cases. What height, according to our rule, should be recorded if the height of our STATLAB mother is perceived as falling exactly midway between 68.3 and 68.4? The usual rule is this: If a more careful inspection cannot indicate which is the nearest point, then record the nearest *even* value, e.g., record 68.4. This means that about half the borderline cases will go up (e.g., those falling between 68.3 and 68.4) and half will go down (e.g., those falling between 68.4 and 68.5). Again we see that we have avoided committing a systematic error.

ROUNDING OFF

The two rules discussed above are useful not only in guiding the scientist's judgments when recording his observations of continuous variables but are also of considerable help to the statistician when he seeks to analyze data—and especially when he finds it necessary, or desirable, to simplify a recorded set of data.

When specifying how a continuous variable is to be recorded, the choice of the discrete scale we are going to use is of great importance. The proper choice depends on the use to which the data will be put, since too fine a scale means wasted effort (e.g., measuring the time to cook a three-minute egg in terms of thousandths of a second) and too crude a scale means essential lost information (e.g., using a watch that has no unit finer than a minute to measure the speed of an Olympic runner doing the hundred-yard dash). Consider three height measurements made one day by a physician: In the morning, he records to the nearest inch the height of an applicant for insurance; in the afternoon, in a study on height loss during aging, he measures the height of a patient to the nearest millimeter; in the evening, he notes for the benefit of the undertaker that the body of an auto-accident victim is $6\frac{1}{2}$ ft long. These measurements differ greatly in fineness of scale, but each is appropriate for its purpose. When in doubt, it is better to err on the side of too fine a scale. If the data are recorded too crudely, the information may be lost forever; but it is always possible and easy to simplify data that were recorded too finely.

The mothers' heights in your sample are recorded to tenths of an inch; for the use we shall make of these heights in our statistical analysis, heights in whole inches will be precise enough. Since that is so, it would only mean unnecessary and laborious work to continue to carry and operate with these values in tenths of an inch. To obviate this useless arithmetical work, you need only *round off* your heights to the nearest whole inch. Thus all nine of the values 67.6, 67.7, . . . , 68.3, 68.4 are rounded to 68 because they are nearer to 68 than to any other whole-inch value. A borderline case like 68.5 presents a problem, since we do not know whether the actual height was nearer 68 or 69 nor can we now attempt to get a better reading. We therefore follow the rule already stated (and for the reasons already given): Round such cases to the nearest *even* value. This means that about half the borderline cases will go up and about half will go down, thus avoiding systematic error. To illustrate these rules, the first six mothers' heights (STATLAB Census page 45) when rounded to the nearest inch become 63, 65, 64, 68, 61, 66; and the last six mothers' heights (STATLAB Census page 52) become 68, 70, 64, 65, 65, 57. (Check these for yourself and correct the erroneous value we have inserted!)

DETECTING ERRORS

Whenever many measurements are made and recorded, it is hard to ensure that the people who will do the measuring will follow the instructions in all cases. Vicissitudes are to be expected whenever we deal with live people and real data. No statistician (except perhaps inside the classroom and with fictitious "illustrative data") works with errorless records. You will therefore find, as you work with STATLAB, that the data recorded in the STATLAB Census will not always be what they should be in a perfect (classroom) world. For example, in the present instance it is quite possible that a technician, told to record the mothers' heights to sixteenths of an inch or tenths of an inch, may have slipped up occasionally and recorded to the nearest whole inch. Or, on a busy and hectic day, he may have forgotten (or neglected) to make the measurement at all and simply inserted an estimated value. When people estimate a height, they usually state their guessed value in whole inches. Experienced statisticians check the quality of a data record by close inspection of the entries themselves. Thus, for example, should either of the preceding errors have occurred in the measurement and recording of the mothers' heights of STATLAB, the statistician would find more values which end in .0 than should be expected by chance.

☐   You are now ready to try out the various rules and ideas we have discussed on the data of your own sample and thereby assess the dependability of your raw data. Then you can simplify your jumble of numbers in preparation for further processing in Unit 4.

WORK TO BE DONE

STEP 1   Retrieve your 40 mothers' heights from Databank, File B-1, and enter them, together with the ID numbers, on the worksheet.

STEP 2   Round the mothers' heights to the nearest inch, according to the methods of this unit.

STEP 3   For the 40 families of your sample, enter on the worksheet the father's height as given in the STATLAB Census.

STEP 4   Round the 40 fathers' heights to the nearest inch.

STEP 5   Enter the mothers' rounded heights and the fathers' heights, both unrounded and rounded, in Databank, File B-1.

STEP 6   Complete the inquiry.

FREQUENCY DISTRIBUTIONS

Your 40 heights of STATLAB mothers were considerably simplified in Unit 3 by rounding them off to the nearest inch, thereby reducing them from three figures to two. We shall now show you how to simplify your data still further by organizing them into a more manageable form. The basic idea is to arrange and tabulate the sample values according to size. The resulting table may then be further simplified by grouping adjacent values. Once you have done that, it will be easier to make several summary statements describing your entire sample.

FREQUENCY TABLES

To make the process clear, we shall use an example similar to the one with which you are working. From the 648 STATLAB girls (Census pages 11 through 36) we drew a sample of 40 and read off the heights of the girls (rounded to the nearest inch). The results are shown in Table 4.

In Table 4 we have listed the values in the random order in which they were drawn. The information contained in the set of

TABLE 4 Forty girls' heights in the order drawn

50	52	56	53	59	55	53	56	55	56	56	52	56	54
59	52	51	53	57	49	54	54	50	48	51	53	51	51
55	57	49	55	52	51	51	53	50	50	61	58		

The heights as recorded in the Census have been rounded to the nearest inch.

data will be more easily seen if the values are rearranged in a mean-ingful order. Let us, in the immortal words of the Yankee baseball manager Casey Stengel, "line them up alphabetically according to size." See Table 5.

You can see how much more visible various features of the sample have become. For example, a glance at Table 5 shows that the observed heights extend from 48 to 61 in, so that our sample has a *range* of $61 - 48 = 13$. The advantages of ordering by size become even clearer when the data are arranged in a *simple frequency distribution* or *simple frequency table*, as is done in the first three columns of Table 6. To construct such a table, it is not necessary first to arrange the values in order as we have done above. Merely inspect the sample to spot that the *minimum* value occurring in the sample is 48 and that the *maximum* value is 61. Then list all the consecutive numbers from 48 to 61 in the x column. (We shall use the symbol x, as is often done in statistics, to refer to the value of a measure or observation.) Finally, tally the observations to find the frequency f with which each value x occurs in the sample. For ex-ample, in Table 6 the entry $f = 5$ opposite $x = 53$ means that there are exactly five girls with height 53 in.

READING A FREQUENCY TABLE

Let us see how easy it is to read off important aspects of a sample once it has been arranged in a frequency table. Table 6 shows that although the observed heights extend from 48 to 61 in, the 40 values are not spread evenly over this range. The seven x values from 50 to 56 inclusive, covering the central portion of the distribution, have between them $4 + 6 + 4 + 5 + 3 + 4 + 5 = 31$ of the sample heights while the other seven x values, constituting the upper and

TABLE 5 Forty girls' heights in order of size

48	49	49	50	50	50	50	51	51	51	51	51	51	52
52	52	52	53	53	53	53	53	54	54	54	55	55	55
55	56	56	56	56	56	57	57	58	59	59	61		

These are the same data that appear in Table 4 but arranged in a systematic (rather than a random) order.

TABLE 6 Frequency distributions for heights of a sample of 40 girls

	Ungrouped				Grouped into class intervals					
	$w=1$				$w=2$			$w=3$		
x	Tally	f	F	Class	f	F		Class	f	F
48	|	1	1	48–49	3	3				
49	||	2	3					48–50	7	7
50	||||	4	7	50–51	10	13				
51	NЖ|	6	13							
52	||||	4	17	52–53	9	22		51–53	15	22
53	NЖ	5	22							
54	|||	3	25	54–55	7	29				
55	||||	4	29					54–56	12	34
56	NЖ	5	34	56–57	7	36				
57	||	2	36							
58	|	1	37	58–59	3	39		57–59	5	39
59	||	2	39							
60		0	39	60–61	1	40				
61	|	1	40					60–62	1	40

The data of Tables 4 and 5 are here arranged in frequency tables of three different class widths.

lower "tails" of the distribution, have only nine of the sample heights. It is the general experience, not only with heights in a human population but also with a great variety of observed quantities, that sample values tend to pile up this way in the central portion of their range.

The frequency table also shows at once that the most popular x value, occurring $f = 6$ times, is $x = 51$. The x value having the highest frequency is called the *mode*. Because of the piling-up tendency, the mode is often found near the center of a distribution. A better idea of the center, however, is obtained by discovering which x value has as many heights above it as below it. This kind of central value is called the *median*.

CUMULATIVE FREQUENCY

For some purposes, such as finding the median, it is convenient to have the frequencies *accumulated* from one end of the distribution. We show in the column headed F (column 4 of Table 6) the frequencies f added up, beginning with the smallest x value. Thus the entry $F = 3$ at $x = 49$ is the sum $1 + 2$, and $F = 7$ at $x = 50$ is the sum $1 + 2 + 4$, and so forth. In contrast with a *simple* frequency distribution, the distribution labeled F is called a *cumulative frequency distribution*.

From the F column it is easy to see that the median is among the values tallied at $x = 53$, since there are 17 heights at x values below 53, and $40 - 22 = 18$ heights at x values above 53. Other characteristics of our distribution are at once visible from the cumulative frequency distribution; for example, one sees that 36 (or 90 percent) of the girls in the sample had a height of 57 in or less, that 33 (or 82 percent) had a height of 51 in or more, etc.

GROUPING THE DATA

In some cases, for reasons to be discussed in a moment, it is desirable to simplify a frequency table still further by grouping consecutive x values into steps of equal width called *class intervals*. We illustrate this procedure in Table 6, first for class intervals of width 2. We have grouped together the heights at $x = 48$ and $x = 49$ to obtain the class labeled 48–49 and so on. (We could of course have phased the grouping differently, using 47–48, 49–50, and so forth. The resulting distribution would be somewhat different in details while giving about the same picture.) The data may also be grouped into intervals of width 3, as shown in the last three columns of Table 6. Thus the class interval 54–56 comprises all the heights 54, 55, and 56.

Grouping simplifies a frequency table in much the same way that rounding off simplifies a set of data. The number of rows in the table is reduced; for example, in Table 6, as we go from an ungrouped distribution to groupings into class intervals of width 2 or 3, the number of rows decreases from 14 to 7 to 5. In consequence, the average number of values tallied per row is increased, and this tends to smooth out some of the minor irregularities. As we shall see in later units, with fewer rows the arithmetical work is considerably reduced. However, it is important not to carry the process of grouping too far. Grouping involves loss of detail and hence of information. For example, knowing that the class interval 48–50 has $f = 7$ does not tell us, as the ungrouped table does, that four of these seven heights are $x = 50$. Experience suggests that it is seldom desirable to use more than about 15 class intervals, and with small samples as few as 7 or 8 intervals may give the best picture.

TRUE CLASS LIMITS

Each integer value x in Table 6 really represents an entire interval of heights. As you will recall from Unit 2, if a technician were reading height to the nearest inch he would record 52 for any height between 51.5 and 52.5. Thus the value $x = 52$ represents the interval from 51.5 to 52.5; the value $x = 53$ represents the interval from 52.5 to 53.5, and so forth. When 52 and 53 are combined into the class 52–53 of width 2, this class truly represents the interval from 51.5 to 53.5. We say that 51.5 and 53.5 are the *true class limits* (lower and upper) of the class 52–53. Similarly the class 51–53 has 50.5 as its true lower limit and 53.5 as its true upper limit. It is convenient to label this class simply 51–53 rather than 50.5–53.5, but the true limits must be considered for graphing (Unit 5) and for calculating the median (Unit 6).

To keep descriptions simple, we may refer to the unit interval represented by 52 as a "class interval of width 1" and call 51.5 and 52.5 its true class limits. This permits us to discuss grouped and ungrouped data together in a unified way.

POPULATION FREQUENCY DISTRIBUTIONS AND THE LAW OF LARGE NUMBERS

The idea of a frequency distribution can be applied not only to random samples but also to any collection of data and even to entire populations. If all the values in the population are known, one can make for them a simple or cumulative frequency table, and this may be done with the data grouped or ungrouped. For example, Table 7 shows the simple and cumulative distributions of the heights of the population of 648 STATLAB boys, grouped in class intervals of width 2. We have given in Table 7 not only the actual counts f and F but have also converted these counts to percentages, indicated by $f\%$ and $F\%$. One merely divides each f or F entry by 648; for example, for class interval 54–55, $f\% = \frac{175}{648} = 27.0$ percent and $F\% = \frac{501}{648} = 77.3$ percent. The comparison of two frequency tables based on different population sizes is made much easier if the percentage scale is used for both. Thus if we were to compare our 648 STATLAB boys with a similar population of 1200 boys, 315 of whom have heights in the interval 54–55, it would not be immediately apparent that 315 and 175 are similar; but when expressed as percentages this becomes obvious (for example, $\frac{315}{1200} = 26.2$ percent and $\frac{175}{648} = 27.0$ percent). Sample frequency tables may also, of course, be expressed in the percentage format, and this is commonly done when the samples are large, for the reason given.

Of course, it is usually not possible to make a frequency table for the population from which you have drawn a random sample.

TABLE 7 Heights of the population of 648 STATLAB boys

Class	f	F	$f\%$	$F\%$
46–47	1	1	0.2	0.2
48–49	26	27	4.0	4.2
50–51	112	139	17.3	21.5
52–53	187	326	28.9	50.3
54–55	175	501	27.0	77.3
56–57	101	602	15.6	92.9
58–59	36	638	5.6	98.5
60–61	9	647	1.4	99.8
62–63	1	648	0.2	100.0

The heights recorded in the Census pages 41 to 66 have been rounded to the nearest inch and then grouped into class intervals of width $w = 2$.

You would hardly take the trouble to sample from a population if you already knew all the population values. The purpose of drawing the sample is precisely to find out something about the population —as the statistician says, to "make an inference about" the population. The basis of such inferences, and hence the basis for most statistical work, is the following fact: A sample frequency distribution will tend to resemble the frequency distribution of the population from which the sample was randomly drawn, at least if the sample is large enough.

To illustrate this fundamental idea, consider the 54–55 class interval of Table 7. This class has 27 percent of all the STATLAB boys. Therefore, whenever a STATLAB boy is chosen *at random* there is a 27 percent chance that his height will be in the 54–55 class interval. Common sense and experience combine to suggest that if we take a reasonably large random sample of the 648 boys, approximately 27 percent of them will have such heights—with similar agreement for the other class intervals. In everyday parlance, this is expressed as the "law of averages."

Statisticians have refined and verified the tendency of large samples to agree with populations, and they call it the *law of large numbers*. Stripped of technicalities, this law states: *As the sample size n is increased, the sample frequency distribution will tend to resemble the population frequency distribution*. More precisely, for each given class interval the $f\%$ of the sample will tend to be close to the $f\%$ of the population. You will have various opportunities to see this law at work as we go along, and also to learn about the *speed* with which the resemblance between the sample frequency distribution and the population frequency distribution is improved as the n of the sample is increased. As we shall see in later units, your sample of $n = 40$ mothers' heights is already large enough to enable you to make credible inferences about the STAT-LAB population of mothers' heights.

Note: In stating the law of large numbers we have said nothing about the size of the population. We have said nothing because of the paradoxical fact that *population size plays almost no role in the law.* A random sample of (say) a thousand people can give equally valid information about the employment status, buying intentions, television program preferences, or political allegiances of a population of ten thousand or of hundreds of millions. Many people find it difficult to understand how a sample constituting so small a fraction can say anything reliable about so large a population. But the law of large numbers says it can—and it does. You will have many opportunities, in your STATLAB work, to check out the validity of this law with your own data.

An example may help you to understand why the size of the population has almost no effect on the behavior of the sample. Candidate Smith hires a pollster to find out his prospects of election. As it happens, 60 percent of the voters favor Smith. This implies that each time the pollster interviews a randomly selected voter, there is a 60 percent chance he will find a supporter of Smith. If he interviews, say, a random sample of 1000 voters, it is very likely that a majority of them will support Smith.

In presenting this example we have not said whether Smith is running for mayor (with 60,000 supporters among 100,000 voters) or for governor (with 6 million supporters among 10 million voters). We have not said which because it does not matter. In either case, there is a 60 percent chance that each randomly selected voter would favor Smith, and it is this chance which governs the behavior of the sample as a whole.

The fact that the characteristics and usefulness of a sample depend hardly at all on population size is essential to the instructional method of STATLAB. You are becoming familiar with the behavior of samples, all drawn from a population of fixed size. If what you are learning were good only for populations of size 1296, this course would be of little value to you. Since in fact population sizes scarcely matter, what you are learning has application to sampling from any large population.

☐ You are now ready to organize your own sample of rounded heights into frequency tables in order to get a better comprehension of the contents of your sample, to permit you to make several initial descriptive statements about it, and to prepare it for further statistical analysis.

WORK TO BE DONE

STEP 1 Scan the 40 mothers' heights (rounded to the nearest inch) in Databank, File B-1, in order to identify your maximum and minimum values.

STEP 2 List in the x column of the worksheet every consecutive integer from this minimum to this maximum, inclusive.

STEP 3 Construct the simple frequency table for your data by tallying your 40 mothers' heights and recording the results in the f column.

STEP 4 Fill in the column headed F by accumulating the f entries, thus making the cumulative frequency table for your data.

STEP 5 Construct grouped frequency tables for your data, using class intervals of width 2 and width 4.

STEP 6 Fill in the columns $f\%$ and $F\%$ for the table with class intervals of width 4.

STEP 7 Enter the three frequency tables you have constructed into Databank, File B-2.

STEP 8 Complete the inquiry.

GRAPHIC REPRESENTATION OF DATA

Your sample is now ready to sit for its portrait. To paraphrase the Chinese, one graph is worth a thousand numbers; for often a large collection of data, even if simplified and grouped into frequency tables, can gain a great deal in comprehensibility by being displayed in pictorial form. Visual scanning of the display gives a good initial sense of the range, shape, and other salient characteristics of the set of data.

If the graphic step is wholly bypassed and the data are fed directly into a calculator for mechanical processing by complex statistical formulas or are given to a statistical clerk to process, we run two dangers:

1   There is the risk that we may never become aware of special, unexpected attributes of the data which are essential for the valid interpretation of outcomes. These attributes, which would be obvious by visual inspection of a good graph, would not be discovered by the machine and might be missed by the clerk because he would not have been told to look for them.

2   Gross errors in arithmetic may go undetected. For example, the

value of a calculated average can be thrown off greatly by misplacing a decimal point and copying down (say) 192 instead of 19.2 for one of the x values. This error may never be noticed in arithmetic calculation, but a visual determination of the center of a graphed distribution may quickly reveal it.

In this unit you will learn some of the guiding principles for making a good graphic display—one which is a valid portrayal of your data, perceptually compelling, and above all informative. After having graphed your data, you will use the graph to check the information you have already extracted from your data in Unit 4 and then go on to gather new information.

THE HISTOGRAM

One of the most useful ways of presenting a set of data graphically is to construct a *histogram*—a series of contiguous rectangles which depict a frequency distribution. The data of Table 6 (in Unit 4) are shown in histogram form in Fig. 2*a*, *b*, and *c* for class intervals of width 1, 2, and 3, respectively. Each class interval is represented by a rectangle (ignore the dotted lines for the moment). The height of each rectangle corresponds to the frequency for that class interval, and the base extends in each class from the lower to the upper true class limit. (We urge you to check the three histograms in this figure against the data in the corresponding three frequency tables. Make certain that you understand how these histograms were constructed before continuing.)

Note, as you run your eyes over the three histograms, how clearly they display the finding that grouping data tends to minimize irregularities and smooths out a distribution. Note also how the mode stands out as the highest point of the distribution and the clear suggestion that the median (the center of the distribution) is greater than the mode. These histograms all agree that although the values tend to pile up in the middle (as we noted in the previous unit), they do not seem to be symmetrically distributed around that middle; the values "tail out"—extend out—farther to the right of the middle area than they do to the left. When a distribution shows such lack of symmetry, we speak of a *skewed* distribution. Finally we can literally see that should anyone report the average of these values to be, say, 56 or more, a considerable arithmetic error will have been made. With continued experience (and increased knowledge about statistics) one can learn a great deal about a collection of data just from looking at its histogram.

THE FREQUENCY POLYGON

The *simple frequency polygon* is quite similar to the histogram. For ungrouped data it is constructed by plotting a series of points

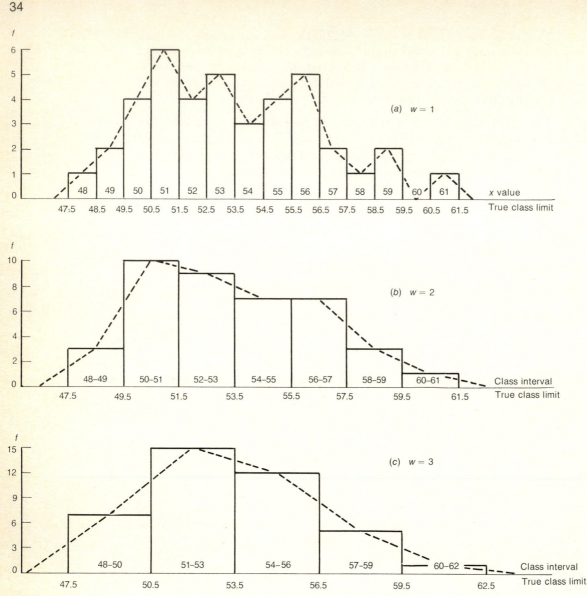

FIGURE 2 Histograms (solid rectangles) and simple frequency polygons (dashed lines) for our sample of 40 girls' heights. The simple frequencies f of Table 6 (Unit 4) are plotted for class intervals of width $w = 1$, 2, and 3.

which correspond to the successive x values on the horizontal axis (sometimes known as the *abscissa* or x axis) and to the frequency for that value on the vertical axis (the *ordinate* or y axis). These points are then connected with a series of straight lines. The first and last points are then connected with line segments to the horizontal axis at the x values preceding the first point and succeeding the last one. This practice represents the fact that the frequency was equal to zero in these two class intervals. (See Fig. 2*a*, where the dashed line represents the frequency polygon.)

For grouped data we must first determine the *midpoint m* for each class interval. The midpoint of a class interval is very simply defined as that value which lies directly midway between the true class limits of the interval. For example, for our class interval 52–53 for which the true class limits are 51.5 and 53.5, the midpoint is $m = 52.5$. Similarly the class interval 51–53, with class limits of 50.5 and 53.5, has midpoint $m = 52.0$. After having determined the midpoint of each interval, we plot a series of points each of which corresponds to the m and f values of that class interval. (See Fig. 2b and c.)

The frequency polygon is most appropriate for the graphic display of a continuous variable; the line which runs unbroken from point to point is supposed to suggest the continuous nature of the variable.

THE CUMULATIVE FREQUENCY POLYGON

When data have been assembled in the form of a cumulative frequency table, the distribution may be depicted in a *cumulative frequency polygon,* as we have done for our sample of girls' heights in Fig. 3. To create a cumulative frequency polygon, points are plotted corresponding to the successive true class limits along the x axis and the *cumulative* frequencies F along the y axis. Thus in Fig. 3b the point at ($x = 53.5$, $y = 22$) represents the fact that $F = 22$ for the class interval 52–53, the upper true limit of which is 53.5. Sometimes it is more convenient to work with $F\%$, and we have shown in Fig. 3 both labelings although normally only one is used. This kind of graph is especially useful in determining quickly the value above or below which a given proportion of a distribution falls. For example: Suppose we wanted to know where the upper and lower 25 percent of the girls' heights fall. To determine the upper 25 percent, all we need to do is construct a horizontal line from 75 on the $F\%$ axis. Where this line intersects the cumulative frequency polygon, drop a perpendicular. Where that perpendicular intersects the x axis, there (approximately) is the value which is the lower limit of the 10 tallest girls. Thus for the ungrouped data it appears that approximately 25 percent of the girls' heights are about 55.7 in *or more*. For the lower 25 percent we construct a similar set of lines starting at the frequency of $F\% = 25$. This gives us (again for the ungrouped data) the estimate that the shortest 25 percent of the girls are about 51.0 in *or less*.

There are several other points to notice from these cumulative frequency polygons: Again we see that the data are not perfectly symmetric, since the upper 25 percent stretches over a wider range (on the x axis) than the lower 25 percent. Widening the class interval, while it smooths out the curve, also induces a change in some of the statistical values you derive from these data, e.g., the point be-

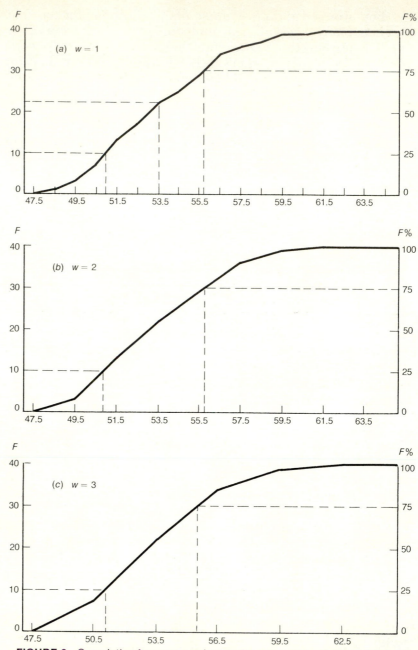

FIGURE 3 Cumulative frequency polygons for our sample of 40 girls' heights. The cumulative frequencies F of Table 6 (Unit 4) are plotted at the upper true limit of height for class intervals of width $w = 1$, 2, and 3. The dashed lines show how certain heights are related to values of F or $F\%$, as explained in the text.

low which 25 percent of the cases fall. Or, to take another specific example, suppose you wanted to know what proportion of the girls in our sample is 53 in or shorter (rounded to the nearest inch). We begin on the x axis at 53.5 in, draw a perpendicular line until it intersects the cumulative frequency polygon, and at that point draw a line to the y axis. And there you can read off the frequency, which is $F = 22$ or $F\% = 55.5$. The value of a good graph has been long appreciated, as is evidenced in Box 2.

FREQUENCY POLYGONS FOR POPULATIONS; FREQUENCY CURVES

As was the case for frequency tables, frequency polygons are useful in portraying entire populations as well as samples. We show the simple frequency polygons of 16 actual populations in Fig. 4. In choosing these examples from among the many thousands that have been published, we have attempted to suggest the almost unlimited variety of applications of statistical methods to every field of human activity. You should at this point study the 16 captions of Fig. 4 to familiarize yourself with the nature of the populations and variables illustrated there. (Figure 4 is on pages 40 to 43.)

A second objective guiding our selection of these examples was to illustrate certain distributional shapes that recur again and again. Turn to Fig. 4 and consider carefully the shapes of Fig. 4a to d. These four figures are drawn from as many areas of study: psychology, demography, animal husbandry, meteorology. Yet the four polygons are quite similar in important respects. Each has a single mode at the center of the distribution. From this mode the polygon falls away in a similar fashion on each side, the two sides being nearly symmetric. The fall is at first rapid, and then slow, as the polygons tail off toward the x axis. This configuration has been called "bell-shaped" from a fancied resemblance to the outline of a bell.

Now turn to Fig. 5a. This curve depicts a mathematical form, corresponding to a specific equation, and is known as a *normal curve*. It can be thought of as a sort of idealization of the common properties we have noticed in the empirical populations depicted in Fig. 4a to d. There is a single mode, from which the curve falls away symmetrically on the two sides, at first rapidly and then slowly as it tails out to approach the x axis. Nature is full of populations which resemble the normal curve to a greater or lesser degree. For example, as you will see, there are several distributions having this shape among the 32 STATLAB variables. The normal curve is, in a sense, at the heart of statistics, and we shall devote an entire unit to its properties.

Figure 4e to h show four empirical populations which resemble the normal shape in some ways but not in others. There is, as

BOX 2

"THERE IS A MAGIC IN GRAPHS"

The history of graphs is a long one. Some see its beginnings among the ancient astronomers and surveyors who employed coordinates to locate points on the earth's surface (the term *ordinate,* for example, was used by early Roman surveyors). There are two somewhat different lines of thought which contributed to the development of the graph as you will come to know it in STATLAB. The first is the mathematical one. Many historians credit the Bishop of Liseux, Nicolas Oresme (1323–1382)—church reformer, scientist, and economist—as the first to use coordinates to express mathematical functions. Admirers of Leonardo da Vinci credit him as the first to have seen the possibility of using coordinates for analyzing quantitative data (in his experiments on gravity, circa 1560). But everyone agrees that it was Descartes who, in 1637, first clearly described the principles on which the modern graph is based. His invention, as wonderfully simple as it was useful, enabled him (and us) to visualize and analyze mathematical formulas by depicting quantitative values as points on a plane marked off by lines drawn perpendicularly to one another.

The second strain in the development of the graph—the descriptive and pictorial one—dates from the eighteenth century. The first person to extend the graph to general use for the portrayal of empirical (and especially economic) data was William Playfair, an English manufacturer, draftsman-engineer (at one time he worked with James Watt), and author (of about a hundred books). At the age of 27, in 1786, Playfair published his *Commercial and Political Atlas,* which he described as representing, "by Means of Stained Copper-Plate Charts, the Progress of the Commerce, Revenue,

The quotations from William Playfair and that from the American Society of Mechanical Engineers are taken from Herbert Arkin and Raymond R. Colton, "Graphs," Harper & Row, New York, 1936. Hubbard's encomium can be found in Willard C. Brinton, "Graphic Presentation," Brinton Associates, New York, 1939.

before, a unique mode from which the two sides fall away, but these shapes lack the symmetry of the normal examples. In each case, one tail falls away more slowly (or is "heavier") than the other. Such asymmetric distributions are said to be skewed in the direction of the heavier tail. Thus Fig. 4*e* and *f* are skewed to the right while 4*g* and *h* are skewed to the left. Skewness may appear to varying degrees. In Fig. 4*e* it is very slight and this curve is nearly normal, whereas in the other three curves the departure from symmetry is quite pronounced. (There is no left tail for Fig. 4*f* because the first

Expenditures, and Debts of England." This book marks the beginning of graphic statistics. The use of graphs in statistics today reflects both the heritage of Playfair and that of Descartes: Graphs are used for presenting empirical data pictorially and for interpreting and analyzing these data mathematically.

From its beginning, graphic statistics has evoked praise from a host of admirers, first among whom was William Playfair himself. In his 1786 volume he promises the reader that with this "new and useful method . . . as much information may be obtained in five minutes as would require whole days to imprint to the memory, in a lasting manner, by a table of figures." About one hundred and thirty years later, the American Society of Mechanical Engineers saw in these virtues a boon to all mankind. In suggesting the formation of a committee on standards for graphic presentation, the society wrote, in 1915: "If simple and convenient standards can be found and made generally known, there will be possible a more universal use of graphic methods, with a consequent gain to mankind because of the greater speed and accuracy with which complex information may be imparted and interpreted." Finally, there is the noble panegyric delivered by Henry D. Hubbard of the National Bureau of Standards (we reproduce here a few choice excerpts):

There is a magic in graphs. The profile of a curve reveals in a flash a whole situation. . . . The curve informs the mind, awakens the imagination, convinces.

Graphs carry the message home. A universal language, graphs convey information directly to the mind. Without complexity there is imaged to the eye a magnitude to be remembered. Words have wings, but graphs interpret. Graphs are pure quantity, stripped of verbal sham, reduced to dimension, vivid, unescapable.

The graphic art depicts magnitudes to the eye. It does more. It compels the seeing of relations. . . . Graphs serve as storm signals for the manager, statesman, engineer; as potent narratives for the actuary, statist, naturalist; and as forceful engines of research for science, technology, and industry. They display results. They disclose new facts and laws. They reveal discoveries as the bud unfolds the flower.

Graphs are dynamic, dramatic. . . . Wherever there are data to record, inferences to draw, or facts to tell, graphs furnish the unrivalled means.

frequency corresponds to the leftmost possible class interval, 1–5. There cannot be a sentence shorter than one word, and it would be misleading to have the frequency polygon come down to zero to the left of $x = 0$. Where the range of a variable is naturally truncated in this way, we use a vertical dotted line to call attention to that fact. See also Fig. 4i, k, and l.)

Figure 5b shows another mathematically defined curve, known as *chi square*, which can be thought to represent an idealized form of a skewed distribution. There is actually a whole family of chi-square curves of varying degrees of skewness. Skewed distributions are also very common, and several STATLAB variables will turn out to have distributional shapes of this kind.

Figure 4*i* to *p* portray eight other populations of various shapes. Figure 4*i* and *j* are rather flat on top, having a roughly constant fre-

FIGURE 4 Simple frequency polygons of 16 actual populations. The nature of each population and its variable *x* is briefly described under each part of the figure, 4*a* through 4*p*, on this and the following pages.

FIGURE 5 Four mathematically defined frequency curves, shown on this and the following pages, which idealize various features seen in the 16 empirical curves of Fig. 4.

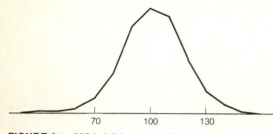

FIGURE 4a 2904 children aged 2 to 18 (the standardization group for the L-M form of the Stanford-Binet test), distributed by IQ.

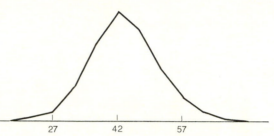

FIGURE 4b 3830 marriages in California with bride of age 40 to 44 (1961), distributed by age of groom.

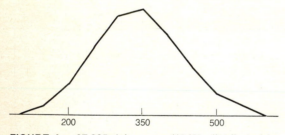

FIGURE 4c 37,885 dairy cows (1949), distributed by yearly pounds of butterfat.

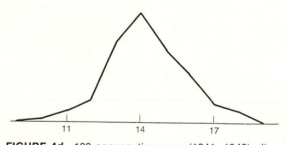

FIGURE 4d 100 consecutive years (1841–1940), distributed by mean June temperature (Celsius) at Stockholm.

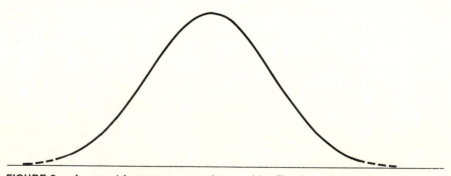

FIGURE 5a A normal frequency curve; it resembles Fig. 4*a* to *d*.

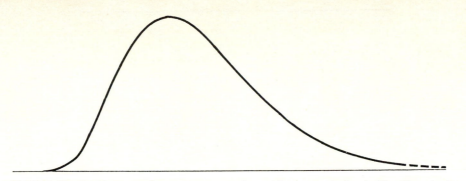

FIGURE 5b A chi-square frequency curve, skewed in a way similar to Fig. 4*f*, or to 4*g* if left and right are reversed.

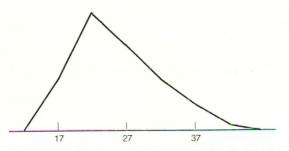

FIGURE 4e 381,172 live births in California (1961), distributed by age of mother.

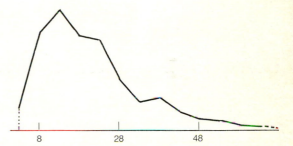

FIGURE 4f 1215 sentences in two essays by Macaulay, distributed by length of sentence in words.

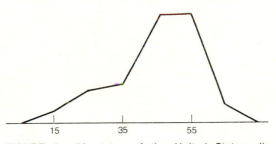

FIGURE 4g 50 states of the United States, distributed by percentage of voting-age population who voted for congressman in 1970.

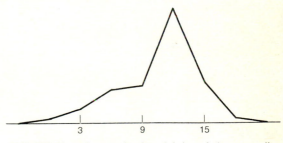

FIGURE 4h 54 stars in the vicinity of the sun, distributed by absolute visual magnitude.

quency over the range of the variable; this property is idealized in the *uniform* frequency curve of Fig. 5*c*. Figure 4*k* and *l* begin at the modal value and fall away on the right side only, with a long, attenuated tail. This shape occurs frequently in physical science and engineering; it is idealized in the *exponential* frequency curve of Fig. 5*d*. Figure 4*m* and *n*, which have two separated high points,

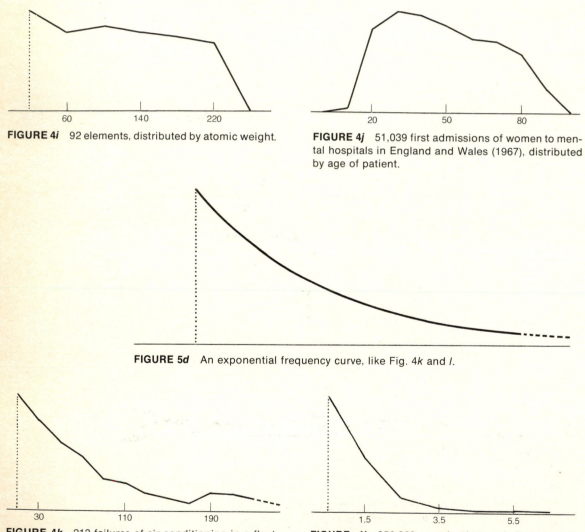

FIGURE 5c A uniform frequency curve, reminiscent of Fig. 4*i* and *j*.

FIGURE 4*i* 92 elements, distributed by atomic weight.

FIGURE 4*j* 51,039 first admissions of women to mental hospitals in England and Wales (1967), distributed by age of patient.

FIGURE 5d An exponential frequency curve, like Fig. 4*k* and *l*.

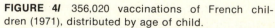

FIGURE 4k 213 failures of air conditioning in a fleet of airplanes, distributed by hours of operation before failure.

FIGURE 4*l* 356,020 vaccinations of French children (1971), distributed by age of child.

43

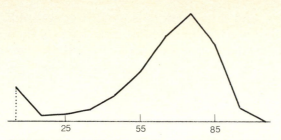

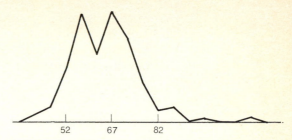

FIGURE 4m 1,701,522 deaths in the United States (1960), distributed by age of the deceased.

FIGURE 4n 136 18-year-olds in the core group of the California Child Development Study, distributed by weight in kilograms.

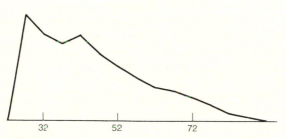

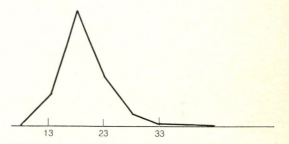

FIGURE 4o 21,404 female members of the American Medical Association (1967), distributed by age.

FIGURE 4p 13,000 patients of the Narcotics Treatment Administration, Washington, D.C. (1970–1973), distributed by reported age of first use of heroin.

are examples of *bimodal* distributions. We might call Fig. 4o triangular. Figure 4p looks at first glance like a normal curve, but on close inspection you will notice the long, thin tail on the right side only, and this unobtrusive but important feature serves to distinguish Fig. 4p from the normal type.

What is the value of idealized frequency curves such as those shown in Fig. 5? Because each of these curves is mathematically defined, it is possible to manipulate it theoretically. That is, one can discover by mathematical operations how a statistical method will work when applied to samples drawn from a population of that particular shape. As we shall see later, differently shaped population frequency curves require different methods of analysis. Theoretical studies therefore give guidance to the working statistician by helping him to choose a method suitable for the sort of population from which he is sampling. To learn to fit the method to the problem is at the heart of the statistician's art, just as it is the essence of good medical practice to fit the drug to the disease or that of jurisprudence to fit the punishment to the crime.

☐ You are now ready to make pictorial representations of your own sample of STATLAB mothers' heights, to read off information from these graphic displays, and thus to increase your understanding of the statistical attributes of your sample—attributes which will later be helpful when you use your sample to estimate the corresponding attributes of the heights of *all* the STATLAB mothers.

WORK TO BE DONE

STEP 1   Using the three (simple) frequency distributions for your sample of mothers' heights, $n = 40$, which you have stored in Databank, File B-2, construct three corresponding histograms (Figs. *a, b,* and *c* on page 1 of the worksheet). Select an appropriate width for the rectangles and write in the height values along the horizontal axis. For your convenience the frequency values are already printed on the vertical axis. Position your second and third histograms exactly below the first one—and cover the same distance on the x axis with each histogram. Label the axes.

STEP 2   In the designated spaces on page 2 of the worksheet, construct a simple frequency polygon for each of the three distributions (Figs. *d, e,* and *f*), again positioning the three polygons so that they lie directly under each other to make comparisons easy. (*Note:* You will first have to determine the midpoints for the class intervals in constructing the second and third frequency polygons.) Again label all axes.

STEP 3   On page 3 of your worksheet construct a cumulative frequency polygon (Fig. *g*) for the ungrouped data only—using the same horizontal spacing as in step 2. Label all axes.

STEP 4   Using the population data of boys' heights (Table 7 in Unit 4), construct a simple (Fig. *h*) and cumulative (Fig. *i*) frequency polygon on page 4 of the worksheet. Label all axes.

STEP 5   Copy in Databank, File B-3, the figures labeled *d, g, h,* and *i* on your worksheets.

STEP 6   Complete the inquiry.

MEASURES OF CENTRAL TENDENCY: THE MEDIAN

We have seen in Units 4 and 5 how a random sample of data may be organized and cast into frequency tables and graphs, thereby simplifying the sample into a form more easily amenable to analysis. Proceeding to the final process of simplification, we shall now see how certain basic features of the sample may be conveyed by a few summary values. Such a summary value is sometimes referred to as a *statistic*, i.e., a value which expresses the end result of statistical processing. Once extracted, the statistic can then be used to estimate the corresponding population *parameter*. A parameter is a summary value describing a key feature of the entire population from which the sample was taken.

Perhaps the statistic most commonly used by the layman as well as by the statistician is the "average." Almost everyone knows what the word *average* means; the difficulty is that it means different things to different people at different times. If you were to hear someone assert that the average net income of the doctors in his town is about $45,000 a year, this could easily be interpreted in at least three ways. It could mean that *most* of the physicians enjoyed such an income, so that it was the "typical" or, to use the

45

statistician's term, the *modal* income; it could refer to the middle income of the town's doctors, there being as many doctors earning more than $45,000 as there were earning less (you will recognize this as the *median* value); and, finally, it could refer to the sum of all the physicians' incomes divided by the number of doctors in that town (the statistician's term for which is *arithmetic mean*). When these three kinds of averages are calculated, it is often found that they differ—and sometimes widely. Which of these three averages is the correct one? The statistician's answer is that they are all correct: Each measures a somewhat different characteristic of the distribution, and deciding which one is the most useful depends upon the question being asked and other complex considerations. It is clear that the measurement of the "average" value of an array of x values is not a simple matter. Three units of this book—6, 7, and 8—are devoted to the study of the many problems which beset its measurement.

The reason why so much time is devoted to these problems is that, for statistical analysis, the most useful thing to know about a sample is its *location*, that is, the place on the scale where the x values tend to be centered. Quite appropriately, the statistics which measure this location are called *measures of central tendency*. For various reasons, some of which we have referred to above, statisticians have developed a number of different measures of central tendency. The two most widely used such measures are the median (which concerns us in this unit) and the arithmetic mean (which is the topic of Unit 7).

In Unit 4 you learned how to approximate a median from a frequency table, and in Unit 5, from a cumulative frequency polygon. The major aims of the present unit are to examine the concept of the median and its arithmetic calculation in some detail, to explore several alternative approximations to the median for your own sample, to point out some of the limitations and ambiguities of such approximations, and to prepare yourself for the next major task: estimating STATLAB population parameters.

THE MEDIAN: DISCRETE DATA

The basic idea of a median is quite simple and intuitively appealing as a measure of central tendency: It divides the sample into upper and lower parts equal in number of observations; it is a "central" value. The median is denoted by placing a tilde (~) above the symbol for the variable concerned. Thus when the variable is denoted by x, the median is written $\tilde{x}$ (read "x tilde"). Although this definition applies to both discrete and continuous data, the methods of determining $\tilde{x}$ for these two kinds of data differ. With discrete data we can determine the median quite directly and with precision because we deal with data that in theory can be measured and recorded without error. This means that no "remedial" action or assumptions will be required to take account of the errors of observation, recording, and rounding which inevitably characterize continuous data.

To examine and illustrate the simple procedures involved in determining $\tilde{x}$ for discrete data for different kinds of samples, we shall look at three commonly occurring varieties. We begin with the simplest type. In the following set of discrete data—sample A: 10, 8, 7, 3, 12—the value of $\tilde{x}$ is easily and accurately determined. First we line up all the observations from the one with the lowest value to the one with the highest to give us the ordered sample: 3, 7, 8, 10, 12. Since there is an odd number of cases, it is clear that the observation third in line divides the sample into two equal parts, each containing two observations. The value of the middle observation thus meets the definition of $\tilde{x}$, and therefore $\tilde{x} = 8$. In this instance the value of an actual observation and the derived statistic $\tilde{x}$ coincide. But not all samples are dealt with that easily.

For our second sample, let us suppose that we have eight observations which, when lined up from lowest to highest, yield the following ordered sample (sample B): $-5, 0, 1, 2, 5, 8, 20, 27$. With an even number of observations there can be no single middle observation; rather we have *two* middle observations, the fourth and fifth ones from the left with the x values of 2 and 5, respectively. The midvalue clearly lies *between* these two values. Since the midway point between 2 and 5 is $\frac{1}{2}(2+5)=3.5$, it is natural to take this value for our median: $\tilde{x} = 3.5$. Again our median divides the sample into two equal parts, with four observations in each part. Note that in this instance the value of $\tilde{x}$ does not appear as an x value in our sample. But this is of no concern because, it will be recalled, $\tilde{x}$ is a derived summary statistic and need not coincide with any actually observed x value.

Now we come to the last type of sample, which may require some comment. This is the type where the middle observation is tied with adjacent ones. First look at sample C: 2, 3, 5, 5, 8, 10. Since the two middle observations are both of value 5, the midvalue is 5 and so $\tilde{x} = 5$. But suppose we had a tie in an odd-numbered sample, as in the following instance (sample D): 2, 3, 5, 5, 8. The middle one of these five observations is again 5, and this time it is tied with its immediately adjacent (but "off center") observation to the right. But this in no way alters the fact that the third observation is the centered observation dividing sample D into an upper part with two observations (5, 8) and a lower part with two observations (2, 3). Thus our definition is met in all details and $\tilde{x} = 5$.

This discussion can be summarized in three simple rules for determining the median for any sample of discrete data:

1   Line up all the observations in order of their x values, from lowest to highest value.
2   For samples with an odd number of observations, the median is determined by giving it the same value as that of the middle observation.
3   For samples with an even number of observations, the median is determined by adding the values of the two middle observations and dividing by 2.

One final word: A median can sometimes depart widely from a commonsensical idea of what a measure of central tendency should be. Consider the following instance (sample E): 1, 1, 1, *1*, 19, 22, 57. Here is a sample of seven numbers, the median of which is *1*. Although it may violate our intuitive feelings of what a proper center for this sample should be, it meets the definition of a median precisely. In fact, there are many real distributions (as opposed to this hypothetical one) where the median behaves in this way. To enable you to see the kinds of medians (sometimes quite unexpected ones) which can meet the definition, we have summarized our illustrative examples (and have added one more) in Table 8. We urge you to examine this table carefully; a searching perusal of these illustrations may suggest to you a number of advantages and disadvantages of the median as a measure of central tendency—a matter we shall discuss in detail in Unit 8. Sir Francis Galton, a founding father of modern statistics, saw some unexpected advantages and uses for the median, as Box 3 shows.

THE MEDIAN: CONTINUOUS DATA

The ideas just presented for discrete data also govern the calculation of the median of a sample of continuous data, but there are differences of detail because the exact values of continuous data cannot be known. We can best explain these details by means of examples which illustrate three different situations.

TABLE 8 Discrete data: hypothetical samples and their medians

Sample	Sample values in order of size	Median ($\tilde{x}$)
A	3, 7, *8*, 10, 12	8
B	−5, 0, 1, *2, 5*, 8, 20, 27	3.5
C	2, 3, *5, 5*, 8, 10	5
D	2, 3, *5*, 5, 8	5
E	1, 1, 1, *1*, 19, 22, 57	1
F	1, 1, 1, *1*, 19, 22, 570	1

For samples of odd size, the median is the central sample value (italicized); for samples of even size, it is the mean of the two central values (italicized).

BOX 3

A LETTER TO THE EDITOR FROM SIR FRANCIS GALTON*

A certain class of problems do not as yet appear to be solved according to scientific rules, though they are of much importance and of frequent recurrence. Two examples will suffice. (1) A jury has to assess damages. (2) The council of a society has to fix on a sum of money, suitable for some particular purpose. Each voter, whether of the jury or of the council, has equal authority with each of his colleagues. How can the right conclusion be reached, considering that there may be as many different estimates as there are members? That conclusion is clearly not the average of all the estimates, which would give a voting power to cranks in proportion to their crankiness. One absurdly large or small estimate would leave a greater impress on the result than one of reasonable amount, and the more an estimate diverges from the bulk of the rest, the more influence would it exert. I wish to point out that the estimate to which least objection can be raised is the middlemost estimate, the number of votes that it is too high being exactly balanced by the number of votes that it is too low. Every other estimate is condemned by a majority of voters as being either too high or too low, the middlemost alone escaping this condemnation. The number of voters may be odd or even. If odd, there is one middlemost value; thus in 11 votes the middlemost is the 6th; in 99 votes the middlemost is the 50th. If the number of voters be even, there are two middlemost values, the mean of which must be taken; thus in 12 votes the middlemost lies between the 6th and the 7th; in 100 votes between the 50th and the 51st. Generally, in $2n - 1$ votes the middlemost is the nth; in $2n$ votes it lies between the nth and the $(n + 1)$th.

I suggest that the process for a jury on their retirement should be (1) to discuss and interchange views; (2) for each juryman to write his own independent estimate on a separate slip of paper; (3) for the foreman to arrange the slips in the order of the values written on them; (4) to take the average of the 6th and 7th as the verdict, which might be finally approved as a substantive proposition. Similarly as regards the resolutions of councils, having regard to the above $(2n - 1)$ and $2n$ remarks.

*This letter, with the title "One Vote, One Value," appeared in *Nature* **75** (1907). Our attention was first called to this letter by its citation in "Introduction to the Statistical Method" by K. R. Hammond, J. E. Householder, and N. J. Castellan, Jr. (Knopf, New York, 1970). As might be suspected, this letter had sequels—see Box 5.

Suppose first that the sample size is odd, say $n = 15$. Just as with discrete data, if the 15 measures are arranged in order of increasing size, the median will be the value of the eighth measure from the lower end. From the sample frequency table it is easy to pick out the class interval that contains this eighth measure. Suppose the frequency table contains two rows with the following entries (as we shall see, only these rows are needed):

Class interval	f	F
20–23	3	6
24–27	4	10

Since there are $F = 6$ measures below the interval 24–27 and $f = 4$ measures in this interval, it is obvious that the eighth measure in order of size will be among the four in the interval 24–27. More precisely, it will be the second in order among the $f = 4$ measures in that interval.

So much is clear from a glance at the table. Unfortunately we cannot know precisely where within the interval the second measure may be. It could be near the lower true limit 23.5, or near the upper true limit 27.5, or anywhere in between. However, common sense suggests that the $f = 4$ measures will probably be scattered throughout the interval. Let us assume that they are evenly spaced throughout the interval, as shown in Fig. 6. Since the interval has width $w = 4$, and since the four measures within it will divide the interval into five equal segments, each segment will be of width $\frac{4}{5} = 0.8$. If we place one of the $f = 4$ measures at each division point, the first measure will be at $23.5 + 0.8 = 24.3$; the second measure will be at $23.5 + 2(0.8) = 25.1$; and so forth. Thus under the equal-spacing assumption, the eighth measure in the whole sample (i.e., the second measure in this interval) falls at 25.1, which is accordingly the estimated or approximated value of the median: $\tilde{x} = 25.1$.

Now let us consider the case where the sample size is even. Then the median is defined to be midway between the two central measures in the ordered sample, just as with discrete data. We can illustrate this case with the same excerpt from a frequency table that we used in the previous case. This time we shall suppose that the sample size is $n = 14$. Then the median is midway between the seventh and eighth measures. Because there are $F = 6$ measures below the interval 24–27, these will be the first and second among the $f = 4$ measures in that interval. As we have seen, the equal-spacing assumption places these at 24.3 and 25.1 (see Fig. 6). Therefore in this case the median is $\frac{1}{2}(24.3 + 25.1) = 24.7 = \tilde{x}$.

Let us summarize the procedure for calculating the median of a continuous sample. First, by inspection of the frequency table, determine the class interval in which the median must lie. Second, divide this interval into $f + 1$ equal segments, where f is the number of measures falling into it: Each segment will have width $w/(f + 1)$. Third, determine the lower true limit of the interval. Finally, add to that limit as many segments as required to bring you to the central value or values of the whole sample.

As the final example let us consider the slightly different cal-

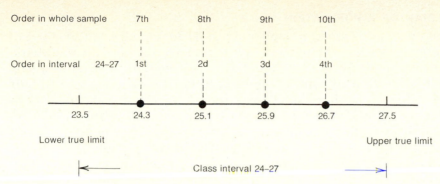

FIGURE 6 Calculation of the median for continuous data. The round dots show the location of $f = 4$ measures within the class interval 24–27 according to the equal-spacing assumption. If $n = 15$, the median is the eighth sample value, located at 25.1. If $n = 14$, the median is midway between the seventh and eighth, or 24.7.

culation required when n is even and there happens to be a value of F exactly equal to $\frac{1}{2}n$. Suppose $n = 40$ and the following entries are found in the frequency table:

x	f	F
52	4	20
53	2	22

Here the interval width is $w = 1$. The median is midway between the twentieth and twenty-first measures in the sample of 40. The twentieth is the fourth and last measure at $x = 52$, that is, in the interval 51.5–52.5. Therefore it is at $51.5 + 4(\frac{1}{5}) = 52.30$ according to the equal-spacing assumption. The twenty-first is the first of the two measures at $x = 53$, and it is at $52.5 + 1(\frac{1}{3}) = 52.83$. Thus $x = \frac{1}{2}(52.30 + 52.83) = 52.56$. (See Fig. 7.) It would be customary practice to round it off and report: $\tilde{x} = 52.6$.

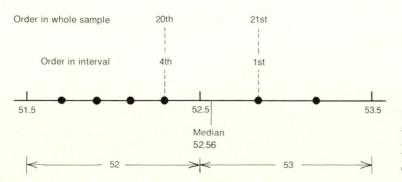

FIGURE 7 Calculation of the median when $F = \frac{1}{2}n$. With $n = 40$ and $F = 20$, the median is midway between the twentieth and twenty-first observations in the whole sample.

MEDIAN OF A POPULATION

The concept of median applies not only to samples but also to entire populations. The median is, in other words, not only a statistic but also a parameter. We shall denote the median of a population by $\tilde{\mu}$; it is the central value of the population distribution, having half the values on either side. (The Greek letter μ is read "mu.") You have seen in Unit 5 how $\tilde{\mu}$ can be determined graphically, and its precise arithmetic calculation from a frequency table is exactly like that for a sample. For instance, the median height of the STAT-LAB boys can be easily calculated from Table 7 (in Unit 4); see inquiry item 6.

The median of a sample, $\tilde{x}$, may be used to *estimate* the median $\tilde{\mu}$ of the population from which the sample was drawn. We shall see how this is done in some detail in Unit 15, but we might anticipate some of the discussion there by making a preliminary observation. As we mentioned in Unit 5, it follows from the law of large numbers that the cumulative frequency polygon of a large random sample will probably be close to the cumulative frequency polygon of the population from which the sample was drawn, the $F\%$ scale being used for both. In consequence, the sample median will probably fall near the population median. This means that the median of a large sample should be a reasonably good estimate of the median of the population. You will shortly have a chance to try this out for yourself.

☐ You are now ready to calculate the $\tilde{x}$ values for your own sample of 40 STATLAB mothers' heights—for class intervals of various widths. With these values (and additional statistics you will collect in the next units) you will be able to compare the virtues and faults of various measures of central tendency and to estimate STATLAB population medians.

WORK TO BE DONE

STEP 1   From Databank, File B-2, retrieve the data necessary to plot on page 1 of the worksheet the cumulative frequency polygon of your sample of 40 mothers' heights, for class intervals of width 2 and 4. Label the x axis in each case.

STEP 2   Determine graphically from each polygon the median height in your sample.

STEP 3   Inspect the *three* frequency tables in Databank, File B-2, in order to locate in each case the two rows of the table needed to calculate the median. Enter these rows on page 2 of the worksheet.

STEP 4   Calculate the three medians.

STEP 5   Round off all five medians (the two obtained via the graphic method, step 2, and the three obtained via calculation, step 4) to the nearest tenth of an inch. Record the rounded value of the calculated median for $w = 1$ in Databank, File B-2.

STEP 6   Record on page 3 of the worksheet the Raven scores of all girls whose mothers were 18 years old at the time of birth. Determine the median score.

STEP 7   Now do the same for the girls with 40-year-old mothers.

STEP 8   Complete the inquiry.

Note: In Unit 8, where you will examine the characteristics and relative effectiveness of the median and arithmetic mean, the comparison of these two measures of central tendency would be facilitated and made more meaningful if you had at hand a collection of medians and means calculated from a *large number of different samples all drawn randomly from the same population.* Fortunately, this desideratum is easily achieved. Your instructor will arrange for the members of your class to combine their values of $\tilde{x}$ for mother's height (as computed from the ungrouped data and then rounded to the nearest tenth of an inch) into a *class distribution* of medians. Each student contributes to this class distribution one value, the value of $\tilde{x}$ obtained from the student's unique sample of 40 mothers' heights. When this class distribution is made available, enter it in Databank, File B-4. (To simplify the later work, the class distribution will be grouped into class intervals of width two-tenths.) After Unit 7, the same will be done for the class distribution of the arithmetic mean.

THE ARITHMETIC MEAN

In this unit you will start to work with one of the simplest and most important of all sample statistics, the *arithmetic mean*. (It is often called just the *mean* and is also popularly known as the "average.") The mean is used more often than any other statistic to specify the center of a sample and consequently plays an essential role in most statistical procedures.

Because of the importance of the mean in statistical theory and practice, every working statistician calculates many means in his time. Ever on the alert to reduce work, statisticians have developed a number of tricks or arithmetic shortcuts which considerably simplify the calculation of a mean. In this unit you will discover how some of the general data-simplifying techniques you have already learned (ordering data into distributions, grouping data, etc.) can help make the calculation of a mean a fairly quick and easy task, even when a large number of cases is involved.

DEFINITION OF THE MEAN

The arithmetic mean of a sample is defined as the sum of the sample values divided by their number. Let us illustrate the definition on the sample of 40 girls' heights introduced in Unit 4. Taking the data in the order they were drawn (i.e., from Table 4), we simply add up the 40 values to get a sum of *2138*:

$$50 + 52 + 56 + \cdots + 50 + 61 + 58 = 2138$$

When we divide this total by the sample size, $n = 40$, we find the arithmetic mean to be $\frac{2138}{40} = 53.45$. It is conventional to denote the arithmetic mean by placing a horizontal bar above the symbol for the variable being averaged. Since x denotes girls' heights here, the arithmetic mean of our sample of girls' heights is written $\bar{x} = 53.45$. The symbol $\bar{x}$ is read "x bar."

Of the two arithmetic processes involved in calculating a mean (summing and dividing), the determination of the sum is usually the more laborious and it is to this problem, therefore, that the calculation shortcuts are addressed. Let us first consider how the construction of a simple, ungrouped frequency table can expedite the calculation of a sample sum.

DIRECT METHOD

Since the value of a sum of terms does not depend on the order in which you add the terms, the same total would be obtained for our sample of girls' heights if we were to add them in order of size (as in Table 5 of Unit 4):

$$48 + (49 + 49) + (50 + 50 + 50 + 50) + \cdots + (59 + 59) + 61 = 2138$$

Here we have bracketed together the equal values. This immediately suggests a shortcut in calculation, i.e.,

$$48 + (2 \times 49) + (4 \times 50) + \cdots + (2 \times 59) + 61 = 2138$$

Instead of adding 40 x's we need merely multiply each different x by its frequency f and add all the resulting products, fx. The mathematical notation for this procedure is Σfx, where the Greek letter Σ (capital sigma) is the conventional symbol meaning "add up the following terms." In this notation, then, the formula for the arithmetic mean for ungrouped data is

$$\bar{x} = \frac{\Sigma fx}{n}$$

FORMULA 1

This method of calculating the arithmetic mean is called the *direct method* and is easily done once your data have been ordered in a simple frequency table, as is illustrated in the first three columns of Table 9.

In calculating $\bar{x} = 53.45$ for our sample of 40 girls' heights, we have treated these heights as if they were exact whole-number values. Of course, as you know, height is a continuous variable, and our 40 values were obtained by rounding the heights as actually measured to the nearest whole inch. For example, the $f = 4$ heights listed in Table 9 as equal to $x = 52$ could originally have had any values in the interval from 51.5 to 52.5 (see Fig. 7). In putting each of these four heights equal to 52, we have in effect assumed that all values in the interval (51.5–52.5) are located at the mid-point 52 of this interval, and that is how we arrived at $fx = 208$ for $x = 52$ (see Table 9).

Let us compare this "midpoint assumption" with the equal-spacing assumption made in Unit 6 when calculating medians of continuous data. If we had spaced our four heights equally, they would have been placed at 51.7, 51.9, 52.1, and 52.3. So placed, their sum would have been $51.7 + 51.9 + 52.1 + 52.3 = 208$. This is exactly equal to $fx = 4 \times 52$, which is the contribution of these four measures to the sample sum when the midpoint assumption is used. A little reflection will convince you that we shall always get the same sum, whether the values in an interval are assumed to be equally spaced or concentrated at the midpoint. Since the midpoint

TABLE 9 Computation of an arithmetic mean and deviations from it: 40 STATLAB girls' heights, ungrouped data

DIRECT METHOD			CODING METHOD, $C = 53$		DEVIATIONS FROM $\bar{x}$	
x	f	fx	x'	fx'	d	fd
48	1	48	−5	−5	−5.45	−5.45
49	2	98	−4	−8	−4.45	−8.90
50	4	200	−3	−12	−3.45	−13.80
51	6	306	−2	−12	−2.45	−14.70
52	4	208	−1	−4	−1.45	−5.80
53	5	265	0	0	−0.45	−2.25
54	3	162	1	3	0.55	1.65
55	4	220	2	8	1.55	6.20
56	5	280	3	15	2.55	12.75
57	2	114	4	8	3.55	7.10
58	1	58	5	5	4.55	4.55
59	2	118	6	12	5.55	11.10
60	0	0	7	0	6.55	0
61	1	61	8	8	7.55	7.55
$40 =$ n		$2138 =$ Σfx		$18 =$ $\Sigma fx'$		$0 =$ Σfd

$$\bar{x} = \frac{2138}{40} = 53.45 \qquad \bar{x}' = \frac{18}{40} = 0.45$$

$$\bar{x} = 53 + 0.45 = 53.45$$

The same data appear in the first four columns of Table 6.

assumption is more convenient in calculating the mean than is the equal-spacing assumption, and since it gives the same results, it makes sense to treat all the observations as if they were at the mid-point.

The same simplification and the same direct method can be used with data grouped into class intervals (although, as always, we can expect to find minor differences in values when the $\bar{x}$'s are computed on data grouped into class intervals of different width). Let us skip ahead for a moment and consider the left half of Table 10, where our 40 girls' heights have been grouped into class intervals of width $w = 2$. We assume that the $f = 3$ observations in the interval 48–49 are at its midpoint $m = 48.5$; the $f = 10$ observations in 50–51 are at 50.5, and so forth. With this assumption the total, calculated by the direct method, is $(3 \times 48.5) + (10 \times 50.5) + \cdots + 1(60.5) = 2136$, so the mean is $\bar{x} = \frac{2136}{40} = 53.4$. The calculation is straightforward: Each midpoint m is multiplied by the frequency f in its interval, the products fm are added, and finally the sum is divided by n. The direct-method formula for grouped data is then conveniently written: $\bar{x} = \Sigma(fm)/n$. It is, of course, identical in meaning to the one for ungrouped data: $\bar{x} = \Sigma(fx)/n$.

CODING

We return now to Table 9. The calculation of Σfx or Σfm is easy on a desk calculator, but it can be slow, tiresome, and subject to many errors when done by hand—especially if the n is large. The *coding method* uses a simple trick that will reduce the work of calculating Σfx or Σfm considerably. This method is illustrated for ungrouped data in the fourth and fifth columns of Table 9. There are four steps involved:

1   You first pick any convenient x value near the *center* of the x distribution. This "guessed mean" is symbolized by the letter C. In Table 9 we have chosen $C = 53$.
2   For each x value in the distribution you now substitute a new "coded" value, labeled x'. The new coded value for each x is obtained by subtracting C from each value of x in turn; that is, $x' = x - C$. In Table 9 these coded values are shown in column 4, the one labeled x'.
3   Next you calculate the arithmetic mean $\bar{x}'$ of this coded x' distribution:

$$\bar{x}' = \frac{\Sigma fx'}{n} = \frac{18}{40} = 0.45$$

4   Finally, you add C to $\bar{x}'$. This step will convert $\bar{x}'$ to $\bar{x}$:

$$\bar{x} = C + \bar{x}' = 53 + 0.45 = 53.45$$

The coded calculation is much easier than the direct method primarily because the numbers fx' are so much smaller than the numbers fx, as can immediately be seen by comparing the corresponding terms in columns 5 and 3 of Table 9.

Why does the trick work? When we subtracted $C = 53$ from each x, we shifted the entire distribution to the left on the number scale, by the constant value C. The arithmetic mean of the x' distribution would therefore be 53 units to the left of the mean of the original distribution. After calculating the mean $\bar{x}'$ of the coded (i.e., shifted) values, we must then add $C = 53$ to shift the mean back 53 units to the right in order to restore the original value of the mean.

CALCULATING THE MEAN FOR GROUPED DATA

By grouping the data, the coding method yields an additional savings in labor, as is illustrated in Table 10. Let us first look at the left side of the table, where the data are grouped into class intervals of width 2.

We take for C the midpoint of any convenient central interval —we have taken $C = 52.5$. For this midpoint the coded value is $x' = 0$. The other values of x' proceed in coded unit steps from midpoint to midpoint, as shown in column 4, Table 10. We then compute the mean of the x' distribution: $\bar{x}' = \Sigma(fx')/n = 0.45$. Next we multiply $\bar{x}'$ by 2, the interval width. The reason why we multiply $\bar{x}'$ by $w = 2$

TABLE 10 Computation of the arithmetic mean: 40 STATLAB girls' heights, grouped data, two widths of class intervals

| | $w = 2$ | | | | | $w = 3$ | | | |
| | $C = 52.5$ | | | | | $C = 55$ | | | |
Class interval	m	f	x'	fx'	Class interval	m	f	x'	fx'
48–49	48.5	3	−2	−6	48–50	49	7	−2	−14
50–51	50.5	10	−1	−10	51–53	52	15	−1	−15
52–53	52.5	9	0	0	54–56	55	12	0	0
54–55	54.5	7	1	7	57–59	58	5	1	5
56–57	56.5	7	2	14	60–62	61	1	2	2
58–59	58.5	3	3	9			40 =		−22 =
60–61	60.5	1	4	4			n		$\Sigma fx'$
		40 =		18 =					
		n		$\Sigma fx'$					

$$\bar{x}' = \tfrac{-22}{40} = -0.55$$
$$x = 55 + 3(-0.55) = 53.35$$

$$\bar{x}' = \tfrac{18}{40} = 0.45$$
$$\bar{x} = 52.5 + 2(0.45) = 53.40$$

The same data appear in the last six columns of Table 6.

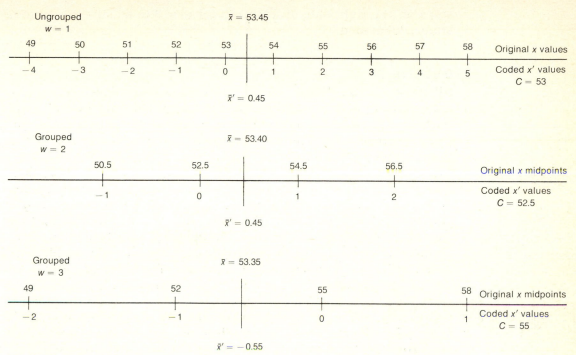

FIGURE 8 Relation between original and coded scales, for the data of Tables 9 and 10.

can best be explained by reference to Fig. 8, where the relations between the unit steps of the original and coded scales for our three examples are depicted. We see there that with our ungrouped data, a one-unit step on the x' scale (say from 1 to 2) corresponds to a one-unit step on the original scale (from 54 to 55 in). There is a one-to-one correspondence in width of unit steps. This is *not* true for grouped data. With class interval $w = 2$, for example, the step from 1 to 2 on the x' scale corresponds to *two* units on the original scale, from 54.5 to 56.5. Since each unit step of the x' scale is equal to two units of the original scale, we must multiply $\bar{x}'$ by 2 to restore the original unit values. Finally, to arrive at $\bar{x}$ we add the constant C as in the case of ungrouped data:

$$\bar{x} = 52.5 + 2(0.45) = 53.40$$

For similar reasons, in calculating the mean for the data grouped into $w = 3$ we must multiply $\bar{x}'$ by 3 before adding C. The general formula for the coded computation of means when data are grouped into intervals of width w is

$$\bar{x} = C + w\bar{x}' \qquad\qquad \textbf{FORMULA 2}$$

We suggest that you verify this formula by checking through the steps on the right side of Table 10 for $w = 3$ to be sure you understand the method.

DISCRETE DATA

All these shortcut methods for calculating the mean are equally applicable to discrete and continuous data. No different assumptions or variations in procedure are required. We might point out, however, that quite often the arithmetic mean of discrete data (no matter which calculation method is used) may come out as a fraction, for instance, 0.001 of a physician per inhabitant. This does not argue against the validity of the calculation method which achieved so homeopathic a healer. A mean is *not* a datum; it is a statistic *derived from* data. We stress this because the notion that we can talk about, for example, the average family with its 2.43 children is an affront to our common sense, and most people balk at accepting such "nonsense." Of course a fractionated child is an arithmetic abstraction: It makes good *mathematical sense* and as such is valid and useful. Nevertheless those who jibe at the statistician's easy reference to fractions of people are performing a salutary job. ("The figure of 2.2 children per adult female was felt to be in some respects absurd, and a Royal Commission suggested that the middle classes be paid money to increase the average to a rounder and more convenient number"—*Punch.*) The statistician needs to be reminded from time to time that he, like all scientists, deals with abstractions of life; and no matter how useful they are, they do not portray life in the full.

DEVIATIONS FROM THE MEAN

Before we continue with our discussion of the mean—its characteristics as a population parameter, the sample mean as an estimate of this parameter, etc.—let us backtrack for a moment and consider a definition of the mean which is somewhat different from the one we gave at the beginning of this unit. This alternative approach will prove useful in succeeding units.

Most values of a sample will deviate from the sample mean in one direction or the other and to a greater or lesser extent, and the arithmetic mean can be defined in terms of these deviations. To do so, however, we have to understand the relation of each deviation to its corresponding x value and the relation of the sum of all the deviations in a sample to the mean of that sample.

To determine for each value x the direction and size of its deviation from the mean (symbolized by d), we merely subtract the mean from the x value: $d = x - \bar{x}$. The sixth column of Table 9 (the column labeled d) shows the deviation corresponding to each x for our sample of 40 girls' heights. Thus for $x = 56$, $d = 56 - 53.45$

= 2.55. Values of x which are smaller than the value of the sample mean yield, of course, negative deviations (i.e., for $x = 48$, $d = 48 - 53.45 = -5.45$).

As you can see from the last column of Table 9, with this sample of 40 girls' heights the *sum* of all the deviations is zero: $\Sigma fd = 0$. But this outcome is not peculiar to this sample; in fact, $\Sigma fd = 0$ *must always be true*.

This can be demonstrated by the following simple algebra. For a single observation we have defined d as

$$d = x - \bar{x}$$

To obtain the equation for the sum of all the deviations in the sample, Σfd, we add the terms of the right side of the equation for *all* the n observations in the sample. Doing this for x, we get Σfx, for $\bar{x}$ we get $n\bar{x}$. Our formula for the sum of the deviations thus becomes

$$\Sigma fd = \Sigma fx - n\bar{x}$$

Since $\bar{x} = \Sigma(fx)/n$, so that $\Sigma fx = n\bar{x}$, the preceding equation becomes

$$\Sigma fd = \Sigma fx - \Sigma fx = 0$$

Using this fact, we can now define the mean in terms of deviations: The mean is that point from which the sum of the deviations is zero. With this definition, $\bar{x}$ is the *center* of the distribution in the following sense: The positive deviation of the sample values from it just balance out the negative deviations from it. If we think of the distribution as a physical system with weight f at point x, the mean is the center of gravity of that system (see Fig. 9). In the next two units we shall see how this conception of the mean helps us to detect important similarities and differences between the mean and the median. Merely to hint why this is so, we might point out that with this new definition of the mean, both $\bar{x}$ and $\tilde{x}$ can now be defined and characterized in the same terms—in terms of their deviations. The sample *median*, as we saw in Unit 6, is that point from which the *numbers* of deviations in the two directions are equal, regardless of the magnitude of the deviations; the sample *mean*, as

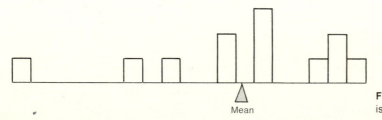

Mean

FIGURE 9 The mean of a distribution is its center of gravity.

we now see, is that point from which the *sums* of the deviations in the two directions are equal, taking the magnitudes into account.

POPULATION MEAN

As we did with the concept of median, we may apply the concept of mean not only to a sample but also to an entire population of x values. The sum of all the x values in a population, divided by the size of the population, is called the *population mean*. It is the parameter most commonly used to specify where the population is centered. The conventional symbol for the population mean is μ.

Various ideas we have discussed for samples carry over in an obvious way to entire populations. The tricks of coding and grouping are even more useful for populations because the frequencies tend to be larger. Table 11 illustrates the calculation of μ for the population of heights of STATLAB boys, grouped into class intervals of width 2, as originally given in Table 7. Note that the boys' mean height, $\mu = 53.60$ in, is quite close to their median height, $\tilde{\mu} = 53.47$ in, found in Unit 6. Here these two measures of central tendency are in close agreement, but as we shall see in Unit 8, this is by no means always true.

SAMPLE MEAN AS AN ESTIMATE

We began this unit by stating that the arithmetic mean of a sample is widely used to specify the center of the x values in that sample. Perhaps even more important is the use of $\bar{x}$ to estimate the mean μ of the population from which the sample was randomly drawn. Indeed, the main reason for drawing a sample is often to make it possible to estimate the parameter μ.

TABLE 11 Mean of the population of boys' heights, $C = 54.5$

Class	f	x'	fx'
46–47	1	−4	−4
48–49	26	−3	−78
50–51	112	−2	−224
52–53	187	−1	−187
54–55	175	0	0
56–57	101	1	101
58–59	36	2	72
60–61	9	3	27
62–63	1	4	4
	648		−289 = $\Sigma fx'$

$$\bar{x}' = \frac{-289}{648} = -0.45$$

$$\bar{x} = 54.5 + 2(-0.45) = 53.60$$

The same data appear in Table 7.

Recall again from Unit 4 the law of large numbers. This law states that if the sample is large, there will be a tendency for the sample distribution to resemble the distribution of the population from which it was randomly drawn. It follows that the $\bar{x}$ of such a sample will probably be close to the μ of the population. We cannot, of course, expect that $\bar{x}$ will coincide exactly with μ. You have already seen in Unit 2 that many quite different samples, each with its own $\bar{x}$, can be randomly drawn from the same population. What one may hope is that $\bar{x}$ will be close enough to μ to provide a useful estimate.

☐ You are now ready to test the efficiency of the various short-cuts and tricks we have discussed by calculating several means of your own sample of STATLAB mothers' heights (for grouped and un-grouped data and for two different sample sizes). With the calculation of these means you will also have two different sets of measures for your sample's location: the median and the mean. In the next unit you will use these two sets of statistics to begin an analytic and comparative investigation of the usefulness of these two measures of central tendency.

WORK TO BE DONE

STEP 1   Retrieve from Databank, File B-1, the heights (rounded to the nearest inch) of the first 10 mothers of your sample, and enter these heights on the worksheet.

STEP 2   Find the sum of these heights, and find their mean. Round $\bar{x}$ to the nearest tenth of an inch.

STEP 3   Retrieve from Databank, File B-2, and enter on the worksheet, the simple frequency table for your 40 mothers' heights, $w = 1$.

STEP 4   Calculate $\bar{x}$ for these 40 heights using the coding method. Round $\bar{x}$ to the nearest tenth of an inch.

STEP 5   Repeat steps 3 and 4 for your 40 mothers' heights grouped into class intervals $w = 2$ and $w = 4$ on page 2 of the worksheet.

STEP 6   Enter in Databank, File B-2, the four means you have computed.

STEP 7   Complete the inquiry.

Note: In order to do the work of the next unit, you must have available the class distribution of $\bar{x}$. Accordingly, the instructor will now arrange to compile into such a distribution the values of $\bar{x}$ (as computed from the ungrouped data and then rounded to the nearest tenth of an inch) obtained by yourself and the other members of the class. When this class distribution is made available, enter it into Databank, File B-4.

MEAN OR MEDIAN?

You have now become acquainted with the two most widely used measures of the center of a sample: the median and the mean. When a statistician is to select a measure of central tendency, his choice will usually lie between these two (the mode is only rarely a good measure to use). Which measure should he choose? Such questions often arise in statistics, since there is usually more than one statistical method available for dealing with a problem. But this does not necessarily mean that all available methods are equally good for a given problem.

The purpose of this unit is to analyze the major issues involved when the statistician makes a choice between the median and the mean. The correct choice will depend on the nature of the sample and the population from which it came—and on the question being asked. In addition to the sample you have already drawn, we shall ask you to draw a sample from a quite different sort of distribution. Then you will be asked to analyze the choice between median and mean for each of your two samples.

POPULATION MEDIAN AND POPULATION MEAN

Since the major use of a sample median or mean is to estimate the population center, we begin the analysis by comparing the two measures viewed as population parameters.

If the population frequency distribution is symmetric (as in Fig. 5a and c, Unit 5), then the population median $\tilde{\mu}$ and population mean μ will always coincide. This may be reasoned as follows. Suppose that the population distribution is symmetric about a point. Then half the population must lie on either side of this point. Therefore this point is the population median $\tilde{\mu}$. Now consider the deviations d from $\tilde{\mu}$. Deviations d at a given distance to the right and to the left of the median will be of equal magnitude but will have opposite signs, and because of the symmetry they will occur with equal frequency f. Therefore they will just cancel each other so that, taking all cases together, the sum of all deviations from $\tilde{\mu}$ will be zero. But recall from Unit 7 that the population mean is the point about which the sum of deviations is zero. We have thus shown that if the population distribution is symmetric, the population median is also the population mean—the two parameters coincide. (In terms of the physical analogy, Fig. 9 in Unit 7, according to which the mean is the center of gravity, it is obvious that a symmetric distribution will balance at its center of symmetry.)

Now suppose that the population frequency curve is skewed to the right, as illustrated by Fig. 10a (which reproduces Fig. 5b of Unit 5). The median splits the population in half, with as many deviations to the right as to the left. The long and heavy right tail will tend to overbalance the short left tail, and this imbalance will tend to pull the mean to the right of the median. By similar reasoning, if the curve is skewed to the left the mean will fall to the left of the median. As Fig. 10a suggests, however, the discrepancy between population median and mean will not typically be very large unless the degree of skewness is extreme. Figure 10b (reproduced from Fig. 5d of Unit 5) represents such an extremely skewed distribution, with a substantial separation of the two parameters.

One may imagine populations which are extreme indeed, for which a single very large value can substantially affect the mean, without of course having much influence on the median. How might the mean and the median family wealth compare on a Greek island with 1000 families—999 fishermen and one shipowner? The statistician faced with the problem of choosing between median and mean of family wealth as a measure of central tendency for the island would want to know the use which would be made of his statistic. If the tax collector wanted to know how much total taxable wealth was available (and assuming no tax loopholes for the shipowner), the statistician would recommend the mean; if, on the other hand, a purveyor of automobiles wanted the information in order

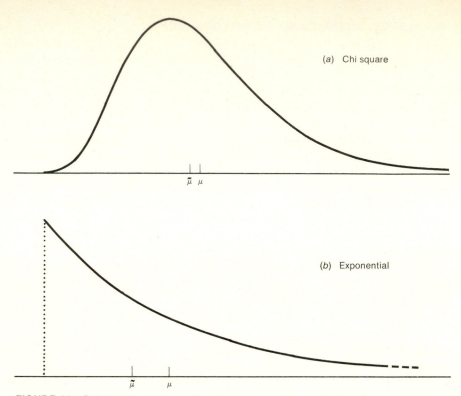

FIGURE 10 Relation between median $\tilde{\mu}$ and mean μ for a moderately skewed and an extremely skewed distribution.

to estimate how many cars he could hope to sell, the statistician would choose the median because it will better reflect the wealth of the generality of families. Both parameters are legitimate measures of central tendency, but they have different uses.

MEDIANS AND MEANS AS ESTIMATES

On many occasions, the main purpose in drawing a random sample is to permit one to estimate the central value of the population. Which of our two statistics, $\bar{x}$ or $\tilde{x}$, viewed as estimates of population center, is preferable? The answer depends on the statistician's purpose and on the nature of the population distribution.

Suppose first that the two population parameters, median and mean, are substantially different, as would tend to be the case with heavily skewed populations. Then one must make up his mind which parameter he wants to use to specify the population center and then choose the corresponding estimate. Obviously it would make no sense to use the sample mean (which by the law of large numbers would tend to be near the population mean) to estimate the population median if the two parameters are far apart.

But now suppose that the population, as is often the case, is near enough to being symmetric that the two parameters are close together. Then *either* sample statistic could be used to estimate *either* population parameter. In such cases it makes sense to choose the statistic that will estimate its own parameter more precisely.

According to the law of large numbers, both our statistics will tend to be close to the corresponding parameters if the sample is large enough. But it may be that a sample of (say) $n = 40$ is large enough to give accurate results with one of the estimates but not with the other. The question of relative accuracy of the two estimates can be studied either mathematically or empirically. We shall give some of the mathematical results in Units 13 and 15. In the work for this unit you can use some sampling results already obtained by your class to bring some empirical evidence to bear on the relative precision of median and mean.

SAMPLE MEDIAN AND SAMPLE MEAN

Sample distributions reflect many of the same features as population distributions. If the sample distribution is symmetric or nearly so, the sample median and sample mean will be close together. If the sample is skewed to the right (or left), the sample mean will tend to fall to the right (or left) of the sample median. In extreme cases, the sample mean can lie at a considerable distance from the bulk of the sample, whereas by definition the sample median can never have more than half the sample on either side. Thus the median will in one sense tend to be more representative than the mean.

Another way in which skewness of a sample can reveal itself is by an asymmetric disposition of the minimum and maximum about the median. With a symmetric sample the median will be midway between the two extremes. On the other hand, if the sample has (say) a long tail to the right and a short one to the left, the maximum will be farther away from the median than is the minimum. (See sample *F*, Table 8, Unit 6.)

An attractive feature of the sample median is its insensitivity to extreme values of the sample. (Again see sample *F* and compare it with sample *E* in the same table.) One must expect that, from time to time, a sample will contain one or more values that are extremely atypical or grossly erroneous. By accident a value will be miscopied, or an extraneous individual will somehow get into the sample, or some major procedural error will occur in an experiment. The chemist preparing a reagent may forget one of the components, for example, leading to an observation (of chemical reactions) of minutes instead of seconds. The result of such a "gross error" is likely to be an improper value far from the rest of the sample. Our two measures of center handle such extreme values quite differently.

For illustration, suppose that in a sample of 10 women, one who is 67 in tall has her height misrecorded as 76 in. When the data are graphed, the suspicious figure is spotted but perhaps the woman is no longer available for a recheck. Experimenters are properly reluctant to "edit" their data. Yet the retention of the false value will boost the sample mean by almost an inch, which is a substantial mistake. The sample median, in contrast, will either be changed not at all by the recording error or else be altered only slightly.

For similar reasons, the sample median has the advantage that it can be computed even if some of the actual sample values are unknown, provided that we know in which tail they lie. Consider for illustration a study of the survival times of a group of cancer patients treated with massive doses of radiation. The clinician, as an investigator, will want to report the average survival time as soon as he can. He will not be able to report the mean survival time until the last patient dies (and this, as a physician, he will seek to postpone as long as possible); the median can be determined as soon as one more than half have died.

As these examples suggest, in some cases the median will be preferable to the mean, either because the latter is untrustworthy or because it cannot be computed. In most situations, however, the choice will depend on the use to which the estimate is to be put or on the precision of the statistic as an estimate of population center.

☐ You are now ready to apply these ideas by considering the medians and means of your own samples, drawn from two quite different STATLAB populations. Using the result obtained by the class as a whole, you can also begin to investigate the key question: Which of the estimates is more precise, and in what circumstances?

WORK TO BE DONE

STEP 1　Draw a fresh sample of $n = 40$ STATLAB families, and record on the worksheet, page 1, the family income at test, rounded to the nearest thousand dollars (e.g., an income printed as 96, meaning $9600, is to be rounded as 10, meaning 10 thousands of dollars. Recall the even rule for borderline cases, Unit 3). Be sure to eliminate duplicate throws, as explained in Unit 2, work step 2.

STEP 2　Make a frequency table of these data on page 3 of the worksheet, using class intervals of width 2: 0–1 (thousands), 2–3 (thousands), etc.

STEP 3　Compute from your frequency table the mean and median incomes of your sample.

STEP 4   Draw a histogram of your frequency table on page 3 of the worksheet.

STEP 5   Retrieve from Databank, File B-4, the class distributions of mean and median mothers' heights, and graph their simple and cumulative frequency polygons on page 2 of the worksheet.

STEP 6   Enter the following in Databank, File C-1:

1   Your new sample data—ID numbers and family incomes rounded to nearest $1000

2   The mean and median incomes for your sample, rounded to nearest $500

3   The frequency table of your sample of family incomes, $w = 2$ (from page 3 of the worksheet)

STEP 7   Complete the inquiry.

Note: In Units 13 and 15 we shall make use of the class distributions of median income and mean income in order to continue the comparative study of these statistics. For these purposes it will be more convenient, and sufficiently precise, to have these statistics rounded to the nearest $500. (For example, a median income of $14,200 is rounded to $14,000 and a median income of $14,300 is rounded to $14,500. Borderline cases such as $14,250 and $14,750 go to their nearest thousand, that is, to $14,000 and $15,000, respectively.) The instructor will arrange for the members of the class to combine these rounded statistics into class distributions of median income and mean income. When these are made available, enter them in Databank, File C-2, for later use.

MEASURES OF DISPERSION

We have stressed, in several of the preceding units, that often the most useful thing to know about a sample is its location. You have now become familiar with two numerical measures, the median and the mean, either of which may serve to specify the center of a sample. To describe a sample solely by its central tendency, how-ever, can be quite misleading. Consider the following two samples (arranged for convenience in order of size and displayed in Fig. 11 as histograms):

$$A:\quad 14\quad 18\quad 19\quad 19\quad 20\quad 21\quad 21\quad 21\quad 23\quad 28$$
$$B:\quad 10\quad 14\quad 17\quad 17\quad 19\quad 22\quad 22\quad 25\quad 29\quad 31$$

Both samples are centered at very nearly the same place (sample A has a mean of 20.4 and B a mean of 20.6; both have a median of 20.5), yet a glance at Fig. 11 shows that these distributions differ considerably in *spread*. The individual values of sample A cluster more closely around the central value than do those of sample B. To describe these distributions so that their important similarities *and* differences will be encompassed, we need a numerical measure

70

of their location and also a numerical measure of the degree of spread of their x values. Not only will such a numerical measure describe and summarize an important feature of a sample but, as we shall see in later units, the degree of spread of a sample tells us a great deal about *the precision of the sample mean as an estimate of the population mean.*

Statisticians have suggested several indices, called *measures of dispersion,* which specify for any distribution the spread of the x values in quantitative terms. As with measures of central tendency, different measures of dispersion are appropriate for different problems. In this unit you will become acquainted with the properties of four measures of dispersion; apply the most important and most widely employed of these measures to your own samples; and (in cooperation with the members of your class) begin an investigation of its virtues, faults, general characteristics, and uses.

THE RANGE

We have already mentioned one natural and intuitive measure of dispersion, the *range*. Recall from Unit 4 that the range is defined as the difference between the maximum and minimum values in the sample. For our present sample A the range is $28 - 14 = 14$, whereas B has range $31 - 10 = 21$. As measured by range, therefore, B is 1.5 times more widely dispersed than A (that is, $\frac{21}{14} = 1.5$).

The range is used when simplicity of calculation is required, but it suffers from the obvious disadvantage that it depends directly on only *two* values, the x value with the maximum positive devia-

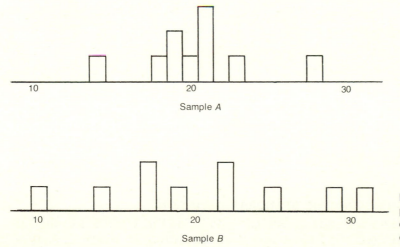

Sample A

Sample B

FIGURE 11 Histograms of two hypothetical samples that differ in dispersion while having the same center.

tion from the mean and that with the maximum negative deviation from the mean. The largest or smallest value in a sample will not infrequently reflect a grossly mismeasured or misrecorded observation.

THE MEAN DEVIATION

A good "global" measure of dispersion ought to reflect the degree to which *all* the observations deviate from the central value. One way to get this is to measure the amount by which each x value deviates from the mean value, add up these deviations, and then find the arithmetic mean of these deviations. Let us return to sample *A*. If we subtract from each value x in sample *A* the sample mean $\bar{x} = 20.4$, we obtain the 10 values $(d = x - \bar{x})$ shown in the second column of Table 12. How large are these deviations on the average? Were we to add the deviations directly and divide by 10, we would not get a very helpful measure of dispersion since we know from Unit 7 that the sum of all the plus and minus deviations from $\bar{x}$, Σfd, will *always* equal zero, no matter what the concentration or spread of the individual x values.

What is wanted is the average of the *magnitudes of the deviations regardless of their signs*, since an x value is equally deviant whether it lies, for example, five units above the mean or five units below. Therefore we must work only with the *absolute values* of the deviations. The mathematical notation for an absolute value is a set of vertical lines enclosing the value. Thus $|d|$ is interpreted to mean the numerical value of d, taken as positive regardless of whether d itself is positive or negative. The sum of the 10 absolute deviations of column 3 of Table 12 is 24.0. Their arithmetic mean, then, is $\frac{24}{10} = 2.4$. This arithmetic mean 2.4 is called the *mean deviation*. The general formula for it, applicable to data arranged in a frequency table, is

$$\text{Mean deviation} = \frac{\Sigma f|d|}{n} \qquad \text{FORMULA 3}$$

You should check for yourself (to be certain that you understand the procedure) that for sample *B* the mean deviation is 5.2. Thus sample *B* is 2.17 times as widely dispersed as sample *A*, as measured by the mean deviation (that is, 5.2/2.4 = 2.17).

THE VARIANCE AND THE STANDARD DEVIATION

The mean deviation is a natural and reasonably simple measure of dispersion, but it is not widely used. It suffers from the fact that absolute values are not mathematically tractable for theoretical

TABLE 12 Sample values and deviations from $\bar{x} = 20.4$ in hypothetical sample A

x values	Deviations d	Absolute deviations $|d|$	Squared deviations d^2
14	−6.4	6.4	40.96
18	−2.4	2.4	5.76
19	−1.4	1.4	1.96
19	−1.4	1.4	1.96
20	−0.4	0.4	0.16
21	0.6	0.6	0.36
21	0.6	0.6	0.36
21	0.6	0.6	0.36
23	2.6	2.6	6.76
28	7.6	7.6	57.76
204	0	24.0	116.40

The same data are depicted in Fig. 11.

studies. This is one reason why mathematicians prefer to turn negative numbers into positive ones in a different way—by squaring them. We show in the last column of Table 12 the square d^2 of each deviation d. The sum of these squared deviations, Σfd^2, is 116.4. The arithmetic mean of the squares of the deviations is thus $116.4/10 = 11.64$. This can also, quite obviously, be considered a measure of dispersion and is called the *variance*. Its general formula is

$$\mathrm{Var} = \frac{\Sigma fd^2}{n} \qquad\qquad \textbf{FORMULA 4}$$

You should check that sample B has variance $386.4/10 = 38.64$.

There is, however, one difficulty with using the variance as a measure of dispersion. Squaring the deviations has the unfortunate feature of changing the units: For example, if d is in inches, d^2 is in square inches. Thus for a sample of observations expressed in inches, we would have its location $\bar{x}$ expressed in inches and its dispersion, variance, in square inches. But this can easily be remedied by taking the square root of the variance and thus restoring the original units. The square root of the variance is called the *standard deviation*, and it is customarily denoted by the letter s with the formula

$$s = \sqrt{\frac{\Sigma fd^2}{n}} \qquad\qquad \textbf{FORMULA 5}$$

The standard deviation, then, is a measure of dispersion that (1) reflects the degree to which *all* the items in the distribution deviate from the central value while (2) at the same time avoiding the difficulties inherent in absolute values, and (3) is expressed in the original units of the sample.

Samples *A* and *B* have respectively the standard deviations $\sqrt{11.64} = 3.41$ and $\sqrt{38.64} = 6.22$. As reflected in the standard deviation, sample *B* is 6.22/3.41 = 1.82 times as widely dispersed as sample *A*. This ratio is intermediate between the ratio 1.5 of the two ranges and the ratio 2.17 of the two mean deviations. As you see, our measures of dispersion do not give quite the same answer to the question: How much more widely spread is sample *B* than sample *A*? Again we are reminded that the statistician must always exercise judgment and choice in committing statistics.

Note: You will find in Appendix B a table of square roots which enables you to obtain such values as $\sqrt{11.64} = 3.41$. The table gives, corresponding to $N = 1.16$, the value $\sqrt{10N} = 3.406$. This means that $\sqrt{11.6} = 3.406$. Similarly, $\sqrt{11.7} = 3.421$. Intermediate values can be found by interpolation. Since 11.64 is four-tenths of the way from 11.6 to 11.7, $\sqrt{11.64}$ will be four-tenths of the way from 3.406 to 3.421, or $\sqrt{11.64} = 3.412$. If two-decimal accuracy suffices, round off 3.412 to 3.41.

The appendix table deals directly with numbers from 1.00 through 99.9. For larger or smaller numbers, just remember that $\sqrt{100} = 10$, that $\sqrt{0.01} = 0.1$, and so forth. *Thus*

$$\sqrt{2830} = \sqrt{28.3 \times 100} = \sqrt{28.3} \times \sqrt{100} = 5.320 \times 10 = 53.20$$

Similarly,

$$\sqrt{0.0179} = \sqrt{1.79 \times 0.01} = \sqrt{1.79} \times \sqrt{0.01} = 1.338 \times 0.1 = 0.1338.$$

COMPUTATIONAL SHORTCUTS FOR THE STANDARD DEVIATION

Because the standard deviation is the most widely used of all measures of dispersion, a number of procedures have been developed to make it easy to compute. To begin with, in order to obtain Σfd^2 one does not need to calculate all the d values and then square each one. It can be shown, by simple algebra, that $\Sigma fd^2 = \Sigma fx^2 - n\bar{x}^2$. This permits us to work directly with the x values rather than first converting each x value into a d. Making the substitution in Formula 5, we get

$$s = \sqrt{\frac{\Sigma fx^2 - n\bar{x}^2}{n}}$$

which can be simplified to yield the following formula:

$$s = \sqrt{\frac{\Sigma fx^2}{n} - \bar{x}^2}$$ FORMULA 6

You should check that, for sample A,

$$\Sigma fx^2 = 14^2 + 18^2 + 2(19)^2 + 20^2 + 3(21)^2 + 23^2 + 28^2 = 4278$$

and since $\bar{x}^2 = (20.4)^2 = 416.16$, therefore

$$s = \sqrt{\frac{4278}{10} - 416.16} = \sqrt{427.8 - 416.16} = \sqrt{11.64} = 3.41$$

This result agrees perfectly with the more laborious direct calculation. Formula 6 not only avoids having to calculate all the d's but also often permits us to work with whole numbers (which fx^2 are likely to be) instead of decimals (which is often the case with fd^2).

Coding will help still more. Let us first discuss the method for ungrouped data (with class intervals of width $w = 1$). First let us see what coding does to the deviation, $d = x - \bar{x}$. As we mentioned in Unit 7, when a number C is subtracted from each x value, the mean $\bar{x}$ gets reduced by the same amount C. As a result, when we code by subtracting C from each x value, there is *no* effect on the deviation: that is, $d' = (x - C) - (\bar{x} - C) = x - \bar{x}$. This means, of course, that the deviation for any given x has the same value in the original and coded units. Since each $d = d'$, each $d^2 = d'^2$ and therefore $\Sigma fd^2 = \Sigma fd'^2$. The net effect is that coding (when $w = 1$) does not change the value of the standard deviation: in other words, $s = s'$.

The fact that the coded x values yield the same standard de-

TABLE 13 Computation of standard deviation for ungrouped data, by coding method, in hypothetical sample A, $C = 20$

x	f	x'	fx'	fx'^2
14	1	−6	−6	36
18	1	−2	−2	4
19	2	−1	−2	2
20	1	0	0	0
21	3	1	3	3
23	1	3	3	9
28	1	8	8	64
	10 =		4 =	118 =
	n		$\Sigma fx'$	$\Sigma fx'^2$

$$\bar{x}' = \tfrac{4}{10} = 0.4$$

$$s' = \sqrt{\tfrac{118}{10} - (0.4)^2} = \sqrt{11.64} = 3.41$$

$$s = s' = 3.41$$

The same data appear in Table 12.

TABLE 14 Computation of standard deviation—ungrouped or width 1: for STATLAB girls' heights, $n = 40$, $C = 53$

x	f	x'	fx'	fx'^2
48	1	−5	−5	25
49	2	−4	−8	32
50	4	−3	−12	36
51	6	−2	−12	24
52	4	−1	−4	4
53	5	0	0	0
54	3	1	3	3
55	4	2	8	16
56	5	3	15	45
57	2	4	8	32
58	1	5	5	25
59	2	6	12	72
61	1	8	8	64
	40 =		18 =	378 =
	n		$\Sigma fx'$	$\Sigma fx'^2$

$$\bar{x}' = \tfrac{18}{40} = 0.45$$
$$\bar{x} = 53 + 0.45 = 53.45$$

$$s = s' = \sqrt{\tfrac{378}{40} - (0.45)^2} = \sqrt{9.2475} = 3.04$$

The same data appear in Table 9.

viation as the original is illustrated in Table 13 for sample *A*. Note that the final column, fx'^2, is just the product of the two preceding columns, x' and fx'. See how much easier the computations have become thanks to our having made the numbers into small integers by the coding process.

A second illustration is given in Table 14, which shows the coded calculation of the standard deviation for the sample of 40 girls' heights already used in Table 9 of Unit 7 to illustrate the coded calculation of the mean. The calculation is straightforward and yields $s = 3.04$, but it is somewhat laborious because of the large number of x values (13) that must be dealt with. This leads to the final trick for simplifying the calculation: grouping.

When the data are grouped into class intervals of width w other than 1, then for the reasons noted in Unit 7 the deviations d of the original variable are w times as large as the corresponding deviations d' of the coded variable: $d = wd'$ (see especially Fig. 8 in Unit 7). This means that after calculating the standard deviation for the coded variable s', we must multiply it by w to restore the original units to obtain s; that is, $s = ws'$. The process is illustrated in Table 15 for the same sample of girls' heights, but now grouped in intervals of width $w = 2$. Thanks to the various tricks, a troublesome calculation has now become easily manageable.

Note that the results depend slightly on the grouping used ($s = 3.04$ for $w = 1$ and $s = 3.03$ for $w = 2$). As we shall make clear in Unit 15, these differences are of no importance. In general the value of s is little affected by the class interval w, provided that w is not much larger than $\frac{1}{2}s$.

Let us conclude this discussion of the computation of standard deviation by mentioning an alternative approach that is often used. Some calculators are designed to make it easy to compute, at the same time, both Σfx and Σfx^2 without first organizing the data into a frequency table. If you have access to one of these machines, you should try it out for yourself. Check that the 40 unorganized sample values of Table 4 in Unit 4 have the sum $2138 = \Sigma fx$ and that the sum of their squares is $114{,}646 = \Sigma fx^2$. When these values are substituted into Formula 1 of Unit 4 and into Formula 6 of this unit, one finds:

$$\bar{x} = \tfrac{2138}{40} = 53.45 \qquad s = \sqrt{114{,}646/40 - (53.45)^2} = 3.04$$

in exact agreement with the results we got by the coding method.

Even though Σfx^2 tends, as in this example, to be a large number, this approach may be tempting because with it one can calculate s without having first to organize the data into a frequency table. In spite of, or even because of, this attractive feature, we do not recommend the method as a general procedure. Recall that one graph is worth a thousand numbers. A statistician who analyzes his data arithmetically, without examining them visually, is depriving himself of the use of his most perceptive sense. If you therefore

TABLE 15 Computation of standard deviation, coding method, data grouped into class intervals of width 2: STAT-LAB girls' heights, $n = 40$, $C = 52.5$

Class	f	x'	fx'	fx'^2
48–49	3	−2	−6	12
50–51	10	−1	−10	10
52–53	9	0	0	0
54–55	7	1	7	7
56–57	7	2	14	28
58–59	3	3	9	27
60–61	1	4	4	16
	40 =		18 =	100 =
	n		Σfx	$\Sigma fx'^2$

$$x' = \tfrac{18}{40} = 0.45$$
$$\bar{x} = 52.5 + 2(0.45) = 53.40$$

$$s' = \sqrt{\tfrac{100}{40} - \left(\tfrac{18}{40}\right)^2} = \sqrt{2.2975} = 1.515$$

$$s = ws' = 2(1.515) = 3.03$$

The same data appear in the first five columns of Table 10.

organize your data to permit effective *graphing,* you will have to construct a frequency table. Once you have a frequency table, the shortcut methods easily give the desired results.

POPULATION DISPERSION

As was the case with measures of central tendency, the various measures of dispersion may be applied not only to samples but also to entire populations. When this is done, we get the population parameters that correspond to the sample statistics.

For example, the population range is defined as the difference between the maximum and minimum values that occur in the population. The concept runs into difficulties with idealized populations that have no maximum or minimum—such as those portrayed in Fig. 5*a, b,* and *d* in Unit 5. However, any actual population will necessarily have a finite range, although it may be large. For example, the STATLAB population has a reported maximum income (at test) equal to $80,000 (family 44–32) and a minimum of $0 (family 11–66). Thus the range is $80,000. (What do you think the figure would be for all the world's families?)

The mean deviation and variance of a population are defined as the arithmetic mean of $|d|$ and d^2, respectively, for all numbers of the population. The population standard deviation is the square root of the population variance. This last parameter is so important that a special symbol is reserved for the population standard deviation: σ, the lowercase Greek sigma.

Let us illustrate these ideas on the STATLAB population of the heights of 648 girls from which we drew the sample of 40 analyzed in Tables 14 and 15. These 648 heights are shown in Table 16 (rounded to the nearest inch) along with the coded calculation of population mean μ, population variance, and population standard deviation σ. The population range is $62-46 = 16$ in; the population standard deviation is $\sigma = 2.61$. (We do not show the work, but the population mean deviation is 2.12.) You may wish to make sure that you understand this technique by verifying that $\sigma = 2.60$ for the 648 STATLAB boys' heights given in Table 7 of Unit 4.

Of course, one does not ordinarily have access to the distributions of the populations one is studying, and thus one must rely on sample statistics as estimates of the unknown population parameters. For our sample of 40 girls' heights, the standard deviation (Table 14) is $s = 3.04$. This value is somewhat above the corresponding population parameter of 2.61. Apparently we happened by chance to get a sample somewhat more spread out than the population. In spite of this, the sample range, 13, is well *below* the population range of 16. There is a good reason for this.

Recall from Unit 4 the law of large numbers, which says that a large sample will tend to have a distribution that resembles the

TABLE 16 Population of 648 STATLAB girls' heights, $C = 54$

x	f	x'	fx'	fx'^2
46	1	−8	−8	64
47	3	−7	−21	147
48	11	−6	−66	396
49	25	−5	−125	625
50	67	−4	−268	1072
51	62	−3	−186	558
52	92	−2	−184	368
53	84	−1	−84	84
54	100	0	0	0
55	73	1	73	73
56	61	2	122	244
57	35	3	105	315
58	19	4	76	304
59	8	5	40	200
60	4	6	24	144
61	1	7	7	49
62	2	8	16	128
	648		−479 =	4771 =
			$\Sigma fx'$	$\Sigma fx'^2$

$$\mu' = \frac{479}{648} = -0.74$$
$$\mu = 54 - 0.74 = 53.26$$
$$\text{Var} = \frac{4771}{648} - (-0.74)^2$$
$$= 6.82$$
$$\sigma = \sqrt{6.82} = 2.61$$

The heights recorded in the Census, pages 11 to 36, have been rounded to the nearest inch.

population distribution. Among other things, this means that for each deviation d, the fraction of the *sample* having that deviation will be close to the fraction of the *population* that has it. It follows that the mean value of d^2 in the sample will tend to be close to the mean value of d^2 in the population, and thus the sample s will tend to be close to the population σ.

In contrast, the law of large numbers does *not* assure us that the sample range will tend to be close to the population range. As is the case with STATLAB income, the population range will always depend only on the extreme population values. Because these values are so rare, it is highly probable that they may escape being caught in a random sample, even in a large sample which is still but a small fraction of the population.

COMPARISON OF THE MEASURES OF DISPERSION

As we pointed out earlier, the sample range is used in applications where a premium is placed on rapid, easy calculation. A good illustration of this application is found in quality control work. There a measure of the dispersion of a small sample from *each* production batch must be plotted on a control chart. Since the shop foreman can easily calculate the range in his head, it is used in preference to a statistic such as mean deviation or standard deviation that requires more extensive calculations. The range is not, however, to be recommended for analysis of important data when time permits a careful analysis. Not only may the range fail to approach the corresponding parameter as the sample gets large, but it is also excessively sensitive to disturbance by gross errors, which tend to fall in the tails of the sample.

Although the mean deviation is a sensible measure of dispersion, the standard deviation has shouldered it out of the statisticians' affections and has taken over the field almost exclusively. This is partly because the standard deviation is so useful in many mathematical analyses. Some of these mathematical analyses have shown that the standard deviation works especially well for samples from normal distributions (see Fig. 5a of Unit 5). Many statisticians think that most actual populations are pretty close to this normal form, and for them the standard deviation has added attractiveness. Unfortunately, the standard deviation does not work well for populations with one or both tails substantially heavier than the normal (for example, Fig. 4f, k, and p of Unit 5).

☐   You will now have a chance to compute the standard deviation of your own samples, check out some of the characteristics of other measures of dispersion, and then combine your results with those of others in the class to see how the statistic s behaves in various circumstances.

WORK TO BE DONE

STEP 1   Retrieve from Databank, File B-2, the ungrouped frequency table for your sample of mothers' heights, and calculate its standard deviation s (by the coding method). Round s to the nearest tenth of an inch.

STEP 2   Do the same for the sample grouped in class intervals of width 2.

STEP 3   Now retrieve from Databank, File C-1, your sample of family incomes. Calculate its s on page 2 of the worksheet. Round s to the nearest $500.

STEP 4   Enter into Databank, File D-1, the standard deviations for your sample of mothers' heights, for both $w = 1$ and $w = 2$, and for your sample of family income, $w = 2$.

STEP 5   Complete the inquiry.

Note: The instructor will arrange for the members of the class to combine their standard deviations of ungrouped mothers' heights (rounded to the nearest tenth of an inch) and of income (rounded to the nearest $500) into class distributions. When these are made available, enter them into Databank, File D-1, for use in Unit 15.

THE NORMAL DISTRIBUTION

When introducing the normal distribution in Unit 5, we mentioned its central role in the theory and practice of statistics. It has been said that the normal distribution enjoys the respect of practical statisticians because they believe it has been established theoretically, and the approval of theoreticians who understand it arises in practice. Both are right—within limits.

Scientists have investigated a wide variety of variables and have found that many (but not all) have frequency polygons that resemble the normal curve. As we saw in Fig. 4a to d of Unit 5, good approximations to the normal curve can be found among such diverse phenomena as IQ scores, age of bridegrooms, butterfat production of cows, and mean June temperatures at Stockholm. As we shall see, several of our STATLAB variables are also very nearly normally distributed.

It is no mere accident of nature that the normal distribution applies to many empirically measured variables. There are good theoretical reasons why this distinctive shape *must* occur under certain circumstances. Indeed, during the eighteenth and nineteenth centuries (the normal curve was described by Abraham De

Moivre about 1730), so impressed were the theoreticians with these reasons that many attempts were made to prove that the mathematics underlying this particular shape governed *all* proper distributions of continuous variables—other shapes were seen as aberrations from this basic form. And it was for this reason that the name *normal* was applied to this distribution. In Unit 13 we shall present, illustrate, and test a theorem which in a sense explains why the normal form is so often found in practice.

The same theoretical considerations which help us to understand why the normal curve is so frequently found in nature also furnish us with powerful aids for statistical analysis, e.g., using samples to estimate population parameters, to test the validity of hypotheses, and to deal with many other statistical problems.

To see how all this is possible, it is first necessary to become acquainted with certain key properties of the normal distribution curve. While this is the primary objective of this unit, you will also have an opportunity to get an indication, by examining population data and by drawing new samples, of how well the normal curve and its properties apply to certain STATLAB data.

PROPERTIES OF THE NORMAL DISTRIBUTION

The term *normal* applies to the *shape* of the distribution and does not restrict its location or dispersion. That is, a normal distribution can have any mean μ and any standard deviation σ.

Figure 12 illustrates three different normal frequency curves: distribution A has $\mu=6$, $\sigma=2$; B has $\mu=16$, $\sigma=3$; C has $\mu=25$, $\sigma=1$. In each case we show the cumulative frequency curve directly below the corresponding simple frequency curve. Each cumulative curve rises to 100 on the $F\%$ scale, reflecting the fact that each distribution accumulates all 100 percent of its values. Correspondingly, each simple frequency curve has under it a total area of 100 percent. For the normal curve, as for all idealized distribution curves, we always use $f\%$ and $F\%$ on the vertical axes rather than f or F. The reason for this is that given in Unit 4: to make comparisons among curves easier. When σ is doubled, as in going from distribution C to distribution A in Fig. 12, the height of the curve is cut in half; since the values are twice as widely spread out, they occur with only half the frequency at any given point on the curve.

Each normal distribution is symmetric, from which it follows (Unit 8) that the mean μ is also the median $\tilde{\mu}$, so that half a normal population lies on each side of the mean. This fact is illustrated in Fig. 12, where each cumulative curve reaches height $F\% = 50$ just at μ. Note also that the normal distribution is unimodal, with its mode at μ. That is, for a normal distribution, mode = median = mean.

In its central portion the simple normal frequency curve is

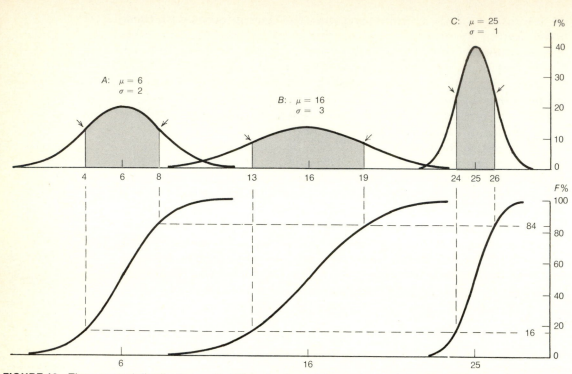

FIGURE 12 Three normal distributions with different values of the mean μ and standard deviation σ. The dashed lines relate the inflection points (arrows) of the simple frequency curve to the points where $F\%$ equals 16 and 84 on the corresponding cumulative frequency curve drawn just below it.

curved downward, while in both tails it is curved upward. The points at which the direction of curvature changes are called *points of inflection;* they are indicated on Fig. 12 by small arrows. An important feature of the normal curve is that its points of inflection always lie at exactly the distance σ from μ; that is, one inflection point occurs at a distance of one σ unit to the left of μ and the other, one σ unit to the right of μ. Thus in Fig. 12 distribution B has its inflections at $\mu - \sigma = 16 - 3 = 13$ and at $\mu + \sigma = 16 + 3 = 19$.

AREA UNDER THE NORMAL CURVE
On each of the three distributions of Fig. 12 we have shaded the portion of the simple frequency curve extending from $\mu - \sigma$ to $\mu + \sigma$. As you can see, the shaded area is about two-thirds of the total: More precisely, it has been mathematically determined to be 68.27 percent of the total area; i.e., this area contains within it 68.27 percent of all the cases in the population.

From the symmetry of the normal distribution certain other areas, or frequencies of measures, can be at once derived from the fact that approximately 68 percent falls between $\mu - \sigma$ and $\mu + \sigma$, as is illustrated in Fig. 13. Half the 68, or 34 percent, must be between μ and $\mu - \sigma$, and another 34 between μ and $\mu + \sigma$. Since there is 68 percent inside the interval from $\mu - \sigma$ to $\mu + \sigma$, there must be 32 percent in the two tails outside this area and hence by symmetry 16 percent in each tail. Thus there is 16 percent to the left of $\mu - \sigma$ and $16 + 68 = 84$ percent to the left of $\mu + \sigma$. These facts are also brought out by the dashed lines of Fig. 12.

Study of Fig. 13 will convince you that most of the frequency of a normal curve will fall within *two* σ units of its mean μ. In fact, between $\mu - 2\sigma$ and $\mu + 2\sigma$ we always find 95.45 percent of the cases. Therefore there must be 4.55 percent outside these limits, 2.28 percent to the left of $\mu - 2\sigma$, 97.72 percent to the left of $\mu + 2\sigma$, and so forth.

One may similarly discuss intervals that extend any given number of σ units on either side of μ. Let us denote the number of σ units away from μ by the letter z. It is possible to compute the area A under a normal distribution that lies within $\mu - z\sigma$ to $\mu + z\sigma$. Here are a few such values:

z	0	0.5	1.0	1.5	2.0	2.5	3.0	3.5	4.0
A	0	0.3829	0.6827	0.8664	0.9545	0.9876	0.9973	0.9995	0.9999

The area 0.8664 corresponding to $z = 1.5$ means that 86.64 percent of the cases fall between $\mu - 1.5\sigma$ and $\mu + 1.5\sigma$. Because of the symmetry of the normal curve, this implies that 43.32 percent of the cases fall between μ and $\mu + 1.5\sigma$. To take another example, since the total area under the curve comprises 100 percent of the cases,

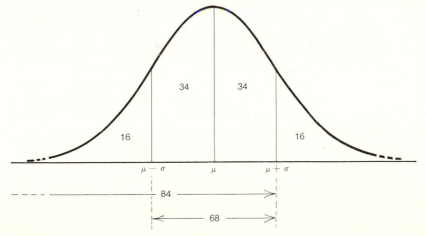

FIGURE 13   Areas (in percent) of a normal curve as related to μ and σ.

the entry $A = 0.9876$ at $z = 2.5$ means that 1.24 percent of the cases fall *outside* $\mu - 2.5\sigma$ and $\mu + 2.5\sigma$. Hence 0.62 percent lie to the *left* of $\mu - 2.5\sigma$. An extensive *normal table* giving areas for various values of z may be found in Appendix B.

As Fig. 13 suggests, the tails of the normal curve extend without limit in both directions, but they are very thin. Both these aspects are unrealistic of actual populations. Any existing population, no matter how large, must have a finite minimum and maximum, so in reality the tails of every population are finite. But however far the tails of an actual population may extend, they are likely to be heavier than the very thin tails of the idealized normal distribution. The ideal normal distribution allows only one case in ten thousand to be as much as 3.7 σ units away from μ, and less than one in a million more than 4.8 σ units away. The real world is more ornery than that. In large populations there are likely to be several individuals who, for one reason or another, are aberrant and have values far from the average. In some cases, these discrepancies are rather minor in relation to the total population, which resembles the normal distribution closely enough to make the discrepancies unimportant.

The areas, as given in the normal table in Appendix B, are of course computed specifically for the normal shape. For distributions close to the normal, however, these areas will be approximately correct, and they give us a useful crude indication even for very nonnormal distributions. Thus for the skewed distribution pictured as Fig. 5b (Unit 5), the area between $\mu + \sigma$ and $\mu - \sigma$ (which we shall write as $\mu \pm \sigma$) is 70 percent rather than the 68 percent of the idealized normal curve. To take more extreme examples of nonnormality, the exponential distribution (Fig. 5c, Unit 5) has 83 percent, and the uniform distribution (Fig. 5d, Unit 5) 58 percent, in the $\mu \pm \sigma$ interval. The great majority of distributions encountered in practice have something like 60 to 80 percent of their cases within the area encompassed by $\mu \pm \sigma$. Similarly, in the interval $\mu \pm 2\sigma$ there will usually be 90 percent or more of the population.

Much the same sort of thing holds true for samples. Thus for samples with distributional shape not too unlike the normal, one typically finds something like 68 percent of the sample between $\bar{x} \pm s$ and something like 95 percent between $\bar{x} \pm 2s$—more or less. You will have a chance to check this out on your own samples.

CHECKING THE NORMALITY OF DISTRIBUTION SHAPE

Many statistical methods of analysis have been specifically designed for samples drawn from normal populations. If you were to use those methods for distributions which are not normal, the results obtained would be questionable to the degree to which your data depart from normality. This means that the honest and wise statistician should first be satisfied that the data are normal (or close enough) before using such methods. How can one judge whether or not the distribution of a population or of a sample has approximately the normal shape? The straightforward way is to plot its simple or cumulative frequency polygon and compare it visually with the normal shapes shown in Fig. 12. This visual comparison (which of course is a subjective and judgmental affair) is best done by people who are familiar with the *kind* of data in question and who know how such data usually behave. Let us illustrate. Recall that the distribution of height of the population of 648 STAT-LAB girls was given in Table 16 of Unit 9. The simple and cumulative frequency polygons are shown in Fig. 14 for this population, grouped into intervals of width $w = 2$ in.

If you study these two polygons and compare them with the curves shown in Fig. 12, you will see that these curves have the normal shapes, at least to a good approximation. In other words, the heights of our girls are very nearly normally distributed. We can have confidence in this impression of normality because, as we shall see in Unit 13, it is reasonable for heights to be normally distributed. In contrast with heights, human weights do *not* tend to be normal, as you will be convinced after a little reflection. For example, consider the women of age 20 to 50 you have met. Their median weight is probably somewhere around 130 lb. You probably know at least one who weighs 100 lb more than that median. However, you probably do not know anyone who is even as much as 50 lb below the median. Clearly the right tail of the weight distribution of such women is much longer than the left.

Figure 15 illustrates this phenomenon for the weights (at the time their pregnancy was first diagnosed) of the population of STAT-LAB mothers. If you study the simple polygon, you will see that although the main part of the distribution resembles the normal more or less, the polygon has a long right tail. This tail extends even farther than shown on the figure: One mother weighed 265 lb—which is 138 lb above the median weight. In contrast, the left tail of the distribution is quite short: The minimum weight in this population (88 lb) is only 39 lb below the median.

Consider now the cumulative polygon of Fig. 15. It also resembles a normal cumulative curve in a general way, but a careful examination shows that it rises to $F\% = 100$ more slowly on the right than it falls to $F\% = 0$ on the left. This is of course a reflection of the long and heavy right tail and short left tail of the simple polygon. The weight of STATLAB mothers is distributed something like a normal curve, but it departs from the normal shape in being skewed to the right. Here again, a visual inspection of the distribution curve, and what we know about weights in general, lead to the

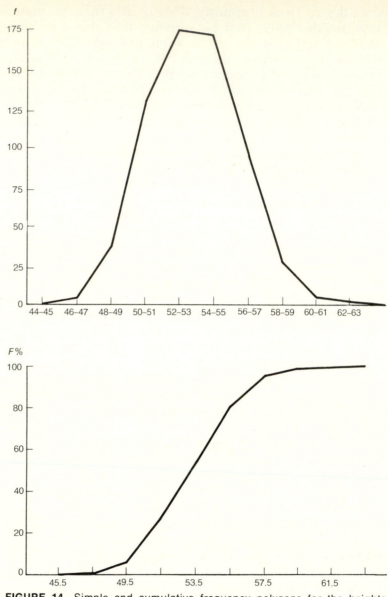

FIGURE 14 Simple and cumulative frequency polygons for the heights (in inches) of the 648 STATLAB girls as given in Table 16.

conclusion that the distribution of our data is not normal. The same is true of weight of STATLAB fathers, but their weight is skewed to a lesser extent. Does your everyday observation support the idea that adult men generally have a more nearly normal weight distribution than do adult women?

Statisticians and medical scientists who work with biometric data soon learn that, generally speaking, heights have a nearly normal distribution whereas weights tend to be skewed to the right. Similarly, those working with data in other fields come to know the distributional shapes that characterize the variables of interest to them. This kind of experience makes it possible to choose statistical tools appropriate for analyzing the data. And all this is important because methods that work very well with nearly normal data may tend to be unsatisfactory for samples from populations with one or two very long tails.

FIGURE 15 Simple and cumulative frequency polygons for the weights (in pounds) of the 1296 StatLab mothers.

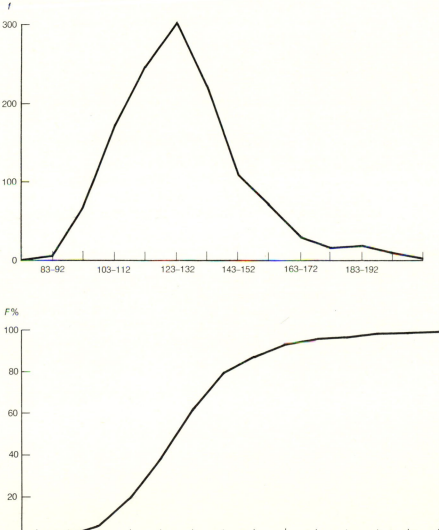

☐ You are now ready to check some of these ideas on two quite different kinds of STATLAB measures: the heights of STATLAB boys and the mental test scores of STATLAB boys and girls. In doing this you will attempt to determine whether heights and mental test scores seem to be distributed in the same way as the fat content of milk from cows and the age of California bridegrooms who married brides who were 40 to 44 years old in 1961.

WORK TO BE DONE

STEP 1   Draw a fresh sample of $n = 40$ STATLAB families, and record on the worksheet, page 1, the child's Peabody and Raven scores.

STEP 2   On page 3 of the worksheet make grouped frequency tables, $w = 5$, for the variables. (Use intervals 0–4, 5–9, 10–14, and so on.)

STEP 3   Compute the mean $\bar{x}$ and standard deviation s for your two grouped samples.

STEP 4   On page 2 of the worksheet plot cumulative frequency polygons of your Peabody and Raven scores, $w = 5$.

STEP 5   From your samples as recorded on page 1 of the worksheet, tally on page 4 of the worksheet the number of Peabody scores that fell within $\bar{x} \pm s$, and also the number that fell within $\bar{x} \pm 2s$. Do the same for the Raven scores.

STEP 6   Enter into Databank,

     File E-1:   ID numbers, Peabody and Raven scores for your new sample (see step 1 above).
     File E-2:   Grouped frequency tables (see step 2 above); $\bar{x}$ and s for Peabody and Raven scores (see step 3 above).
     File E-3:   Cumulative frequency polygons (see step 4 above).

STEP 7   Complete the inquiry.

GRAPHIC METHODS RELATED TO THE NORMAL DISTRIBUTION

Because of the central importance of the normal distribution in the theory and practice of statistics, much effort has gone into devising statistical methods appropriate for normal data. In this unit we shall show you two graphic methods that are very helpful in handling populations and data distributed in something like a normal way. Most data are close enough to the normal to permit these methods to be useful, so they are well worth learning. The techniques you will learn in this unit are among the most useful tools in the armamentarium of the applied statistician.

NORMAL PROBABILITY PAPER

As you have seen in Unit 10, one can make a fairly good judgment about the degree of normality of a distribution by carefully studying its cumulative frequency polygon and comparing it with the normal cumulative curve (Fig. 12, Unit 10). It is, however, easy to be deceived when judging the precise agreement of two curves by eye. But by constructing your frequency polygon on *normal probability paper* a visual inspection can be made highly precise.

91

Normal probability paper is a graph paper that has an ordinary scale on the horizontal or x axis but a special scale for $F\%$ on the vertical axis. This special scale has been arranged in such a manner that it *straightens out* a normal cumulative curve. That is, the normal cumulative curve that appears as an S-shaped curve when plotted on the usual graph paper becomes a perfectly straight line when plotted on normal probability graph paper. (You will see, in a moment, how this is done.) Accordingly, if a cumulative polygon is nearly normal its plot will become nearly straight. By plotting in this way we gain a great deal, because we can judge quite easily and with great confidence, by eye, whether or not a line is straight. Further, if we see that the line is not straight we can easily see the direction and extent of its departure from a straight line.

Figure 16 shows the cumulative frequency polygon of the heights of the population of STATLAB girls. The polygon is plotted on ordinary graph paper on the left (as was already done in Fig. 14 of Unit 10) and on normal probability paper on the right. A glance at the right-hand plot is enough to assure us that the graph is very nearly a straight line—and hence that the height of these 648 girls has a very nearly normal distribution. (The linearity is so good that it takes a careful look to see a slight curvature in the downward direction. Lay a straightedge alongside the plot to make the curvature easier to see.)

FIGURE 16 Cumulative frequency polygons for the heights (in inches) of the 648 STATLAB girls. On the left, the population distribution is plotted with $F\%$ on the ordinary scale, as was already done in Fig. 14; on the right, the same distribution is shown with $F\%$ on the normal probability scale.

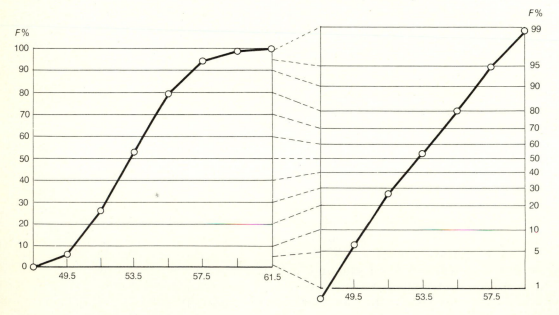

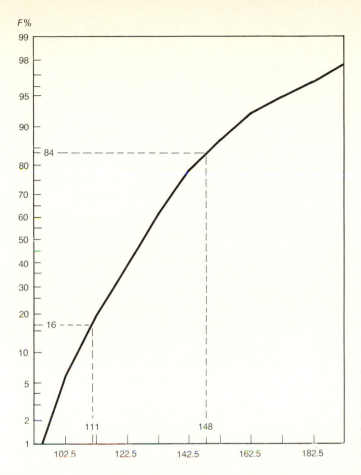

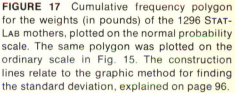

FIGURE 17 Cumulative frequency polygon for the weights (in pounds) of the 1296 STAT-LAB mothers, plotted on the normal probability scale. The same polygon was plotted on the ordinary scale in Fig. 15. The construction lines relate to the graphic method for finding the standard deviation, explained on page 96.

The situation is quite different with mothers' weights, shown in Fig. 17. As we noted in Fig. 15 in Unit 10, the plot of the cumulative polygon on ordinary paper looks a good deal like a normal cumulative curve, but it shows on careful examination some nonnormal features. The graph rises to $F\% = 100$ more slowly than it falls to $F\% = 0$, reflecting the long right tail of the weight distribution. But see how much more compellingly this phenomenon is shown on the normal probability paper graph. The plot is at once seen to be not at all straight but strongly curved downward. And this curvature holds not just for the upper part of the curve, corresponding to the few very heavy mothers, but is visible throughout the entire distribution.

The normal probability scale used on the vertical axis of normal paper does not reach up to $F\% = 100$ or down to $F\% = 0$ because the tails of the idealized normal curve are infinite. As you will recall from Unit 5, on an ordinary plot of a cumulative polygon the leftmost segment does go down to $F\% = 0$ while the rightmost does go up to $F\% = 100$. These two extreme segments cannot be shown on normal paper, and in practice one often omits all values above (say) 99 percent or below 1 percent.

Figures 16 and 17 show the plots on normal probability paper of entire populations. The paper is also useful for plotting samples. Of course, when a small sample is plotted there are bound to be some departures from linearity, even if the population itself is normal, because of sampling fluctuations. To help give you a feeling for the magnitude of such departures, we show in Fig. 18 the normal probability paper plots of 19 samples of size $n = 40$, drawn from the population of girls' heights, which is plotted in the upper left-hand corner of the figure. If you spend some time studying these figures, you will be better prepared to judge intelligently whether one of your own samples is so far from the normal shape as to justify the conclusion that it probably came from a nonnormal population.

Normal probability paper can be bought in an engineering supply store. However, you will find printed on the inside back cover of this book two normal probability scales of different sizes which you can use to convert, easily and quickly, any ordinary (and cheaper) sheet of graph paper into normal probability paper.

GRAPHIC METHOD FOR THE STANDARD DEVIATION

Even when one takes advantage of such shortcut devices as grouping and coding, it is a fair amount of work, as you have seen, to compute the standard deviation s of a sample or σ of a population. If the distribution is reasonably close to the normal shape, there is a very easy way to determine the standard deviation graphically, at least to a good approximation. As a by-product, this graphic method also yields an approximate value for the mean.

Let us state the method first for the σ of an entire normal population. The method relies on the facts, mentioned before (see Fig. 13, Unit 10), that 16 percent of a normal population lies to the left of $\mu - \sigma$ and 84 percent lies to the left of $\mu + \sigma$. By spotting the points where $F\% = 16$ and $F\% = 84$, one can read off from the graph of the cumulative curve the points on the horizontal axis equal to $\mu - \sigma$ and $\mu + \sigma$. The mean μ is located midway between these points; σ is equal to the distance between μ and $\mu + \sigma$.

The graphic method is illustrated in Fig. 19 for the population of 648 girls' heights. Although the method can be used with a plot on ordinary graph paper, the necessary visual interpolations are usually more accurate when applied to the nearly linear plot on normal probability paper. We have marked on the normal probability paper graph the points corresponding to $F\% = 16$ and $F\% = 84$. These points are respectively above $x = 50.7 = \mu - \sigma$ and $x = 55.9$

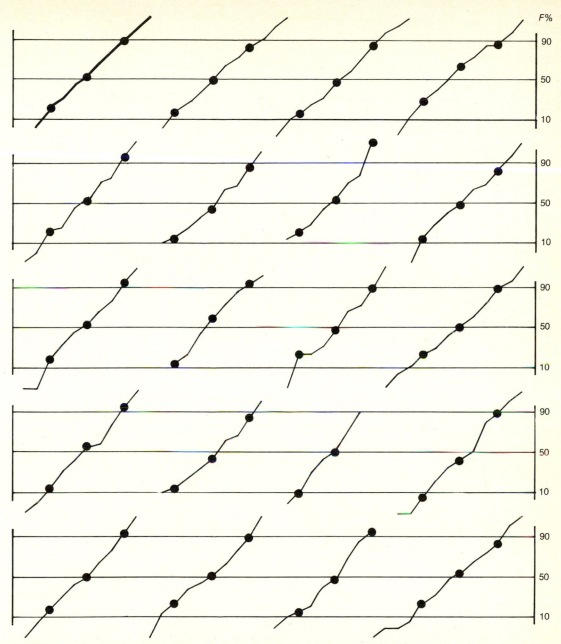

FIGURE 18 Cumulative frequency polygons, plotted on the normal probability scale, for 19 samples from the same population. In the upper left is the population distribution of the 648 STATLAB girls' heights (already shown in Fig. 16). The other polygons represent 19 samples, each of size 40, drawn from this population. The dots correspond to the true class limits 50.5, 53.5, and 56.5 in.

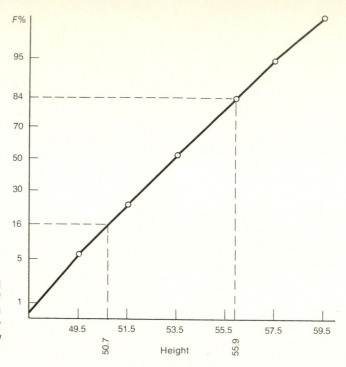

FIGURE 19 The graphic method for standard deviation, illustrated on the population of 648 STATLAB girls' heights plotted on the normal probability scale. The dashed lines show how $F\% = 84$ corresponds to height 55.9 in and how $F\% = 16$ corresponds to height 50.7 in.

$= \mu + \sigma$ on the x axis. The midpoint of these two x values yields the mean, that is, $\mu = (55.9 + 50.7)/2 = 53.3$; the distance between $\mu = 53.3$ and $\mu + \sigma = 55.9$ is $55.9 - 53.3 = 2.6 = \sigma$. These results, attained with so little effort, are quite close to the values $\mu = 53.26$ and $\sigma = 2.61$ calculated by the coding method (Table 16 in Unit 9). They are accurate enough for most practical purposes and in any case provide a very easy check on the arithmetic.

Although the graphic method for σ and μ is derived from properties of the normal distribution, it works well enough to give a useful check for many population distributions that are quite non-normal. In particular, it will give reasonably good results with most skewed distributions. You can check this phenomenon in Fig. 17 for the highly skewed distribution of mothers' weights. You should construct the appropriate lines on the normal probability paper plot and get $\mu + \sigma = 148$ and $\mu - \sigma = 111$, leading to $\mu = 129.5$ and $\sigma = 18.5$. The true values are $\mu = 130.8$ and $\sigma = 22.8$. The agreement is close enough for the check to be useful.

The graphic method can also, of course, be applied to a sample if its distribution is shaped something like a normal. Its cumulative frequency polygon will reach $F\% = 16$ somewhere near $\bar{x} - s$ and reach $F\% = 84$ somewhere near $\bar{x} + s$. From the sample cumulative frequency plot (preferably on normal probability paper) one can read off $\bar{x} + s$ and $\bar{x} - s$, and then at once compute $\bar{x}$ and s. For an example (where for a change we shall use the cumulative frequency curve plotted on ordinary graph paper), turn to Fig. 3a in Unit 5. For our sample of 40 girls' heights, it appears that $F\%$ reaches 16 at $x = 50.6$ and 84 at $x = 56.4$. Therefore, approximately, $\bar{x} - s = 50.6$ and $\bar{x} + s = 56.4$. It follows that $\bar{x} = 53.5$ and $s = 2.9$. (The computed values are $\bar{x} = 53.45$ and $s = 3.04$.)

Even if you intend to compute mean and standard deviation arithmetically, you should first get the graphic values. As we have repeatedly urged, the data should always be graphed. Once this has been done, it takes only a few seconds to read off the graphic values of $\bar{x}$ and s. Graphic values give reassuring checks on the numerical results.

☐ You are now ready to see for yourself how adequately these graphic methods can determine the degree of normality of the distributions yielded by the heights of boys in the STATLAB population and that of your samples of family incomes and children's mental test scores. You will also be able to check out, with your own data, the speed and precision with which graphic methods can estimate means and especially standard deviations for various distributions. Once you have learned to do this easily, you will find it quite feasible to get an approximate value of s even for rough-and-ready analyses.

WORK TO BE DONE

STEP 1 Plot the cumulative frequency polygon of the heights of the population of STATLAB boys (given as Table 7 in Unit 4) on the normal probability paper provided on the worksheet. Read off the values of μ and σ by the graphic method.

STEP 2 Using the frequency table of your sample of 40 incomes (see Databank, File C-1), plot this distribution in the space provided in the worksheet. Read off the values of $\bar{x}$ and s by the graphic method.

STEP 3 Plot your sample of Peabody scores, $w = 5$, from the cumulative frequency table recorded in Databank, File E-2. Read off the values of $\bar{x}$ and s by the graphic method.

STEP 4 Complete the inquiry.

CONVERTED SCORES: RANKS, PERCENTILES, STANDARD SCORES

This unit applies some of our findings about the nature of x distributions, graphic methods, standard deviations, etc., to two major practical problems. The first is how to interpret a *single x* value. To illustrate: The school psychologist must evaluate the mental test score of a *specific* 8-year-old; the manager of a chain of grocery stores worries about the earnings of a *particular* store; the doctor wants to assess the performance of *his* 82-year-old patient's heart. The second practical problem of concern is how to form a *composite score* made up of several x values taken from widely differing distributions. A moment's thought will impress you with the fact that most of our mental tests, indices of social status, economic activity, etc., are composite scores based on seemingly quite disparate measures. Thus an index of a nation's economic welfare may reflect, in a single value, the number of railroad car loadings, tons of steel produced, new buildings constructed, and bankruptcies filed.

The objectives of this unit are to examine these problems, consider some of the solutions proposed, and then test these solutions in deriving a composite mental test score for your sample of StatLab children and evaluating their performances on it.

STANDARDS, REFERENCE GROUPS, COMMON SCALES

To assess an individual's performance is to make a comparison with some relevant standard. Thus in judging a specific 8-year-old child, the psychologist uses as a standard other 8-year-old children (rather than, say, 17-year-olds). The manager's worries will be allayed (or intensified) only after he compares his ailing branch store's earnings with the other grocery stores in his chain (or similar chains)—not with the earnings of hardware stores in Istanbul. And the doctor asks whether his 82-year-old patient's heart is pumping as much blood as an octogenarian's heart should. In other words, the standard against which to assess an individual's performance is the performances of the group to which the individual belongs, his *reference group*.

In deriving composite scores we are often faced with the problem of adding and comparing incompatibles. Consider our economist: He must somehow express tons of steel and number of bankruptcies in units of comparable magnitudes lest in his composite score the hundreds of millions of tons of steel manufactured entirely swamp the effect of the equally important few thousands of bankruptcies reported. This necessity has led to the following suggestion: In constructing a composite score we must first convert x values of widely differing scale units into scores using a common scale.

RANKS

The simplest way to transform x values of varying scale units into scores on a common scale is to arrange the original values in ranks. Thus by assigning the highest value of each x distribution a rank of 1, the second highest 2, and so on down to the lowest value, we bypass the varying magnitudes of the original distributions. An example can make the consequences and uses of ranking clear. In Table 17 we have listed the heights (rounded to the inch) and weights, together with the ranks, of the first six boys of the last page of the STATLAB Census.

TABLE 17   Heights, weights, and ranks of six STATLAB boys

ID no.	Height (inches)	Height (ranks)	Weight (pounds)	Weight (ranks)	Mean ranks
66-11	51	4	61	3	3.5
66-12	53	2	93	1	1.5
66-13	54	1	66	2	1.5
66-14	49	5.5	56	5	5.25
66-15	49	5.5	53	6	5.75
66-16	52	3	58	4	3.5

The heights have been rounded to the nearest inch.

Note in Table 17 that the actual magnitude of the original values and the degree of spread among them play almost no role in the ranks. The difference in weight between boy 66-12, for example, and boy 66-13 is 27 lb while the difference between this second boy and boy 66-11 is only 5 lb, yet the ranks are equally spaced: They receive ranks 1, 2, 3 (column 5).

Next note that ranks permit us to compare scores from different distributions which in their original units are not comparable. Thus a height of *54 in* is equal (in ranks) to a weight of *93 lb*.

Can ranks be used to derive composite scores? The answer is a weak "maybe." Suppose that we wanted to assess our boys with respect to a "body build index" reflecting in equal degree their heights and weights. It would appear that all we need do is to average the two corresponding ranks for any one boy. Thus boy 66-12, who was first in weight and second in height, would receive the same index score as boy 66-13, who was first in height and second in weight (column 6). This seemingly reasonable tradeoff ($1 + 2 = 2 + 1$), however, illustrates the questionable validity of using ranks for composite scores. By giving equal weight to equal rank differences derived from original values which are widely different in magnitude, we may be falling into the error of the cook who, with a recipe calling for a stew made of one-half horse meat and one-half rabbit meat, used one horse and one rabbit. That is, the difference between ranks 1 and 2 on height might *not* be equal to the difference between ranks 1 and 2 on weight—as we shall see on page 105.

Finally, note from Table 17 how *ties* in x values are dealt with. Each of the tied x values is assigned the average of the ranks which would be encompassed by these values. Thus boys 66-14 and 66-15, being tied for *the lowest* of the six height values, must cover between them the two lowest ranks, 5 and 6. Each of these boys is therefore assigned the *mean* of those two ranks: $(5 + 6)/2 = 5.5$ (column 3).

Rank scores have several virtues: They are simple to determine; their meaning is easily grasped; they permit comparison of the performance of one individual on several different tests or of several individuals on one test; and they do yield a rough-and-ready composite score.

However, rank has its disadvantages. By ignoring the variations in magnitude between adjacent ranks on any one measure, a great deal of information is lost; moreover, frequently a conversion into ranks results in so large a number of ties that ranking loses all value; and finally, the method of ranks is convenient only when you are dealing with relatively few observations. When you are dealing with large numbers of observations, you can make use of an elaborated variant of the ranking procedure: the percentile rank.

PERCENTILES

It is often useful to convert the original x distribution into a new scale made up of units of which each contains an equal fraction of the total number of cases. We then assign each x value its new scale value. In fact, we have already used such a scale. The median can be considered as the value which divides an x distribution into a two-unit scale. This permits only the crudest of scaling, of course, since it specifies merely whether a particular x value falls in the upper or the lower half of the scale.

A slightly more differentiated scale can be produced by dividing the distribution into four equal quarters. The values of the three dividing points are called *quartiles*. The first quartile (designated Q_1) separates the bottom quarter of the cases from the upper three-quarters. The second quartile (Q_2) separates the second and the third quarters—and is, of course, equal to the median. The third quartile (Q_3) separates the top quarter from the rest. With this four-unit scale we can specify the quarter within which a particular value lies.

These scales can be of whatever sized units are appropriate to our purposes. Occasionally, we wish to fractionate the distribution into tenths, and here the 9 dividing points among the 10 segments are called *deciles*. Much more commonly, we divide the distribution into 100 equal parts. The values of the 99 dividing points in such a scale are called *percentiles*. The percentiles are designated by the capital letter P and a subscript. For example, P_{15} is the fifteenth percentile, marking off the bottom 15 percent of the cases from the top 85 percent. It should be clear that the following equivalences hold for the preceding scales:

$$Q_1 = P_{25}$$
$$Q_2 = P_{50} = \text{median}$$
$$Q_3 = P_{75}$$

A convenient graphic method for finding percentiles has already been presented in Unit 5. For example, on page 35 we used the cumulative frequency polygon of our sample of 40 girls' heights and found that P_{25} is about 51.0 in. The use of an approximate method for computing percentiles is especially appropriate because there seldom exists an exact answer for a percentile calculation. Suppose, for example, that a statistician wishes to compute the second decile P_{20} for a sample of $n = 47$ x values. Ideally, P_{20} should be a point with 20 percent of the values to its left and 80 percent to its right. If he chooses (in his sample of 47) for P_{20} any point between the ninth and tenth smallest values, there will be $\frac{9}{47}$ or 19.1 percent to the left. If he chooses for P_{20} any point between the tenth and eleventh, there will be $\frac{10}{47}$ or 21.3 percent to the left. The best compromise is to take the tenth observation itself as P_{20}. With this choice, there are 9 to the left and 37 to the right; of these $9+37=46$ values, $\frac{9}{46}$ or 19.6 percent are to the left. That is as close to 20 percent as it is possible to get. The example illustrates the usual situation: One can seldom find a point with exactly a specified proportion of a distribution on either side, though of course if the sample is large one can usually come quite close. (Can you see why, for the special case of the median P_{50}, an exact solution does always exist?)

We can use the percentile scale to specify where a particular value is *ranked* relative to all the cases in the distribution. The *percentile rank* is defined as the percentage of cases in the original distribution which falls *below* a specified value. Suppose that a person's original x value outranks 95 percent of his fellows on the given variable. His x value would be accorded a percentile rank of 95. A person excelling only 2 percent of the group would have a percentile rank of 2.

Percentile ranks are customarily stated in whole numbers, and a common practice is to *round downward*. For example, percentile ranks calculated to be greater than 38 but less than 39 would all be reported as 38 (just as age in years is rounded downward to the last birthday). Percentile ranks must therefore range from a minimum of zero to a maximum of 99. (No value can have a percentile rank of 100 since it cannot exceed *all* values, including itself.)

The simplest way to obtain percentile ranks is to use the cumulative frequency polygon. Thus if we wish to determine the percentile rank of a girl 58 in tall in our population of STATLAB girls, we can use Fig. 14 of Unit 10. We erect a vertical line from the 58-in value on the x axis up to the cumulative curve and then draw a horizontal line over to the percentile scale. At that point we read off the percentile rank, rounding downward. The result is that the girl 58 in tall has a percentile rank of 91. Check to see that a girl of 49 in would have a percentile rank of 2.

Although they are somewhat more complicated to obtain, percentile ranks, like other rank scores, are easily understood. Moreover, they permit the comparison of single individuals on several tests or several individuals on a single test; and—keeping in mind the same reservations made about ranks—they may sometimes be useful for obtaining a quick combined score from two disparate distributions.

Percentile ranks are often applied to individuals who were not members of the original sample on which the percentile values were established. On the college admission test, for example, once a distribution of scores has been obtained by testing a large sample of high school students, any other student's score can then be converted into a percentile rank with respect to that distribution. These percentile ranks, of course, are determined on a *sample* rather than on the entire population. Hence they are *estimates* of the true population percentile ranks.

One of the most effective uses of percentile ranks is in providing a *profile* of an individual's standing on a whole set of quite different variables. Thus the varying percentile ranks earned by a student on a battery of achievement tests will give an informative picture of his comparative strengths and weaknesses. And, of course, his profile can be directly compared and contrasted with similar profiles of other students. We illustrate this by comparing the profiles of two StatLab children on a variety of variables. Arbitrarily we select the first and the last children (11-11 and 66-66) for comparison. For each child the following nine different values (representing physical characteristics of the child and parents, mental attributes of the child, and economic standing of the family) have been converted to percentile ranks, using the entire StatLab population as a reference group: birth length and weight, height and weight 10 years later, score on Peabody and Raven mental tests, mother's age at birth of child, father's age at birth of child, family income at time of birth. The profiles for the two families are depicted in Fig. 20. Note how the similarities and differences of the two children stand out.

For a converted score which has most of the advantages of percentiles and at the same time avoids shortcomings which percentiles share with simple rank scores, we can use *standard scores*.

STANDARD SCORES

Standard scores, or *z* scores as they are usually designated, make it possible to compare individual values from disparate distributions while preserving the degree of spread among the original values. A *z* score is determined by three values: the individual's own *x* value and the two key parameters of his reference population, μ and σ. To obtain a *z* score involves two steps. We first locate the individual in his group by calculating how far, and in what direction, he lies from the center of his reference group, μ, and we then express this individual location in terms of the reference group's variability as measured by σ. This two-step process is summarized by the basic formula

$$z = \frac{x - \mu}{\sigma}$$

FORMULA 7

In effect, *z* scores express deviations in σ units (see Unit 10).

In most cases, of course, we must use $\bar{x}$ and s as estimates for μ and σ, and the formula for *z* scores becomes

$$z = \frac{x - \bar{x}}{s}$$

FORMULA 8

To investigate and illustrate the attributes and uses of *z* scores,

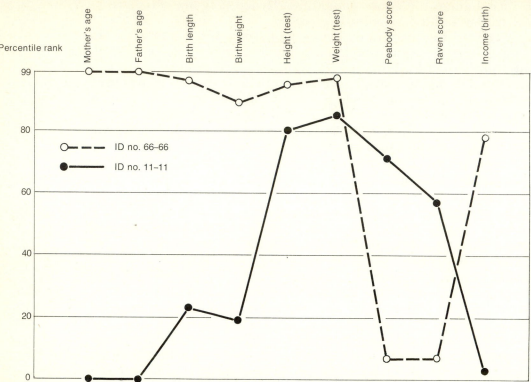

FIGURE 20 The percentile-rank profiles of the first and last children listed in the STATLAB Census, for nine variables.

let us return to our six STATLAB boys of Table 17 and convert their original heights and weights into z scores. In this instance, since we know the relevant STATLAB population values for heights and weights, we shall use μ and σ.

The height of STATLAB boys has a mean of 53.60 in and a standard deviation of 2.60 in; the mean weight is 70.9 lb and the standard deviation is 12.8 lb. To illustrate with boy 66-11, who is 51 in tall: His standard score for height becomes

$$z = \frac{51 - 53.60}{2.60} = -1.00$$

and that for his weight of 61 lb becomes

$$z = \frac{61 - 70.9}{12.8} = -0.77$$

In the same manner the z scores for the remaining five boys can be derived, and Table 18 presents them all.

Many of the advantages of standard scores can be illustrated from Table 18. First note that some z scores are positive and some are negative. This difference in sign carries an immediate and clear meaning: A negative sign means that the original value is *below* the mean of its group, and a positive sign means that it is above. An original value exactly on the mean would, of course, have a z score of zero. If you look further down on page 66 of the STATLAB Census for boy 66-42, you will find that he was 53.6 in tall, giving him a

$$z = \frac{53.6 - 53.60}{2.60} = 0$$

Second, the z score expresses x values in units of a common scale—standard deviations.

Third, with composites made up of z scores we avoid the horse and rabbit stew error of combining ranks because information about the relative distances between original values is preserved. For example: Recall that boys 66-12, 66-13, and 66-11, who were at varying distances from each other in their original weights, were lined up at equal distances from each other through the ranking transformation (Table 17). With the z-score transformation, however, the distance of 27 lb between boys 66-12 and 66-13 becomes 2.11 σ units and the 5-lb difference between boys 66-13 and 66-11 becomes 0.39 σ units (Table 18). Let us see how this is achieved— using our height and weight data as illustrations.

In deriving our z score we first obtained "distance scores" for both heights and weights: $x - \bar{x}$ (or $x - \mu$). Thus, for height, boy 66-14 has a distance score of $49 - 53.60 = -4.60$ and, for weight, boy 66-13 has a distance score of $66 - 70.9 = -4.9$. Can we then say that boy 66-14 is about as far from the mean height as boy 66-13 is from the mean weight? In terms of numerical values the answer is obviously yes (-4.60 is about the same as -4.9). But consider now what happens when we take the second step and divide the distance scores by their appropriate standard deviations. The 4.9-lb distance in weight is only 0.38 of these σ units below the population mean ($-4.9/12.8 = 0.38$) while the 4.60-in distance in height is 1.77 σ units below the population height ($-4.60/2.60 = -1.77$). A boy who is 0.38 of a σ unit below his group mean is really not very far from that mean—more than one-third of his fellows weigh even less than he does (as we can quickly determine by looking up the B

TABLE 18 z scores for six STATLAB boys

ID no.	Height z score	Weight z score	Mean of height and weight z scores
66-11	−1.00	−0.77	−0.88
66-12	−0.23	1.73	0.75
66-13	0.15	−0.38	−0.12
66-14	−1.77	−1.16	−1.46
66-15	−1.77	−1.40	−1.58
66-16	−0.62	−1.01	−0.82

The data of Table 17 are here transformed into standard scores.

values for a z of 0.38 in the normal table—assuming that the population of boys' weights is sufficiently normal). On the other hand, a boy who is 1.77 σ units below the mean of his group's height is a fairly extreme case—only about 3.8 percent of his comrades are shorter than he is. It is clear now that the distance score of −4.60 in is located at almost five times as great a distance from *its* mean than is −4.9 lb from *its* mean! Obviously z scores are excellently designed to permit us (1) to make fair comparisons among scores from different variables—giving due weight to each score in terms of the location of the original x value in its own population—and (2) to add z scores from different variables, since the original values are now expressed in generalized relative scores.

Finally, when the z-score standard is based on a good random sample it can be used to derive a z score for any individual who belongs to the population sampled, including those who were not in the sample itself. For all the preceding reasons, whenever we deal with normal or nearly normal distributions the z score or standard score is the converted score of choice with which to assess individual performances and form composite scores.

☐ You are now ready to test these ideas in evaluating the mental test performances of some of the STATLAB children in your sample. In this investigation we shall assume that a composite score reflecting in equal degree a child's performance on the Peabody test (which requires some verbal facility) and the Raven test (a nonverbal spatial relations test) will yield a more useful approximation to the child's "general" mental capacity than either test alone.

WORK TO BE DONE

STEP 1   Retrieve from Databank, File E-1, the ID numbers and scores on two mental tests of the first 10 children *in the order drawn* in the work of Unit 10. Enter on the worksheet.

STEP 2   Determine and record the 10 children's ranks on each test.

STEP 3   Determine and record the percentile ranks for these 10 children on each test. In determining the percentile ranks, apply the graphic method to the cumulative frequency curves in Databank, File E-3. (These curves for the Peabody and Raven scores are based on your own sample, $n = 40$.)

STEP 4   Compute and record the z scores for these 10 children on each test. For computing the z scores, use your sample statistics, $\bar{x}$ and s, for the Peabody and Raven tests from Databank, File E-2.

STEP 5   For these 10 children compute and record the three composite scores: based on ranks, on percentile ranks, and on z scores.

STEP 6   Complete the inquiry.

Before going on to the next major area of statistical analysis, the inferential, it would be useful to check out your command of the material on descriptive statistics already covered. To aid you in this, we have devised this exercise which is at once both a self-test *and a* review. *The following summary statements can serve you as test items with key words or phrases italicized. You are urged to exploit these statements to the full in probing your familiarity and comprehension of the statistical logic, the formulas, the concepts, and the technical terminology. Unit references are provided so that whenever a statement reveals a weak spot in your understanding, you can go back and review the appropriate unit.*

This is the first of four such self-test reviews. Quite aside from the study value of these exercises, we believe that a reading of even the bare-bones outlines at the completion of each major area will be convincing and perhaps even pleasantly surprising testimony of how far you have come in your study of the art and science of statistics.

☐ The major objectives of statistics can be listed as (1) *description*, (2) *estimation*, (3) *testing*, and (4) *correlation*. (UNIT 1)

☐ Almost every statistical principle relating to inference which we shall discuss assumes that the *n* events being analyzed were drawn in such a way as to meet the basic guarantee demanded by a *randomly drawn sample.* (UNIT 2)

☐ Data may be either *discrete* or *continuous.* If they are continuous, the values must be recorded as though they were discrete, thus making error unavoidable. To minimize unavoidable errors and to avoid *systematic errors*, statisticians recommend the *nearest-point rule* for recording continuous values and the *even-number rule* for rounding them. (UNIT 3)

☐ Organizing *x* values into *simple* and *cumulative frequency distributions*, with *class intervals* of varying *widths* and specified *true class limits*, facilitates statistical description and analysis. Attributes of *population frequency distributions* can be inferred from *sample frequency distributions* as specified by the *law of large numbers.* (UNIT 4)

☐ By constructing and examining *histograms, simple frequency polygons*, and *cumulative frequency polygons*, you increase the probability of alerting yourself to special or unexpected characteristics of the data as well as to gross arithmetic and recording errors which may have been committed in processing the data. Frequency polygons also make possible a graphic determination of the values of certain statistics. Certain *dis-*

tributional shapes recur again and again in frequency polygons depicting quite different *empirical populations.* For some of these shapes there exist *idealized curves* generated by specified mathematical formulas, thus permitting theoretical analysis of such distributions. (UNIT 5)

☐ An important measure of location of a distribution is the *median.* Although the definition of and the methods for calculating either the median *parameter* $\tilde{\mu}$ or the *statistic* $\tilde{x}$ are essentially similar for both discrete and continuous data, we introduce the *assumption of even spacing of the values* for calculating medians from continuous data. (UNIT 6)

☐ The most widely used measure of location is the *arithmetic mean* (μ and $\bar{x}$). A number of shortcuts simplify the calculation of the arithmetic mean and are embodied in the formulas $\bar{x} = \Sigma(fx)/n$ and $\bar{x} = C + w\bar{x}'$. In calculating $\bar{x}$ for grouped data the *midpoint assumption* is used. The following alternative to the common definition of the mean helps us to detect important differences and similarities between the mean and median: The mean is that point from which $\Sigma fd = 0$. (UNIT 7)

☐ We estimate the parameters μ and $\tilde{\mu}$ from the statistics $\bar{x}$ and $\tilde{x}$. In comparing the mean and median, the following can be said: (1) In *symmetric distributions,* $\mu = \tilde{\mu}$; (2) in *skewed distributions,* the relation between the loci of the mean and median is systematically determined by the *direction of skewness;* (3) the mean and median differ greatly in their sensitivity to *extreme values;* and (4) a mean can be computed only when we know all the x values, although this restriction does not hold for the computation of a median. (UNIT 8)

☐ In addition to the two measures of location of a distribution, we have considered four *measures of dispersion.* The *range* is the difference between the maximum and minimum x values; the *mean deviation,* $\Sigma(f|d|)/n$; *variance,* $\Sigma(fd^2)/n$; and the *standard deviation,* $\sqrt{\Sigma(fd^2)/n}$. This last measure is the most preferred (for various reasons). The symbol σ is used to denote the standard deviation of a population while the letter s is the symbol for a standard deviation of a sample. The standard deviation can be calculated by shortcut methods as indicated by the formulas $s = \sqrt{[\Sigma(fx^2)/n] - \bar{x}^2}$ and $s = ws'$ where $s' = \sqrt{[\Sigma(fx'^2)/n] - (\bar{x}')^2}$. As is true of most statistics and parameters, we infer σ from s but must do so with caution where the *distribution tails* are substantially heavier than *normal.* (UNIT 9)

☐   The term *normal,* which describes a mathematical curve central to statistical theory and practice, applies only to its shape and does not specify its μ or σ. The shape of the normal curve is symmetric and *unimodal,* and therefore mode = median = mean. In its central portion it curves downward; in both tails, upward. These two *points of inflection* occur at a distance of one σ unit from μ. In all normal distributions a constant percentage of the *area of the curve* falls within the interval $\mu \pm z\sigma$. Advantage has been taken of this fact in constructing the most useful of statistical tables, the *normal table,* for various values of z. Many statistical methods of analysis have been specifically designed for samples drawn from normal populations. To use such samples for nonnormal distributions can yield questionable results. It is therefore important to determine the normality of a distribution prior to using such analytic methods. Some of the graphic procedures already discussed can help in such determination. (UNIT 10)

☐   Because of the changed shape of a normal cumulative distribution when plotted on *normal probability paper,* the reliability of visual inspection to determine the normality of a distribution can be greatly increased. Construction of such cumulative distributions also permits the graphic determination of good approximations to standard deviations and means—even for nonnormal and skewed distributions. (UNIT 11)

☐   To assess single x values in terms of an appropriate *reference group* and to construct *composite scores* made up of x values from disparate distributions, we find it useful to cast the x values into *converted scores.* The most commonly used converted scores are *ranks, percentile ranks* (easily determined graphically), and *standard scores* (z *scores*) as determined by the formulas $z = (x - \mu)/\sigma$ or $z = (x - \bar{x})/s$. (UNIT 12)

CENTRAL LIMIT THEOREM, STANDARD ERRORS, AND CONFIDENCE INTERVALS

You are now familiar with the idea that a sample statistic may be used to estimate the corresponding population parameter. For example, the median income in your sample of 40 estimates the median income of the 1296 StatLab families, and your sample mean of Peabody scores estimates the mean of all the Peabody scores earned by the StatLab children. Furthermore, according to the law of large numbers such estimates are likely to be accurate if the sample is large enough. But how accurate are your estimates with the sample size you used? Indeed, by what criterion should the accuracy of an estimate be judged when the true population value may never be known with certainty?

In this unit we shall see that there is available a key mathematical theorem that goes a long way toward giving concrete, *numerical* answers to these basic questions. With the help of the data and statistics you and your classmates have already collected and calculated, you will not only be able to see how this theorem helps to answer these questions but you will also have an opportunity to make an empirical check on the validity of the theorem—to see whether the results of its mathematical reasoning hold up in practice. Then, applying the results of the theorem to your own continuing statistical analysis of STATLAB, you will be able to make further estimates, with *specified degrees of confidence,* of certain physical, economic, and mental attributes of the STATLAB population.

The theorem deals with a set of concepts already familiar to you (normal distributions, $\bar{x}$'s and μ's and σ's) and one new concept: the frequency distribution of a statistic.

THE FREQUENCY DISTRIBUTION OF A STATISTIC

We have seen how a frequency distribution can make it easier to comprehend the way a *variable* (such as height or income) is distributed among the members of a population. The notion of a frequency distribution can also be usefully applied to the way a *statistic* (such as the mean height or the median income) is distributed. The distribution of a complete collection of mean heights—one mean for every different sample of a given size that can possibly be drawn from the same population—would be an example of a distribution of a statistic, in this case the $\bar{x}$ distribution. In the same way, the frequency distribution of a complete collection of median incomes—one median for every sample of a given size that can possibly be drawn from the same population—would be an $\tilde{x}$ distribution. In fact you already have, in your Databank, samples drawn from these very distributions, only we have been calling them *class distributions*. Thus you have a collection of $\bar{x}$'s of STATLAB mothers' heights, where each $\bar{x}$ was calculated from a different sample of $n = 40$ drawn from STATLAB, and a collection of $\tilde{x}$'s of STATLAB family incomes similarly drawn.

These class distributions, however, are far from constituting a complete collection of $\bar{x}$'s or $\tilde{x}$'s. Consider all the possible distinct samples of $n = 40$ that might be drawn from the 1296 STATLAB families. The number of all such samples is almost incredibly large, but it is finite. Actually there are 1.74×10^{124} different samples of $n = 40$ which can be drawn from STATLAB. To write out this number you would have to write 174 followed by 122 zeros! Next imagine that for *each* of these samples we have computed the $\bar{x}$ of the heights of the 40 mothers in that sample. We now have a collection of 1.74×10^{124} $\bar{x}$'s. (They are not, of course, all distinct numbers, since two different samples may well yield the same $\bar{x}$.) Finally, suppose that these 1.74×10^{124} $\bar{x}$'s are organized into a frequency distribution. *This* distribution is what is meant by an $\bar{x}$ distribution. Of course an $\bar{x}$ distribution (or an $\tilde{x}$ distribution) will have all the attributes of any distribution, e.g., there will be a *mean* of the $\bar{x}$ distribution, a *standard deviation* of the $\bar{x}$ distribution, and a *shape* to characterize the $\bar{x}$ distribution.

The number 1.74×10^{124} is so huge that it would not be possible to carry out the steps described above, and nobody could in practice construct the frequency table for all these values of $\bar{x}$. And yet it is essential to know a lot about the attributes of an $\bar{x}$ distribution (its mean, its standard deviation, its shape), if we are to provide the answers to the questions posed at the beginning of this unit. It is here that we receive help from the key mathematical theorem we have referred to—a theorem which does, by mathematical reasoning, what would be impossible by empirical investigation.

THE CENTRAL LIMIT THEOREM

Our theorem describes what happens to the mean, the standard deviation, and the shape of an $\bar{x}$ distribution as the size n of the samples from which the $\bar{x}$'s are calculated becomes larger and larger (or, as the mathematicians phrase it, "as the sample size n is increased without limit"). This theorem is so central to the theory of probability that it is called the *central limit theorem*. The first versions of the theorem were discovered in the eighteenth century, and mathematicians have spent two hundred years extending and improving it. The mathematical proof is very difficult, suitable only for graduate courses in probability theory. However, the conclusions of the theorem, what it says about the $\bar{x}$ distribution, are easy to understand. Moreover, as we have indicated, you already have the information which will enable you to make an empirical check on these conclusions. The theorem can be considered to consist of three main parts.

Part 1: The mean of the $\bar{x}$ distribution coincides with the mean μ of the population. It is fortunate indeed that it can be proved mathematically that the mean of the 1.74×10^{124} $\bar{x}$'s, for example, is equal to the mean of the STATLAB population. For if the distribution of $\bar{x}$ values had a center that was far from μ, then we could hardly use a single $\bar{x}$ from this off-center $\bar{x}$ distribution as an estimate of the true STATLAB center, μ.

Part 2: The standard deviation of the $\bar{x}$ distribution is equal to $\sigma/\sqrt{n}$. Here n is the sample size and σ is the standard deviation of the population from which the sample was drawn. This formula for the standard deviation of the $\bar{x}$ distribution indicates two things clearly:

1 When the numerator (in this case, σ) is increased or decreased, the value of the entire expression must be correspondingly increased or decreased: The formula then asserts that the standard deviation of the $\bar{x}$ distribution, $\sigma/\sqrt{n}$, is proportional to the standard deviation of the population, σ.

2 As the denominator (in this case, $\sqrt{n}$) is increased the value of the entire expression must be decreased: The formula tells us that with a sufficiently large sample size, the standard deviation of the $\bar{x}$ distribution will be small.

Part 3: As the sample size n is increased, the $\bar{x}$ distribution tends to have the normal shape regardless of the shape of the population distribution. This is the most surprising part of the central limit theorem. One might be prepared to find that the $\bar{x}$ distribution would be normally distributed if the samples were drawn from a normal population. It is harder to believe that the $\bar{x}$ distribution will tend to be normal if the population itself has (say) a uniform (rectangular) shape like Fig. 5c in Unit 5. But this is what the theorem says, and this can be demonstrated mathematically; you will soon have an opportunity to check out the validity of this conclusion yourself. And it is this third part of the central limit theorem which, as Box 4 argues, accounts for so much that is normal in nature.

There is, however, one proviso, or caution, which must be kept in mind concerning all this. Like the law of large numbers, this third part of the central limit theorem speaks of a *tendency* that becomes more and more apparent as n increases. This tendency is much faster when the population distribution is symmetric than when it is heavily skewed. For example, even if the population has a very *non*normal-looking symmetric distribution like that of Fig. 5c, the $\bar{x}$ distribution derived from it will be very nearly normal with a sample n as small as 3 (see Fig. 21). When the population is skewed, however, as depicted in Fig. 5b, the $\bar{x}$ distribution will lose its skewness rather slowly as n is increased. If the population is heavily skewed, one can expect traces of skewness to be visible in the $\bar{x}$ distribution where the sample is as large as $n = 40$. This, too, you will be able to check out with your data.

FIGURE 21 Simple frequency curve for the arithmetic mean of $n = 3$ observations drawn from a population that has the uniform distribution.

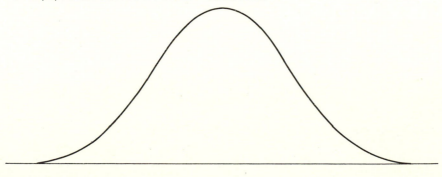

BOX 4

THE CENTRAL LIMIT THEOREM, MULTIDETERMINATION, AND NORMALITY

The central limit theorem tells us, among other matters, that a distribution of means —for any measurable variable—will always be nearly normal provided that each mean is based on a large sample. This mathematical discovery provides a possible explanation for the curious but well-established fact that a host of quite dissimilar biological, physical, social, and psychological phenomena, when measured, yield similar, nearly normal distributions. Let us see how the central limit theorem explanation works out for a very simple measurable attribute of people—the heights of adults.

Your height is the result of both genetic and environmental factors. It is polygenic, i.e., influenced by many different genes rather than by a single gene pair. Similarly, many environmental factors, such as nutrition and childhood diseases, can affect height. Some factors tend to make you taller, some shorter; and your eventual height is a sort of average of the specific "sample" of the many genetic and environmental factors you have drawn by your lot from the large population of possible "height factors." And the same is true of every other person: Each individual's height can therefore be considered to be an average of a different sample of many height-determining factors. Therefore, in view of the central limit theorem it is not surprising to find that height has a normal distribution.

Not all is normal in nature. The weights of people, for example, are not normally distributed. But why does not the same argument "prove" that weight should be normal? An answer which might be given is that weight is not just the cumulative result of many small factors but, especially for the obese, results also in part from one or two dominant causes (e.g., a glandular disturbance). Similarly, a family's income may be very large as the result of one cause, say inherited wealth. Weight and income, not being in many cases the average of numerous factors of comparable importance, do not come fully under the sway of the central limit theorem. (Of course, height too can be primarily determined by a single dominant factor such as dwarfism or acromegaly. But instances of this are rare, whereas obesity and affluence are common enough to affect the population in a major way.)

Most attributes of people, animals, and other natural objects—their abilities, propensities, sizes, weights, shapes, etc. —are multidetermined, and it is because of this that so many phenomena in nature have a nearly normal distribution. The reverse of the argument is also reasonable: If a variable is observed to be normally distributed, it is a good guess that the magnitudes of this variable are multidetermined. Because mental test scores show a near-normal distribution, most psychologists believe that these scores reflect many determinants: genetic factors; long-term environmental conditions and temporary factors.

We can summarize the three parts of the central limit theorem with the following statement: The $\bar{x}$ distribution tends to be normal, it has a standard deviation of $\sigma/\sqrt{n}$, and its mean is equal to the true mean μ of the population from which it has been derived. We shall soon see that in this statement, together with what we already know about the normal distribution, we have almost everything we need to assess the credibility of any single $\bar{x}$ as an estimate of μ. It is probably no exaggeration to say that without the insights provided by the central limit theorem, much of the structure of statistics concerned with the problem of inference would be without sure foundation.

We now turn to an application of the findings of the central limit theorem to the problems of estimating population parameters from sample statistics.

STANDARD ERROR OF THE MEAN

As the first of several statistical applications of the central limit theorem, let us take one of the questions raised at the beginning of this unit: By what criterion should the accuracy of an estimate be judged? Let us discuss this question for $\bar{x}$ as an estimate for μ.

By the first part of the central limit theorem, the $\bar{x}$ distribution is centered at μ. Since the standard deviation is a measure of dispersion of any distribution, this means that if an $\bar{x}$ distribution has a small standard deviation, most of the individual $\bar{x}$'s will cluster close to the central value μ. Therefore the chance will be relatively good that any single $\bar{x}$, computed from a single sample, will lie close to μ. In this case we may say that $\bar{x}$ provides an accurate estimate of the true value of the population mean. Conversely, if the $\bar{x}$ distribution has a large standard deviation, the various $\bar{x}$'s comprising it will be widely dispersed and hence there is a relatively poor chance that any single $\bar{x}$ will fall close to μ; we can therefore characterize $\bar{x}$ as an inaccurate estimator of μ. In other words, the value of the standard deviation of the $\bar{x}$ distribution, $\sigma/\sqrt{n}$, provides us with a quantitative criterion against which to judge the accuracy of $\bar{x}$ as an estimate of μ: The smaller the value of $\sigma/\sqrt{n}$, the greater is the precision of $\bar{x}$ as an estimate of μ.

As an illustration, let us recall that the population of the 648 STATLAB girls' heights has the standard deviation $\sigma = 2.61$ (Table 16 in Unit 9). Therefore the standard deviation for an $\bar{x}$ distribution of samples, $n = 40$, drawn from this population would be

$$\frac{2.61}{\sqrt{40}} = 0.41 \text{ in}$$

This can be written in symbols as follows: $SD_{\bar{x}} = 0.41$. The subscript $\bar{x}$ on SD reminds us that it is the statistic $\bar{x}$ whose standard deviation (SD) is under consideration. The formula for the standard deviation of the statistic $\bar{x}$ would then be

$$SD_{\bar{x}} = \frac{\sigma}{\sqrt{n}}$$

FORMULA 9

Although $SD_{\bar{x}}$ provides all we need for a criterion with which to judge the accuracy of $\bar{x}$ as an estimate of μ, it unfortunately demands more than we can usually supply. In practical work the value of σ, the standard deviation of the entire population, is usually not known and therefore we cannot derive the value $SD_{\bar{x}}$. What we can do, however, is use s as an approximation for σ, as explained in Unit 9. Accordingly, in practice the statistician uses $s/\sqrt{n}$ in place of $\sigma/\sqrt{n}$. This value $s/\sqrt{n}$ has been given a very revealing name. Since $\bar{x}$ is used to estimate μ, the degree to which $\bar{x}$ departs from μ may be thought of as an "error of estimate." Because $s/\sqrt{n}$ is a measure of the deviations of $\bar{x}$ from μ, the value $s/\sqrt{n}$ is called the *standard error of $\bar{x}$*. Expressed in conventional symbols, its formula is

$$SE_{\bar{x}} = \frac{s}{\sqrt{n}}$$

FORMULA 10

Let us return to our illustration. In Unit 4 we drew a sample of $n = 40$ from the population of STATLAB girls. From this sample we computed the statistics $\bar{x} = 53.45$ (Unit 7) and $s = 3.04$ (Unit 9). We can now compute the standard error of $\bar{x}$:

$$SE_{\bar{x}} = \frac{s}{\sqrt{n}} = \frac{3.04}{\sqrt{40}} = 0.48 \text{ in}$$

The standard error of 0.48 in, as we see, is a fairly good approximation for the standard deviation of the $\bar{x}$ distribution as determined by the $SD_{\bar{x}}$ formula, which yielded 0.41 in. (The approximation is a little large, reflecting the fact that our sample of 40 girls' heights happened to be somewhat more widely dispersed than the "average" sample would be.) In any event, we now have a feasible way of arriving at a quantitative answer to the question: What criterion can I use to judge the accuracy of using my sample mean $\bar{x}$ as an estimate of the true population mean μ? The answer is: Calculate the standard error of the mean, $SE_{\bar{x}}$; the lower the value of the standard error of your mean, the better is your $\bar{x}$ as an estimator of μ. It should be emphasized here that everything you need for calculating $SE_{\bar{x}}$ you can get from your own sample.

STANDARD ERROR—SOME GENERAL REMARKS

Although the central limit theorem makes it possible to compute the standard error of a sample mean only, statisticians have worked out ways of getting the standard error for many other statistics, and we shall give several examples in later units. Whenever you report the value of any statistic—mean, median, etc.—it is always desirable to give the standard error of that statistic as well so that the reader of your report may have some idea of the accuracy of your statistic as an estimate for the population value. To fail to report a standard error is to ask your audience to buy a pig in a poke.

A word of caution is in order here. As we have noted in Unit 9, s tends to be a rather poor and unreliable estimate for σ when sampling from a population with a heavy tail or tails. If s is a poor approximation for σ, then $s/\sqrt{n}$ will be a poor approximation for $\sigma/\sqrt{n}$. In such an event, the use of the standard error of $\bar{x}$, $\mathrm{SE}_{\bar{x}}$, as a criterion of the quality of $\bar{x}$ as an estimate of μ should be undertaken with great caution. Nevertheless, even a crude standard error is better than none at all. (We shall be able to say more about this important point in Unit 15.)

There is a conventional format for reporting a standard error along with a statistic, namely:

Statistic $\pm$ standard error of the statistic

In this format, we would say (on the basis of our sample of 40) that "the mean height of all the STATLAB girls is 53.45 ± 0.48 in."

The major reason why it is so essential to provide the standard error of any statistic, when you use that statistic to make an inference about a population parameter, is that the standard error can tell you, *in quantitative terms,* the degree of *precision* of the estimate. This, of course, is what we have been after in the last several units, and it is the payoff of the entire analysis undertaken in the present unit. We are now ready to turn to the last step of that analysis and see what such expressions as 53.45 ± 0.48 in really mean.

CONFIDENCE INTERVALS

We can now use the standard error of the mean $\mathrm{SE}_{\bar{x}}$ to give us a quantitative answer to the question: How accurate *is* our estimate? In so doing, the standard error of the mean will acquire deeper significance.

Recall from our summary of the central limit theorem that if n is sufficiently large, the $\bar{x}$ distribution will be normal, with mean centered at μ and a standard deviation of $\mathrm{SD}_{\bar{x}} = \sigma/\sqrt{n}$. Now recall from Unit 10 that in all normal distributions a constant percentage of the cases will fall within the interval $\mu \pm z\sigma$. Thus 68 percent of the cases lie within one standard deviation of the mean in either direction, 95 percent lie within two standard deviations, etc. All this applies, of course, to the $\bar{x}$ distribution. Accordingly, there is a 68 percent chance that any randomly chosen $\bar{x}$ will fall somewhere within the distance $SD_{\bar{x}}$ of μ (in one direction or the other). Suppose we happen to get an $\bar{x}$ which does in fact lie within $SD_{\bar{x}}$ of μ; then, of course, μ must lie within $SD_{\bar{x}}$ of our $\bar{x}$. (This is saying no more than that if A lies within a mile of B, then B lies within a mile of A.) Let us write this more succinctly: If $\bar{x}$ lies within $\mu \pm SD_{\bar{x}}$, then μ lies within $\bar{x} \pm SD_{\bar{x}}$. We can now generalize this beyond our example of a particular value of $\bar{x}$: Since we know that the probability of a randomly chosen $\bar{x}$ lying within $\mu \pm SD_{\bar{x}}$ is 68 percent, then there is also a 68 percent chance that μ will lie within the interval $\bar{x} \pm SD_{\bar{x}}$. This interval $\bar{x} \pm SD_{\bar{x}}$, which has a 68 percent chance of containing μ, is called the 68 percent *confidence interval* for μ.

Suppose you wanted an interval with a higher degree of confidence. That can easily be arranged. Since 95 percent of the cases in a normal distribution fall within two standard deviations of the mean, using the same reasoning as above, the interval $\bar{x} \pm 2SD_{\bar{x}}$ is the 95 percent confidence interval for μ. Note, however, that as you moved from a 68 percent confidence interval to a 95 percent confidence interval, you have been forced to trade off precision for confidence. That is, with the wider confidence interval you can say that there is a 95 percent chance that the true population mean lies somewhere in the interval $\bar{x} \pm two\ SD_{\bar{x}}$, whereas with the 68 percent confidence interval you can *narrow* the interval within which μ lies to half that range, making it $\bar{x} \pm one\ SD_{\bar{x}}$. In statistics, as elsewhere in this world of limited options, the sad necessity of making choices cannot be escaped. In any event, for any desired degree of confidence merely look in the normal table in Appendix B to find the appropriate z; $\bar{x} \pm z\sigma/\sqrt{n}$ is then the desired interval.

Again we remind you that in practical work one does not usually know the value of σ, and we must replace it by the estimate s, just as we did in getting the standard error. Thus $\bar{x} \pm s/\sqrt{n}$ will be an interval that has *approximately* the probability of 68 percent of covering μ and will be an approximate 68 percent confidence interval. The estimate s may not be a very good estimate of σ, but fortunately the accuracy of the approximation is favored by a compensation of errors. Sometimes s will underestimate σ by a good deal, causing the confidence interval to be too narrow and hence less likely to cover μ with the advertised degree of confidence; sometimes, in partial compensation, s will overestimate σ and a larger chance of coverage will result. If, therefore, we compute a goodly number of confidence limits in our statistical lives, we shall find that we cover the true μ fairly close to 68 percent of the time by using the $\bar{x} \pm s/\sqrt{n}$ interval. Another way of saying this, in the context of your class work with this unit, is: We would expect about 68 percent of your class to cover the STATLAB μ for any of the quantitative variables (e.g., height or weight) when each member uses $\bar{x} \pm s/\sqrt{n}$ as his estimating interval.

We return to our illustration of girls' heights. From our sample of 40 girls' heights, the interval $x \pm \mathrm{SE}_{\bar{x}} = 53.45 \pm 0.48$ is approximately a 68 percent confidence interval for the mean height of all 648 STATLAB girls. In this case, the interval (which extends from 52.97 to 53.93 in) does cover the true μ, which we know is 53.26 in.

We should note what is truly a great achievement: We can draw a single sample (say of $n = 25$, or 30, or 40, and so on) and from that *one* sample we can estimate not only the μ for the population (size without limit) but we can also determine the standard error of our estimate. Then by use of the normal table we can determine the degree of confidence with which we can assert that the population μ lies within such-and-such a range. In your work you will be given an opportunity to demonstrate all this for yourself.

YOUR WORK IN THIS UNIT

Because your work in this unit deals with matters of basic importance for the rest of the course, it is advisable to discuss its rationale a bit. There are two major objectives which can be gained from your work here:

1 An empirical check on the conclusions of the central limit theorem—a check which should also deepen your understanding of the theorem
2 An empirical check on the statistical implications of the theorem as they apply to the determination of the degree of precision and confidence with which you can estimate a population μ from a single sample $\bar{x}$

We have said that by using the class distribution of $\bar{x}$'s (e.g., for mothers' heights) you could check out the validity of the central limit theorem. This merits further discussion. When in Unit 2 you drew a random sample of 40 STATLAB mothers' heights, you chose it in such a way that *each of all the possible samples* was equally likely to be drawn. Accordingly, the computed value of your $\bar{x}$ is randomly chosen from the 1.74×10^{124} $\bar{x}$'s which form the $\bar{x}$ distribution. Similarly, every other member of your class has drawn a random $\bar{x}$ from this $\bar{x}$ distribution, so that the class has collectively a *random sample from the $\bar{x}$ distribution*. If, for example, your class has 50 students in it, then in Databank, File B-4, there is recorded the distribution of a sample of size $n = 50$ from the *total* $\bar{x}$ distribution of mothers' heights.

But can such a sample of only $n = 50$ tell us anything useful about so huge a collection? Yes: Because as we have already emphasized, the size of a population plays almost no role in the law of large numbers. *Only the sample size matters*. And a *random* sample of 50 is large enough to tell you quite a lot about the $\bar{x}$ distribution. It is therefore perfectly feasible to use your class distributions to check out the predictions of the central limit theorem about the behavior of the $\bar{x}$ distribution. You should be able to answer such questions as the following: As measured by your class sample, is the $\bar{x}$ distribution of mothers' heights normal? Is the $\bar{x}$ distribution of family incomes normal? Are your answers in accord with what the central limit theorem would demand? Do your class samples of the $\bar{x}$ distributions appear to have the center and dispersion that are predicted theoretically?

For the second part of your work you can use your own unique samples of mothers' heights, family income, and mental test scores to obtain standard errors of the means and hence confidence intervals. When these results are discussed in class, the instructor will reveal the true parameter values and you will be able to see whether and to what degree your sample inferences about all of **StatLab** were valid, whether your confidence intervals succeeded in covering the true means, and whether the predicted percentage of your classmates succeeded in catching the true population means in their confidence intervals.

WORK TO BE DONE

STEP 1   Retrieve the class distribution of $\bar{x}$ for mothers' heights from Databank, File B-4, and plot it on the worksheet in the space provided. Label the horizontal axis.

STEP 2   Read off the mean and standard deviation of this class distribution.

STEP 3   Now do the same for family income (Databank, File C-2).

STEP 4   Use the spaces provided on the worksheet to calculate the standard errors for your sample means of mothers' heights, family income, and Peabody score. The necessary s values can be retrieved from Databank, Files D-1 and E-2.

STEP 5   Complete the inquiry.

PROPORTIONS AND THEIR STANDARD ERRORS

The statistical methods we have been considering so far were devised to deal with variables that are numerical or quantitative. As we explained in Unit 3, some such variables are continuous, answering the question "How much?": time, length, weight. Others are discrete, answering the question "How many?": number of points showing on a die, number of fingers on a hand, number of cigarettes smoked last week. But in either case the answer is a numerical quantity, which we have conventionally denoted by x.

We now turn to the analysis of data which arise from questions that are not quantitative but *qualitative* or *categorical*: What is the child's blood type? (O, A, B, AB) What is the color of the flower? (red, yellow, pink, etc.) What is the father's occupation? (professional, laborer, managerial, clerical, etc.) In this unit we shall take up the simplest case, in which the individual must fall into one of only *two* categories. In this simplest case it is always possible to formulate the question so that it may be answered Yes or No. Is the child female? Has the mother ever smoked? Does the patient have tuberculosis? Will the voter support school bonds? Does the electric motor meet the manufacturer's specifications? Is the man employed this week? Is the television set turned to Channel 5?

121

As these examples show, the range of applications of Yes or No data is very great. Furthermore, variables that are originally quantitative and continuous, such as age, are often converted to a categorical form: The registrar of voters asks, "Will you be 18 on the first Tuesday after the first Monday of November?" and the Social Security office asks, "Are you 65 or over?" Similarly, a discrete variable, such as number of points scored on a test, can be made categorical: "Did you pass the test?" A quantitative variable that has been reduced to a Yes or No form is said to be *dichotomized* (the Greek roots mean "cut into two").

In this unit we shall show you how the central limit theorem and the concepts related to the accuracy of an estimate (standard error, confidence intervals) can be extended from quantitative variables to qualitative data that fall into two categories. (Later, in Unit 22, we shall take up some methods for qualitative data that fall into more than two categories.)

Once having done that, we shall be able to deal with another important question, one which must be faced at the start of every sampling investigation: How large a sample is necessary in order to achieve any desired degree of accuracy? In the work of this unit you will have the chance to try out these methods and ideas on the analysis of the occupations of STATLAB fathers and the mental test scores of STATLAB children.

PROPORTIONS AS POPULATION PARAMETERS

If the individuals of a population are classified into two groups—those who answer Yes and those who answer No to a question of interest to us—we shall usually want to know how many fall into each group or (equivalently) what *proportion* of the total falls into each group. Let us denote by P the proportion who answer Yes and by Q the proportion who answer No. It is obvious that $P + Q = 1$. If N is the total size of the population, then the *number* of people who answer Yes is NP and the number who answer No is NQ. Clearly, since each individual falls into one group or the other, $NP + NQ = N$.

To illustrate these notations, consider the question: Was a STATLAB child right-eye dominant on his first test? If you will recall the code for laterality (page 319), you can see that, for the first eight girls listed on Census page 11, the answers were respectively: Yes, No, Yes, No, Yes, No, Yes, and Yes. Continuing in this way, one can eventually find that of the $N = 1296$ children in STATLAB, $NP = 780$ were right-eyed and $NQ = 516$ were left-eyed at first test. Thus for this illustration, $P = \frac{780}{1296} = 0.602$, $Q = \frac{516}{1296} = 0.398$, and of course $P + Q = 0.602 + 0.398 = 1$.

Sometimes we find it more convenient to use *percentages*

instead of proportions. To convert a proportion into a percentage we need merely multiply the proportion by 100. Thus in the preceding illustration, P and Q become 60.2 percent and 39.8 percent.

Now P and Q are parameters of the population, and ordinarily their values will not be known. We shall be interested in estimating these parameters from the data of a sample. By a simple device, the central limit theorem can be applied to this problem. Let us first encode the Yes and No answers into numerical x values. We can do this quite simply by defining the variable x to be equal to 1 for any individual who answers Yes and equal to zero for any individual who answers No. (The variable x which encodes the answers in this way is called an *indicator variable* because it indicates a Yes answer by taking on the value 1.) In the eyedness illustration, $x = 1$ for the first, third, fifth, seventh, and eighth girls; $x = 0$ for the second, fourth, and sixth. By encoding the Yes or No answers in this way, a *qualitative* categorization is made *quantitative* and thus amenable to the methods of Unit 13. In other words, we can now determine for our indicator variable an arithmetic mean and a standard deviation.

Let us see how we can calculate the population mean μ for our indicator variable x. Since $x = 0$ for each of the NQ individuals in the population who answer No, the sum of their x values is also zero. Since $x = 1$ for the NP individuals who answer Yes, each individual contributes 1 to the sum, so that the sum of their x values is NP. Thus the sum of all the x values in the population is $0 + NP = NP$. Dividing this sum by the number N in the population gives, of course, the population mean of the x variable. The result is $\mu = (NP)/N = P$. We have shown that, for an indicator variable, *the population arithmetic mean μ is identical with the population proportion of Yes answers*. (In the eyedness illustration, $\mu = P = 0.602$.)

We are also going to need the population standard deviation, which (see Unit 9) is the square root of the variance. Recall that the variance is the mean of the squares of all the deviations from the population mean μ. In the present case, $\mu = P$. There are NQ individuals who give the answer No and get $x = 0$. For each of them, the deviation from the population mean is $d = x - \mu = 0 - P = -P$, so that the square of the deviation is $d^2 = (-P)^2 = P^2$. Similarly, the NP individuals answering Yes get $x = 1$ and thus have deviation $d = x - \mu = 1 - P = Q$, so that $d^2 = Q^2$. (Here we have used the fact that $P + Q = 1$.) In summary, there are NQ individuals, each of whom has $d^2 = P^2$; and there are NP individuals, each of whom has $d^2 = Q^2$. Therefore the sum of the squares of the deviations for all members of the population is

$$(NQ) \times P^2 + (NP) \times Q^2 = N(QP^2 + PQ^2) = (NPQ)(P + Q) = NPQ$$

To get the variance we need only divide that sum by N:

$$\mathrm{Var} = \frac{(NPQ)}{N} = PQ$$

Thus, finally, the population standard deviation is

$$\sigma = \sqrt{\mathrm{Var}} = \sqrt{PQ} \qquad \text{FORMULA 11}$$

Turning again to the eyedness illustration, there we found that $P = 0.602$, so in this case $\sigma = \sqrt{0.602 \times 0.398} = 0.489$.

SAMPLE PROPORTIONS AS ESTIMATES

If you are interested in a population proportion and have drawn a sample from that population, it seems natural to use the corresponding sample proportion as an estimate. Let us denote the proportion of Yes individuals in the sample by p and the proportion of No individuals by q. Using the same reasoning that we employed above, out of a sample of size n there are np Yes individuals and nq No individuals, and $np + nq = n$. Of course, $p + q = 1$.

In a similar fashion we can compute the sample mean $\bar{x}$ of our indicator variable x. Each of the nq No individuals has $x = 0$; each of the np Yes individuals has $x = 1$; and therefore the sum of all x values in the sample is equal to np. Division by n gives the mean of the sample x values: $\bar{x} = (np)/n = p$. We have shown that, for an indicator variable, the sample mean is identical with the proportion of Yes answers in the sample.

We are now ready to apply the central limit theorem to proportions. Recall from Unit 13 that, for any variable x, the mean $\bar{x}$ of a sample of size n has approximately the normal distribution with mean μ and standard deviation $\sigma/\sqrt{n}$ (provided that n is large enough). Let us apply that general fact to our particular indicator variable. We conclude that p has, for a large sample, approximately the normal distribution with mean $\mu = P$ and with standard deviation

$$\mathrm{SD}_p = \frac{\sigma}{\sqrt{n}} = \frac{\sqrt{PQ}}{\sqrt{n}} = \sqrt{\frac{PQ}{n}} \qquad \text{FORMULA 12}$$

Of course, since P will in practice be unknown, we must replace it by its estimate p in Formula 12. In that way we get the standard error of p:

$$\mathrm{SE}_p = \sqrt{\frac{pq}{n}} \qquad \text{FORMULA 13}$$

We now have everything we need in order to use the statistics of *quantitative variables* to analyze dichotomous—and originally nonquantitative—variables. More specifically, we can now use the sample p to make inferences about the population P and thus obtain standard errors and confidence levels. For the last time, consider the eyedness illustration. We drew a new random sample of $n = 50$ StatLab children and found that 35 of them were right-eyed at first test. That is, in our sample $p = \frac{35}{50} = 0.7$ while $q = \frac{15}{50} = 0.3$. The standard error of p is $SE_p = \sqrt{(0.7 \times 0.3)/50} = 0.065$. We would report: On the first test, the proportion of the StatLab children who were right-eye dominant was 0.700 ± 0.065.

This report can of course also be interpreted to mean that the 68 percent confidence interval for P runs from $0.700 - 0.065 = 0.635$ to $0.700 + 0.065 = 0.765$. As it happens, this interval *fails* to cover the true value, $P = 0.602$. One must of course expect the 68 percent interval to fail to cover the parameter from time to time—in fact, it should fail about $100 - 68 = 32$ percent of the time. (Note, however, that in this example the 95 percent confidence interval *does* cover the true P.)

As with other cases, the central limit theorem merely asserts that there is a *tendency* for p to have a distribution of the normal shape as n gets larger and larger. How large must n be for such confidence intervals to be of practical use? In discussing this question in Unit 13, we said that the normal tendency works faster for symmetric populations than for skewed ones. When you are dealing with proportions, the population is symmetric when $P = Q = \frac{1}{2}$, for in that case the mean is at 0.5 and half the individuals are at equal distances to the left and to the right of the mean. Accordingly, it is not surprising that p has a nearly normal distribution for quite small sample sizes provided that P and Q are nearly $\frac{1}{2}$. If they are far from $\frac{1}{2}$, however, with P near 0 and Q near 1 or vice versa, then n has to be very large for p to be safely treated as normal. A rule of thumb is to require both np and nq to exceed 5 (or preferably 10) before placing much faith in the confidence interval for P, based as it is on the assumption that p is normally distributed about P.

PLANNING THE SIZE OF A SAMPLE

You have now seen how the data collected in a sample may be used to estimate a population mean (Unit 13) or proportion (this unit). We must still, however, deal with a question that faces any investigator planning to draw a sample: How large a sample should I draw? If the sample is made too small, it will not provide enough information and the inferences based on it will be insufficiently reliable or accurate. On the other hand, it is a waste of time, money, and effort to draw a sample larger than necessary.

When inferences are to be based on the central limit theorem, as in this unit and Unit 13, then some light is thrown on the question by the fact that the estimate $\bar{x}$ has the standard deviation $\text{SD}_{\bar{x}} = \sigma/\sqrt{n}$. For any given value of σ, this formula tells us how large n must be to cause the estimate $\bar{x}$ to have the degree of accuracy specified by the experimenter. (As we saw in Unit 13, the standard deviation of $\bar{x}$ is a good measure of the accuracy of $\bar{x}$ as an estimate.)

Unfortunately, the value of σ is in practice seldom known. Once the sample has been drawn, σ can be estimated by s—but the investigation must be planned *before* the sample is drawn. In general, there are two lines of approach to the planning of sample size.

One possibility is to try to guess what the value of σ is likely to be, using past experience with similar populations. For example, the biometrician knows that, in populations of adult males, the standard deviation of height is usually about 3 in or a little less. If, therefore, he is planning to estimate the mean height of StatLab fathers and he wishes the estimate to have a standard deviation of 0.25 in, he could calculate the necessary sample size by equating the formula $\sigma/\sqrt{n}$ for $\text{SD}_{\bar{x}}$ to the desired value: that is, $\sigma/\sqrt{n} = 0.25$. Replacing σ by the guessed value 3 gives $3/\sqrt{n} = 0.25$. It is easy using this equation to find $\sqrt{n} = 3/0.25 = 12$ and hence $n = 12^2 = 144$.

A second possible approach is to draw the sample in two stages. The biometrician realizes that the StatLab fathers may have σ larger than 3 in, so that $n = 144$ would prove to be insufficient; or their σ may be smaller than 3 in, so that $n = 144$ would be wastefully large. He may decide to draw a first sample of, say, $n = 100$. From the s of this sample he can estimate σ and use that estimated value to decide how large to make the second sample so that the total sample size will be reasonable. Such two-stage sampling is often an efficient way to proceed in cases where there is no practical barrier to taking the data in two stages.

So far we have been discussing the general estimate $\bar{x}$ of Unit 13. When we turn specifically to proportions, matters get simpler. Recall Formula 11, which shows how the standard deviation $\sigma = \sqrt{PQ}$ of an indicator variable can be calculated for any given value of P. For example, when $P = 0.5$, then $\sigma = \sqrt{0.5 \times 0.5} = 0.5$; when $P = 0.4$, then $\sigma = \sqrt{0.4 \times 0.6} = 0.49$; and the same value holds when $P = 0.6$. Figure 22 shows how σ varies as P is changed. You will note that σ reaches its peak at $P = 0.5$ and that σ can never be larger than 0.5, no matter what the population proportion P may happen to be. It follows that the maximum value of the standard deviation $\text{SD}_p = \sqrt{PQ/n}$ of p is $0.5/\sqrt{n}$.

An investigator who pessimistically assumes that σ has its maximum possible value $\sigma = 0.5$, so that SD_p has its maximum possible value $0.5/\sqrt{n}$, and calculates the necessary sample size on this basis, is sure to have a sample that is large enough. Furthermore, since the values in the central portion of the curve of Fig. 22

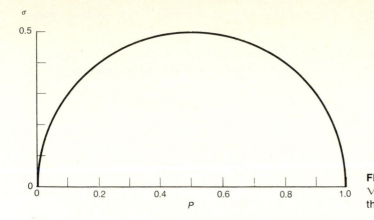

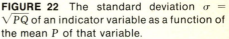

FIGURE 22 The standard deviation $\sigma = \sqrt{PQ}$ of an indicator variable as a function of the mean P of that variable.

are all near the maximum value, his sample size will not be wastefully large unless P should turn out to have a value close to zero or to 1.

For example, suppose that you want to estimate the proportion of STATLAB mothers who were smokers at the time of test and wish to have the standard deviation of this estimate not larger than 0.1. You need only equate this desired value to $0.5/\sqrt{n}$ and calculate from $0.1 = 0.5/\sqrt{n}$ that $\sqrt{n} = 0.5/0.1 = 5$ and hence that $n = 5^2 = 25$. This value of n is sure to be large enough. Furthermore, if P has any value reasonably near 0.5 you will not have taken many more observations than needed. (For example, you would need a sample of $n = 21$ or more for any value of P between 0.3 and 0.7.)

☐ You are now ready to see for yourself how all this works. We shall ask you to compute estimates of P, the standard errors of those estimates, and the related confidence intervals, for your own two-category data. Among other things, you will attempt to estimate the proportion of STATLAB fathers who are in the upper employment categories and the proportion of STATLAB children whose Peabody scores are greater than 80. The work will illustrate not only data that are naturally qualitative but also numerical data made qualitative by dichotomization. We shall also give you a chance to plan the appropriate size of one of your samples.

WORK TO BE DONE

STEP 1 You are going to investigate by drawing a sample the proportion P of STATLAB fathers who had at time of test any one of the following professional-managerial occupations: codes 0 (professional), 1 (teacher/counselor), or 2 (manager-official). Calculate on the worksheet the size of sample needed to ensure that the standard deviation of your estimate will not exceed 0.07.

STEP 2 Now draw a sample of this size, and encode the results on the worksheet. Put $x = 1$ for any father with occupation code 0, 1, or 2; put $x = 0$ for all other cases.
(*Note:* The worksheet provides more blanks than are necessary.)

STEP 3 Calculate on the spaces provided your estimate p for P and its standard error.

STEP 4 Next turn to Databank, File E-1, and count the children in your Peabody sample who got a score greater than 80. Calculate from this datum your estimate of the population proportion of such children and the standard error of that estimate.

STEP 5 Enter into Databank, File F-1, the ID numbers of your new sample (step 2); the x values will not be needed. Also record the results of step 3.

STEP 6 Complete the inquiry.

STANDARD ERRORS OF s AND $\tilde{x}$

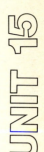

The central limit theorem of Unit 13 tells us that the sample mean $\bar{x}$ has approximately a normal distribution, at least if n is large enough. Furthermore, this normal distribution is centered at the population mean μ. Similarly, we found in Unit 14 that the sample proportion p has a nearly normal distribution, centered at the population proportion P.

These are examples of a general phenomenon: Most statistics computed from large samples have approximately normal distributions, centered at the corresponding population parameters. This is true, for example, for median, mean deviation, variance, and standard deviation.

To use such a statistic for calculating a confidence interval for the corresponding population parameter, we need one more ingredient: the standard error of that statistic. For $\bar{x}$ this is provided by Formula 10 and for p by Formula 13. As these examples suggest, a somewhat different approach is needed for each standard error. In this unit we shall develop standard errors for two more statistics: the sample median and the sample standard deviation.

As in the case of $\bar{x}$, we cannot show you the mathematical proofs of the limit theorems which underlie the results. We can, however, do something at least as important by helping you to find out how the theorems work with actual data. You already have in your Databank class distributions that will verify and delineate the meaning of the theoretical conclusions. You will also be asked to compute standard errors for s and $\tilde{x}$, using samples you have previously drawn.

STANDARD ERROR OF THE STANDARD DEVIATION

We have given in Unit 9 the reasons why the standard deviation σ is the most popular measure of the dispersion of a population. Accordingly, the sample standard deviation s, as the estimate for σ, is one of the most frequently computed statistics. Like any other statistic, s has its distribution, which consists of the s values obtainable from every possible sample of size n drawn from a given population. We would expect the sample statistic s to have a distribution centered near the corresponding population parameter σ. Furthermore, the shape of the s distribution should be nearly normal, at least if the sample size is large enough. Unfortunately, the tendency of s to approach a normal distribution is rather slow if the population is heavily skewed. (You may recall that the same was true for $\bar{x}$: See Part 3 of the central limit theorem in Unit 13.) We will give you a chance to check the normality of s for two different populations.

As with any statistic, it is desirable to attach to a computed value of s its standard error, in order to indicate its reliability as an estimate of σ. The parameter σ is not only important in its own right. It also enters into the formula for the standard deviation of $\bar{x}$, $\mathrm{SD}_{\bar{x}} = \sigma/\sqrt{n}$ (and, as we shall soon see, for the standard deviation formula of other statistics). It is therefore necessary that s be an accurate estimate for σ so that the standard error of $\bar{x}$, $\mathrm{SE}_{\bar{x}} = s/\sqrt{n}$, will be reliable. One might say that s, by providing the standard error of $\bar{x}$, is the custodian of the virtue of $\bar{x}$, but how virtuous is s itself? We need to ask Juvenal's question: *Quis custodiet ipsos custodes?*

As mentioned in Unit 10, it used to be the general opinion that nearly all natural populations were normal. It is not surprising, therefore, that the nineteenth century mathematicians who first studied the behavior of s assumed that their samples came from normal populations. A formula for the frequency curve of s for a sample from a normal population has been known for a hundred years, and extensive tables of its distributions have been published. Our needs will be met by the following simple approximation:

If n is not too small (say n greater than 10), then

$$SD_s = \frac{\sigma}{\sqrt{2n}} \qquad \qquad \textbf{FORMULA 14}$$

This result verifies and refines the assertion of the law of large numbers—that s is a reasonable estimate for σ, at least for normal samples. That is, the distribution of s is centered at σ; and if n is large enough, the standard deviation $\sigma/\sqrt{2n}$ will be small. Thus the distribution of s will be tightly concentrated near its center, and the estimate s is likely to be close to σ.

Of course, in practice you will not know σ, and in the formula $\sigma/\sqrt{2n}$, σ must be replaced by its estimate s. This gives the formula for the standard error of s for a sample from a normal population:

$$SE_s = \frac{s}{\sqrt{2n}} \qquad \qquad \textbf{FORMULA 15}$$

Thus from our sample of $n = 40$ girls' heights (that variable being itself nearly normal), for which we found (Table 15 in Unit 9) $s = 3.04$, we would calculate $SE_s = 3.04/\sqrt{80} = 0.34$ and report: We estimate the standard deviation of the population of STATLAB girls' heights to be 3.04 ± 0.34 in. Because s is nearly normally distributed about σ, we may interpret this as a confidence interval. (As it happens, this 68 percent confidence interval *fails* to cover the true $\sigma = 2.61$, given in Table 16 in Unit 9. Check whether the 95 percent confidence interval does cover σ.)

Now what about populations that are not normally distributed? Can we continue to use Formula 15? Unfortunately not, because (as we hinted in Unit 9) the behavior of s is heavily influenced by the precise shape of the tails of the population. It will be recalled that in calculating the standard deviation we square each deviation from the mean (Formula 5, Unit 9). Consider two values with deviations from the mean of $d = 1$ (which is close to the mean) and $d = 5$ (farther toward the tail). When squared these become, respectively, $d^2 = 1$ and $d^2 = 25$. If these two d^2's were to be averaged, the second would dominate the first. Thus when one averages the values of d^2 in Formula 5, $s = \sqrt{\Sigma(fd^2)/n}$, a few individuals far out in the two tails can have a major effect on the result. Suppose now that our population is basically like a normal one, but that a few percent of the cases are moved from the center rather far out into one or both tails. Since there are so few such cases in the population, many samples of moderate size will fail to catch any of them and s will behave as if computed from a normal sample. Many other samples, however, will contain one or two of these tail-dwellers, and then the computed s would be much larger. The net effect is that we would have substantially more variability among all the sample s's than for s's drawn from a normal population. This means, of course, that for samples drawn from populations with a heavy tail or tails, the standard error of the mean $\bar{x}$ (which depends on the sample s) is a questionable criterion of the accuracy with which $\bar{x}$ can estimate μ. That is why we issued the caveat in Unit 13 when discussing $SE_{\bar{x}}$.

As an illustration, consider Fig. 5b of Unit 5. Populations with something like this shape are fairly common in practical work. Because this distribution has a precise mathematical definition, it is possible to work out the standard deviation of s for large samples drawn from it. It turns out the SD_s in this case is 1.48 times as large as the value given by Formula 14. If you relied on that formula when sampling from a population like Fig. 5b, you would be seriously misled into thinking s to be much more accurate than it really is. And this illustration is by no means an extreme one. If you were to compute s for a large sample from the exponential distribution (Fig. 5d in Unit 5), its true standard deviation would be 2.65 times the value given by Formula 14.

In principle it would be possible to work out an appropriate formula like Formula 14 for each shape of distribution. Unfortunately the result would depend a great deal on the precise shape of the population tails, and in practice there is no good way to get this information from the sample unless we draw a sample of many hundreds of observations.

In our judgment, statisticians have not yet found a satisfactory way to attach a standard error to an estimate of the standard deviation, if the population is substantially nonnormal. Nevertheless, most statisticians would recommend that you compute SE_s from Formula 15, because they think most populations are close enough to the normal for this value to be useful. However, we urge you to remember that the computed standard error may be overoptimistic in case one or both tails of the population are heavier than with the normal shape.

STANDARD ERROR OF THE MEDIAN

We now turn to a discussion of $\tilde{x}$. As with the other statistics we have studied, theoretical considerations lead us to expect that the distribution of $\tilde{x}$ will be centered near the true population $\tilde{\mu}$, and that it will have a nearly normal distribution. For reasons to be given below, the median $\tilde{x}$ tends to be normal even for a sample of modest size drawn from a heavily skewed population.

Knowledge of $SD_{\tilde{x}}$ will as in the other cases lead to a method for attaching a standard error to the estimate $\tilde{x}$. Let us again begin by supposing that the sample comes from a normal population that has standard deviation σ. Then it can be shown mathematically that the normal distribution of $\tilde{x}$ has (approximately) the standard deviation

$$\mathrm{SD}_{\tilde{x}} = \frac{1.25\sigma}{\sqrt{n}} \qquad \textbf{FORMULA 16}$$

But what if the population is not normal? Suppose, for example, that it has a distribution like the solid curve of Fig. 23. Here both tails are longer and heavier than would be the case with a normal distribution, especially the right tail. To emphasize this fact, we show by the horizontal arrows how much the extreme values in this population would have to be pulled back toward the center to make the population normal (the dashed straight line). Let us call the population represented by the straight line (the population created by these pullbacks) the *hypothetical (normal) population.* Although there is only a small proportion of these extreme values in the real population that have to be pulled back, they make a substantial contribution to the standard deviation σ of this real population, for the same reason that a few extreme values in a sample make up most of the sample standard deviation s. Consequently, the hypothetical (normal) population of Fig. 23 will have a standard deviation—call it σ_h—that is substantially smaller than the σ of the real population.

Now let us see what the pullbacks indicated by the horizontal arrows on Fig. 23 would do to a sample. By the law of large numbers, a large sample drawn from the original heavy-tailed population would have a sample cumulative frequency polygon like the solid curve for the population itself. This means that the central values of the sample would have come from the central portion of the population, as you would anyway expect to be the case. These central population values are scarcely affected at all by the pullback. Recall that the sample median $\tilde{x}$ is the central value of the sample. The changes required to convert the original heavy-tailed population to the normal population of Fig. 23 would therefore alter the value of the median $\tilde{x}$ of the sample scarcely at all. In Fig. 23, the pullbacks do not alter the population median $\tilde{\mu}$ at all. Thus the median of the hypothetical population (dashed line) is also the median $\tilde{\mu}$ of the real population (solid line).

Since the pullbacks shown on Fig. 23 will leave the median of nearly all samples drawn from the population unchanged, the distribution of $\tilde{x}$ would be (nearly) the same whether we drew the sample from the original heavy-tailed distribution or from the hypothetical normal distribution into which it has been changed. By the earlier results (see Formula 16), the median $\tilde{x}$ of a sample from this hypothetical normal population has a (nearly) normal distribution, centered at $\tilde{\mu}$, and has standard deviation

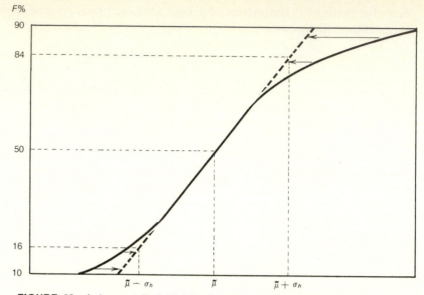

FIGURE 23 A heavy-tailed distribution and the hypothetical (normal) distribution fitted to it. The solid curve depicts the cumulative frequency polygon of a population with median $\tilde{\mu}$, plotted on normal probability paper. This population has heavy tails, especially on the right, and its standard deviation σ is correspondingly large. The dashed line depicts the hypothetical (normal) population that agrees with the actual population in its central part and therefore has the same median $\tilde{\mu}$. The standard deviation σ_h of the hypothetical population is, however, much smaller than the standard deviation σ of the actual population.

$$\mathrm{SD}_{\tilde{x}} = \frac{1.25\sigma_h}{\sqrt{n}}$$

FORMULA 17

It therefore follows that this formula also gives us the standard deviation of the median of a large sample from the original heavy-tailed distribution.

We can summarize this discussion in the following way. Suppose that you have a population *the central portion of which* is distributed like a normal population with standard deviation σ_h. Then the median $\tilde{x}$ of a large sample will have a nearly normal distribution, centered at the population median $\tilde{\mu}$, and with standard deviation given by Formula 17.

Of course, in practice a value of σ_h would have to be obtained from the sample itself. If we denote by s_h the sample estimate for σ_h, we can write the formula for the standard error of $\tilde{x}$ as

$$\mathrm{SE}_{\tilde{x}} = \frac{1.25s_h}{\sqrt{n}}.$$

FORMULA 18

To use this formula, all we need know is the value of s_h for the hypothetical (normal) distribution. To show how easily this is done, we drew a sample of $n = 50$ STATLAB mothers and recorded their weights at the time pregnancy was diagnosed. The sample median turned out to be $\tilde{x} = 129.5$ lb, and we wanted to attach a standard error to this estimate for $\tilde{\mu}$. We organized these 50 weights into a frequency table with class intervals of width 10. The results are plotted on normal probability paper in Fig. 24. As you see, the sam-

FIGURE 24 Finding the standard error $SE_{\tilde{x}}$ of the median $\tilde{x}$ of a sample. The cumulative frequency polygon of our sample of 50 STATLAB mothers' weights is plotted on normal probability paper as the solid line. The dashed line depicts a hypothetical (normal) distribution fitted by eye to the central part of the sample. Since this hypothetical cumulative distribution reaches $F\% = 16$ at 109 lb and reaches $F\% = 84$ at 149 lb, its standard deviation is $s_h = (149 - 109)/2 = 20$ lb.

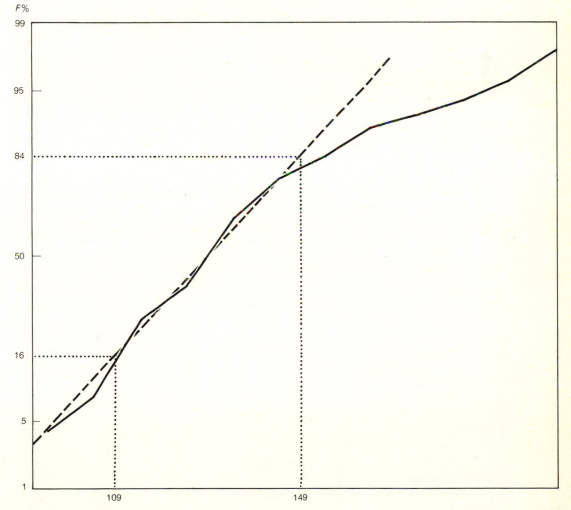

ple cumulative has a heavy right tail, reflecting the skewness only to be expected in a distribution of adult weights. The central portion of the plot, however, is reasonably straight, and we have drawn the straight dashed line that appears to follow as nearly as possible the trend of the sample cumulative polygon in its central portion. (The line is said to be "fitted by eye" to the central portion of the polygon.) This straight dashed line represents a hypothetical normal distribution that, on the basis of the sample, resembles the central portion of the population of STATLAB mothers' weights. By the graphic method of Unit 11 we find that the fitted normal distribution has standard deviation $s_h = 20$. We find in this way, by Formula 18, that $\mathrm{SE}_{\bar{x}} = (1.25 \times 20)/\sqrt{50} = 3.5$. We would report: On the basis of our sample, the STATLAB mothers had median weight 129.5 ± 3.5 lb. (Check with Fig. 17 in Unit 11 to see whether our 68 percent confidence interval did cover the true $\tilde{\mu}$.)

It is instructive to compare $\bar{x}$ and s with respect to their standard errors for nonnormal populations. As explained in the previous section, the value of s is heavily dependent on the few extreme values in the sample, and accordingly the behavior of s is heavily influenced by the tails of the population. Since only a small proportion of the population is in the tails, a sample of even fairly large size will contain but few extreme values. It is therefore hard to infer from the sample what the population tails are like. In consequence, it is hard to infer much from the sample about the behavior of s and especially hard to get a reliable standard error.

The situation of $\bar{x}$ is just the opposite. The value of $\bar{x}$ depends only on the center of the sample, and hence its behavior depends primarily on the center of the population. Most of the population is in its center, so there is a fair amount of information about the center in a sample of even quite modest size. In consequence, it is relatively easy to infer from the sample how $\bar{x}$ behaves, and the methods presented above readily provide a usable standard error for $\bar{x}$. Most of the continuous variables encountered in practical work have a distribution the central portion of which is enough like the central portion of a normal distribution that this method may be used. For even so nonnormal-looking a distribution as the exponential (Fig. 5d, Unit 5), this method will work well enough. We shall give you a chance to try it out with one of your own samples.

L'ENVOI: MEAN AND MEDIAN COMPARED

In Unit 8 we compared the two key measures of central tendency, $\bar{x}$ and $\tilde{x}$, from several points of view. Formula 17 permits us to round out that discussion by comparing $\tilde{x}$ and $\bar{x}$ with regard to accuracy. (And speaking of rounding out our mean and median comparison, we offer the discussion in Box 5.)

Recall, from Unit 13, Formula 9 for the standard deviation of $\bar{x}$: $\mathrm{SD}_{\bar{x}} = \sigma/\sqrt{n}$. It must be remembered that in this formula σ is the standard deviation of the *actual* population from which the sample came, whereas in Formula 17, σ_h is the standard deviation of the hypothetical normal distribution, the one that most closely fits the center of the population. Let us see what difference this difference makes in the relative accuracies of $\bar{x}$ and $\tilde{x}$.

Let us first consider samples drawn from a normal population. In this case, the hypothetical normal distribution is identical with the actual population distribution, so $\sigma_h = \sigma$. Comparison of Formulas 9 and 17 shows at once that $\bar{x}$ is more accurate than $\tilde{x}$ in this case. Indeed, $\tilde{x}$ is 25 percent more widely dispersed than $\bar{x}$, as dispersion is measured by the standard deviations of the estimates. You have in fact already checked that $\bar{x}$ is more accurate than $\tilde{x}$ for samples from a normal population, using the sampling results of your own class (Unit 8, item 9 of inquiry, recorded in Data-bank, File C-2). There you compared the dispersions of $\bar{x}$ and $\tilde{x}$ for samples from the very nearly normal distribution of mothers' heights.

But now consider a population like that shown by the solid curve of Fig. 23. As we mentioned above, σ is larger than σ_h in such cases and can be substantially larger. It would not be surprising to find σ 25 percent larger than σ_h—or even more, in case the population has a heavy tail or tails. If so, the accuracy advantage of $\bar{x}$ would be eliminated or even reversed.

In summary, the sample mean is a more accurate estimate than the sample median for samples drawn from populations that are normal or nearly so. Many statisticians consider that most populations are nearly normal and they accordingly prefer $\bar{x}$. If one or both tails of the distribution are substantially longer and heavier than they would be for a normal distribution, however, the advantage of $\bar{x}$ over $\tilde{x}$ is reduced. It is possible that the two estimates will then be about equally accurate or even that the median will be the better.

As an illustration, consider once more the weights of STATLAB mothers. By fitting a straight line to the central portion of Fig. 17 in Unit 11, you can easily read off $\sigma_h = 17$ lb for this population. It follows from Formula 17 that the median of a sample of n such weights has a standard deviation of

$$\mathrm{SD}_{\tilde{x}} = \frac{1.25 \times 17}{\sqrt{n}} = \frac{21.2}{\sqrt{n}} \text{ lb}$$

Now the actual standard deviation of the weights is $\sigma = 22.9$: This is substantially larger than σ_h, reflecting the heavy right tail of the distribution. It follows that the mean of the sample of n has the $\mathrm{SD}_{\bar{x}} = 22.9/\sqrt{n}$. We see that there is little to choose between $\tilde{x}$ and

BOX 5

THE MIDDLEMOST ESTIMATE EXPRESSES THE *VOX POPULI*

Galton's letter on the uses of the median (see Box 3, page 49) was followed, in the best tradition of England's letter to the editor columns, by several weeks of comment and countercomment by the readers of *Nature.* To begin with, Galton himself followed up his letter with a brief article in which he extended the political interpretation of the median. His article was titled "Vox Populi," excerpts of which follow:

> In these democratic times, any investigation into the trustworthiness and peculiarities of popular judgments is of interest. The material about to be discussed refers to a small matter, but is much to the point.
>
> A weight-judging competition was carried on at the annual show of the West of England Fat Stock and Poultry Exhibition. A fat ox having been selected, competitors bought stamped and numbered cards, for 6*d.* each, on which to inscribe their respective names, addresses, and estimates of what the ox would weigh after it had been slaughtered and "dressed." Those who guessed most successfully received prizes. About 800 tickets were issued, which were kindly lent me for examination after they had fulfilled their immediate purpose. These afforded excellent material. The judgments were unbiased by passion and uninfluenced by oratory and the like. The sixpenny fee deterred practical joking, and the hope of a prize and the joy of competition prompted each competitor to do his best. . . . According to the democratic principle of "one vote, one value" the middlemost estimate [median] expresses the *vox populi.* Now the middlemost estimate is 1207 lb., and the weight of the dressed ox proved to be 1198 lb., so the *vox populi* was in this case 9 lb., or 0.8 per cent of the whole

weight too high. . . . This result is, I think, more creditable to the trustworthiness of a democratic judgment than might have been expected.

Two weeks later, *Nature* published a demurrer to Galton's thesis on democracy, credibility, and the median. Under the title "Mean or Median," R. H. Hooker took a dim view of the allegedly unbiased nature of the cattlemen's judgment, and of Galton's statistical analysis. Here, in part, is Hooker's letter to the editor:

> Galton's letter raises several interesting points as to the theoretical treatment of statistical data, to two of which I would like to allude.
>
> In the first place, as to bias . . . it might be expected that buyers would have an instinctive tendency to underestimate the weights of animals; and similarly farmers (sellers) might be expected to overestimate. . . . The second and more important point . . . is the use of the median in this connection. . . . I should, in fact, like to strike a note of hesitation in regard to the too general use of the median in preference to the mean. . . . I am not sure that Mr. Galton is quite right in regarding the present instance as a case of "vox populi" at all. . . . The judgment of buyer and seller as to how much beef there is in a given ox is really much more a matter of skill than of popular judgment. . . . In such circumstances, is the median a nearer approximation to the truth than the mean? Here the question could be answered by calculating the arithmetic mean [and] judging from the data in Mr. Galton's article, the mean would seem to be approximately 1196 lb., which is much closer to the ascertained weight (1198 lb.) than the median (1207 lb.).

I should accordingly like to ask Mr. Galton whether he would indicate what, in his opinion, are the chief considerations to be taken into account in giving preference to the mean or the median as a better measure of the "average"? It is a point on which there is considerable difference of opinion.

Within a week Galton published a rejoinder which restated his case for the median as the only central measure which the ballot box can arrive at, whether we wish it or not. Galton refused, however, to accept the challenge with which Hooker had ended his letter: "I had no intention," writes Galton, "of trespassing into the technical and much discussed question of the relative merits of the Median and the several kinds of Mean, and beg to be excused from not doing so now."

In the very next issue of *Nature* appeared a letter by a new protagonist, the statistician G. Udny Yule. This letter sought to set Galton straight, to clear matters up for Hooker, and to put an end to the whole affair—and all done in statistical reasoning that will sound familiar to the reader of StatLab. Borrowing Hooker's no-nonsense title "Mean or Median" (a title, by the way, that StatLab, too, has borrowed for Unit 8), Yule wrote:

The two applications of the *median* suggested in Mr. Galton's letter . . . seem to me to be somewhat distinct. In the case of a jury or committee voting as to a sum of money to be given, there is no question of truth, but only of expediency. If any amount be proposed and put to a vote, the proposition will . . . be defeated so long as that amount is above the median; the process of voting tends, therefore, to give an amount *not greater than* the median. Mr. Galton's suggested procedure is in this case, it seems to me, quite correct. . . . The case of averaging a series of estimates with the view of arriving at objective truth appears to be on a different footing. If there is a considerable sprinkling of fools or knaves amongst estimators, or of persons with a tendency to bias . . . according to the suggestion of Mr. Hooker . . . the question as to the choice of means is one that is difficult to answer. The important question is in fact, not the "probable error" but the probable bias for the whole frequency distribution may centre round an entirely erroneous value. [Here Yule is making the important point that if your original data are erroneous there is very little use in applying sophisticated statistical analysis—or any other kind.] If, on the other hand, the observers are honest and unbiased, the choice of average turns on the form of the frequency distribution. . . . For the normal distribution . . . the probable error of the median is greater than that of the mean in the ratio of 1.25:1 approximately. [Compare our Formulas 9 and 16.] For a flatter topped curve with more curtate [shortened] tails the ratio of probable errors is greater than 1.25:1, and accordingly for all such distributions the arithmetic mean is the better form of average. But for a curve with a high central peak and long tails, the probable error of the median may be less than that of the mean, and it will be the more stable form of average.

Yule concludes his little statistical lecture with the following stern conversation-stopper: "In the absence of definite knowledge as to the frequency distribution of estimates in any specific case, it does not seem to me that any confident judgment as to choice of means can be given."

If you like this sort of thing, you might find it of interest to read this exchange in its entirety (including a note on *vox expertorum* by F. H. Perry-Coste). All the material can be found in the 1907 section of volume 75 of *Nature* under the following dates and on the following pages: 28 February, page 414; 7 March, page 450; 21 March, page 487; 28 March, page 509; 4 April, page 534.

$\bar{x}$ in this example, at least as far as accuracy is concerned, and indeed $\tilde{x}$ is slightly the more accurate. The choice between $\tilde{x}$ and $\bar{x}$ would presumably be made on other grounds, as in Unit 8.

Of course, in practice you would not have available the population distribution and could not calculate the two standard deviations as we have done in this illustration. But you would usually have past experience with the variable in question, and this should give some guidance in making the choice. Again it appears that the statistician who is familiar with the behavior of the variables with which he is working has a great advantage.

One final comment: We have given two formulas for $SD_{\tilde{x}}$, Formulas 16 and 17, and you might think that it would be necessary to decide which one to use as the basis for the standard errors of $\tilde{x}$. In fact, only the second formula is ever called for in practical work. Formula 16, although it is necessary in the course of our development of the general result, is valid only if the sample comes from a normal population. But if you know that your population is normal, you would have no occasion to compute $\tilde{x}$. In a normal population $\tilde{\mu} = \mu$, so the statistics $\tilde{x}$ and $\bar{x}$ are estimates of the same thing. Since in the normal case $\tilde{x}$ is 25 percent more widely dispersed than $\bar{x}$, in these circumstances only $\bar{x}$ would ever be used. Even if you are interested in estimating $\tilde{\mu}$, it would be reasonable to compute $\tilde{x}$ only if you thought the population was nonnormal or that it might be so. And in these circumstances, only Formula 17 can be relied upon.

☐ You are now ready to try out these ideas and methods. To complete the inquiry you will need to recall that the STATLAB population of mothers' heights has a very nearly normal distribution, with mean 64.43 in and standard deviation 2.50 in. The STATLAB family income at test has a highly skewed distribution, with mean $15,640 and standard deviation $6840.

WORK TO BE DONE

STEP 1   Retrieve from Databank, File D-1, the class distributions of s for mothers' heights and family income. Plot these distributions on the worksheet, page 1.

STEP 2   For your own samples of $n = 40$ mothers' heights ($w = 1$), and $n = 40$ family incomes, retrieve from Databank, File D-1, the sample standard deviations s, and record them on the worksheet, page 2.

STEP 3   Retrieve from Databank, Files B-4 and C-2, the class distributions of $\tilde{x}$ for mothers' heights and family income. Plot these distributions on the worksheet, page 3.

STEP 4   For your own sample of $n = 40$ family incomes at test, retrieve
from Databank, File C-1, the cumulative frequency table. Plot it
on the worksheet, page 4. Fit by eye a normal distribution to the
central portion of the distribution. Retrieve from Databank, File C-1,
the median income $\tilde{x}$ of your sample, and record it on the worksheet, page 4.

STEP 5   Complete the inquiry.

STANDARD ERROR OF A DIFFERENCE

In Unit 13 we explained why it is always desirable to give the standard error of a statistic when it is used to estimate a population value. Armed with a standard error, the user of your report can judge the accuracy of your estimate and can obtain confidence intervals for the parameter estimated. There we also saw how, thanks to the central limit theorem, one can determine the standard error of a mean. In Units 14 and 15 we extended that reasoning to provide us with techniques for obtaining standard errors for proportions, medians, and standard deviations. In this unit we round out our study by considering a special type of standard error, one which is equally applicable to means, proportions, medians, and standard deviations, as well as to all other statistics. This standard error and its associated confidence interval assess the accuracy with which *two* statistics can estimate the *difference* between the corresponding parameters of two populations.

The need to know the accuracy of estimates of differences arises very frequently in industry, scientific experimentation, public opinion polls, economic studies, etc. How much more variability is there in the iron content of ore *A* than ore *B*? Here we compare the σ's of the iron content of ores A and B as estimated, of course, by sample s's of A and B. How much does the median family income of Sweden exceed that of Italy? The difference between the two $\tilde{\mu}$'s will be inferred, of course, from the two corresponding $\tilde{x}$'s. How much greater is the mean brain weight of rats who live in a stimulating environment than that of rats isolated in laboratory cages? We are dealing here with differences in μ's as estimated by differences in $\bar{x}$'s. How much greater is the proportion who support tax exemption for church property among churchgoers than among atheists? This is a question about differences in P's deduced from examining differences in p's. In all of these—and similar questions —we want to know, of course, not only the size of the estimated difference but also the accuracy of that estimate. In other words, to every difference we estimate we should attach the standard error of that difference.

STANDARD ERROR OF A DIFFERENCE

Let us first discuss the standard error of a difference in the context of comparing population means μ. Let us indicate the mean of one population by μ_1 and the mean of a second population by μ_2. (Whenever we deal with two or more populations we specify the population to which each parameter—or statistic—belongs by attaching some distinguishing subscript to the parameter or statistic.) We want to estimate $\mu_1 - \mu_2$, the difference between the two. The method is obvious: Take a sample from the first population and calculate its mean $\bar{x}_1$, and take a sample from the second population and obtain its mean $\bar{x}_2$. If the samples are randomly chosen and their respective n's (n_1 and n_2) are large enough, then we can use $\bar{x}_1$ and $\bar{x}_2$ as estimates of μ_1 and μ_2. It would then follow that $\bar{x}_1 - \bar{x}_2$ is a reasonable estimate for $\mu_1 - \mu_2$. What is the standard error of this estimate?

We begin with the observation that the difference between two means, $\bar{x}_1 - \bar{x}_2$, has like all statistics a frequency distribution. This distribution consists of all the values obtained by subtracting every possible sample mean of the second population from every sample mean of the first population. We already know that to obtain a useful standard error of such a statistic we must know three things about its distribution: the *mean* of the $\bar{x}_1 - \bar{x}_2$ distribution, the *shape* of the distribution, and the *standard deviation* of the distribution. Fortunately, probability theory provides answers to all three questions:

1   The distribution of $\bar{x}_1 - \bar{x}_2$ is centered at $\mu_1 - \mu_2$.
2   The distribution is normal if the separate $\bar{x}_1$ and $\bar{x}_2$ distributions are normal and, as you know, this is usually the case if the sample sizes are not too small.

3 There is a probability theorem which proves that if two statistics arise from the two separately drawn samples, the variance of their difference is equal to the sum of their separate variances:

$$\text{Var}_{\bar{x}_1 - \bar{x}_2} = \text{Var}_{\bar{x}_1} + \text{Var}_{\bar{x}_2}$$

Let us see how, in three easy steps, this can yield the standard error of the difference between means.

We know from Unit 9 that the variance of any distribution is equal to the square of the standard deviation of that distribution. This permits our rewriting the theorem we have just stated as follows:

$$(\text{SD}_{\bar{x}_1 - \bar{x}_2})^2 = (\text{SD}_{\bar{x}_1})^2 + (\text{SD}_{\bar{x}_2})^2$$

If we now take the square root of both sides, we get

$$\text{SD}_{\bar{x}_1 - \bar{x}_2} = \sqrt{(\text{SD}_{\bar{x}_1})^2 + (\text{SD}_{\bar{x}_2})^2}$$

Since we know (Formula 9) that $\text{SD}_{\bar{x}} = \sigma/\sqrt{n}$, it then follows that $(\text{SD}_{\bar{x}})^2 = \sigma^2/n$. Making this substitution we now write

$$\text{SD}_{\bar{x}_1 - \bar{x}_2} = \sqrt{\frac{\sigma_1^2}{n_1} + \frac{\sigma_2^2}{n_2}}$$

FORMULA 19

And now comes our usual "of course": Of course in practice we do not know the population deviations σ_1 and σ_2, so we substitute for them their sample estimates s_1 and s_2 to obtain the final expression for the formula for a standard error of a difference between two means:

$$\text{SE}_{\bar{x}_1 - \bar{x}_2} = \sqrt{\frac{s_1^2}{n_1} + \frac{s_2^2}{n_2}}$$

FORMULA 20

We know that the distribution of differences between means, $\bar{x}_1 - \bar{x}_2$, is centered at the real difference, $\mu_1 - \mu_2$, and is (approximately) normally distributed. We can therefore use the standard error of the difference between means to construct appropriate confidence intervals which will have the same interpretations as confidence intervals for any other normally distributed statistic. Thus if we report the difference between means and the standard error of the difference as

$$(\bar{x}_1 - \bar{x}_2) \pm \text{SE}_{\bar{x}_1 - \bar{x}_2}$$

we are saying that there is a 68 percent chance that the true difference, $\mu_1 - \mu_2$, will be found somewhere between the estimated difference $(\bar{x}_1 - \bar{x}_2)$ *minus* $\mathrm{SE}_{\bar{x}_1 - \bar{x}_2}$ and $(\bar{x}_1 - \bar{x}_2)$ *plus* $\mathrm{SE}_{\bar{x}_1 - \bar{x}_2}$. Similarly, there is a 95 percent chance that the true difference will lie within the estimated $2\mathrm{SE}_{\bar{x}_1 - \bar{x}_2}$ of the estimated difference $\bar{x}_1 - \bar{x}_2$ and so forth. Again we may use the normal table to construct any confidence interval we wish for the true difference between means.

To illustrate: Suppose that we wanted to know how much, on the average, the STATLAB boys differed from the STATLAB girls in their Peabody mental test scores. Call the STATLAB boys "population 1" and the girls "population 2." We must first draw two samples, one for the boys and one for the girls. The two samples need not, of course, have the same n—but they must be drawn separately, independently, and of course randomly. Let us detail the most efficient way of doing this with our STATLAB Census. We shall, say, want a sample of $n_2 = 16$ for the girls and $n_1 = 13$ for the boys. As you know, the STATLAB Census is so constructed that Census pages 11 through 36 contain the records of the girls and pages 41 through 66 are devoted to the boys. This means that on each drawing of a page (the first throw of the dice for each draw), every time the red die shows a 1, 2, or 3 we have been directed to a girls' page and when it shows a 4, 5, or 6 a boy is indicated. The second throw specifies an individual on that page whose Peabody score we record in either the boys' or girls' column as appropriate. We continue drawing our individuals until one of the samples is completed. (Let us assume that the boys' sample is completed first.) We continue throwing the dice, but *discard* all "page drawings" where the red die shows either a 4, 5, or 6, until the second sample is also complete. In this way we shall have drawn *separate* and *random* samples with a minimum of superfluous (discarded) throws. We stress this point because the probability theorem basic to the derivation of the standard error of the difference,

$$\mathrm{Var}_{\bar{x}_1 - \bar{x}_2} = \mathrm{Var}_{\bar{x}_1} + \mathrm{Var}_{\bar{x}_2}$$

demands that the two samples be separately and independently drawn.

Table 19 presents the samples drawn in accordance with these procedures. These data will yield the following statistics: for the boys: $\bar{x}_1 = 83.00$, $s_1 = 7.07$; for the girls: $\bar{x}_2 = 80.94$, $s_2 = 8.12$. Combining the information from the two samples, we estimate $\mu_1 - \mu_2$ by the statistic $\bar{x}_1 - \bar{x}_2 = 83.00 - 80.94 = 2.06$. In other words, our sample statistics estimate that on the average the STATLAB boys surpass the girls by 2.06 points on the Peabody test. Using Formula 20 we find the standard error of that estimated difference to be

TABLE 19 Random samples (in order drawn), Peabody scores, STATLAB boys and girls

Boys ($n_1 = 13$)		Girls ($n_2 = 16$)	
ID no.	Score	ID no.	Score
65-41	77	26-21	80
65-21	79	36-64	70
52-64	91	26-36	81
64-61	83	21-55	93
56-43	87	26-65	87
63-64	81	36-51	84
43-54	70	31-22	76
53-26	81	14-55	92
63-56	84	12-13	88
65-22	99	25-56	88
41-33	85	25-62	85
56-61	87	35-21	78
53-53	75	12-41	66
		31-54	84
		15-13	66
		26-14	77

In drawing these samples, we happened to complete the boys' sample with ID number 53-53 when only 14 girls had been drawn. The next throw gave 46; since 46 is a boys' page, this throw was ignored. On our next throws we obtained 15-13 (recorded as the fifteenth girl); then 62 (ignored); then 55 (ignored); and finally 26-14 (recorded as the last girl).

$$\sqrt{\frac{(7.07)^2}{13} + \frac{(8.12)^2}{16}} = \sqrt{3.84 + 4.12} = 2.82$$

We are now prepared to make the usual statement: The mean Peabody score of the STATLAB boys exceeds that of the STATLAB girls by 2.06 ± 2.82 points.

A reader of this report, noticing that the standard error 2.82 is large relative to the estimated difference 2.06 itself, could reasonably conclude that our estimate is so inaccurate that we have not really even shown that there is any difference between the boys' and girls' scores. This conclusion would be strengthened when that reader determined the confidence limits for the true difference. Using the found difference and its standard error, he could say: "There is a 68 percent chance that the true difference between the average scores of the STATLAB boys and girls lies in the interval 2.06 ± 2.82, or somewhere between -0.76 and 4.88 points. In other words, there is a 68 percent chance that the true difference is as high as 4.88 points, or as low as zero, or even 0.76 points *in favor of the girls*—or anything in between." (A minus value, of course, means that $\bar{x}_2$ is greater than $\bar{x}_1$.) "And what is more," your now very critical reader can assert, "there is a 95 percent chance that the girls exceed the boys by as many as 3.58 points!" [2.06 − 2(2.82)] The prudent statistician would be justified, after considering the alleged sex difference in Peabody scores, to render the Scottish verdict of "Not Proven."

DIFFERENCES BETWEEN PROPORTIONS

A proportion, as we have seen in Unit 14, can be thought of as a special case of the mean, and we can end our presentation of the standard error of the difference between means by presenting the formula for the standard error of the difference between proportions.

We know from Formula 12, Unit 14, that the standard deviation of a proportion is $SD_p = \sqrt{PQ/n}$. If we now consider p as a special case of $\bar{x}$, then everything we have said here about $\bar{x}$ applies to proportions and it immediately becomes apparent why the formula for the standard deviation of the difference between two proportions becomes

$$SD_{p_1-p_2} = \sqrt{\frac{P_1 Q_1}{n_1} + \frac{P_2 Q_2}{n_2}}$$

FORMULA 21

Substitution of estimates for the parameters gives

$$SE_{p_1-p_2} = \sqrt{\frac{p_1 q_1}{n_1} + \frac{p_2 q_2}{n_2}}$$

FORMULA 22

STANDARD ERRORS OF DIFFERENCES BETWEEN OTHER MEASURES

The basic probability theorem that gave us our standard error of the difference between means is a generally applicable one, for it asserts that the variance of the difference between *any* two statistics is equal to the sum of their separate variances. This is an intuitively appealing conclusion since it says, in effect, that the degree of error of an estimate of a difference between two measures is the sum of the errors with which each individual measure was estimated to begin with. Accordingly we can see why, by exactly analogous reasoning, we can determine standard errors of the estimates of the differences between medians, or standard deviations, or any other parameters. Thus for the median and standard deviation, Formulas 23 and 24 apply:

$$SE_{\tilde{x}_1 - \tilde{x}_2} = \sqrt{\frac{(1.25 \, s_{h_1})^2}{n_1} + \frac{(1.25 \, s_{h_2})^2}{n_2}}$$

FORMULA 23

$$SE_{s_1 - s_2} = \sqrt{\frac{s_1^2}{2n_1} + \frac{s_2^2}{2n_2}}$$

FORMULA 24

Two final words of caution are in order. First, the formula for the standard error of the difference between s_1 and s_2 (Formula 24) works only if both populations are normal. You will recall that in Unit 15 we came to the conclusion that for nonnormal distributions we could not arrive at a satisfactory standard error of s. If in nonnormal cases we cannot measure the accuracy of any single s, it is apparent that we would be hard put to it to measure the accuracy of the difference between two such s values. Second, we must repeat that our formulas for the standard error of a difference are applicable only for two separately and independently drawn samples. When the two statistics are computed from the same sample, so that the statistics are thereby related to each other, a different approach is needed.

☐ You are now ready to use and to check out one of the major findings of probability theory as it applies to the central problem of determining the accuracy of estimates. In your work you will have as your immediate objective the estimate of the true differences between boy and girl STATLAB infants in certain physical and biological attributes: birthweights and blood types. The success or failure with which you and your classmates can make such estimates will be some indication of the validity of the reasoning behind the concept of the standard error of a difference.

WORK TO BE DONE

STEP 1   Draw samples of sizes $n_1 = 20$ of boy infants and $n_2 = 15$ of girl infants. Record the ID numbers on page 1 or page 2 of the worksheet, as appropriate. Record, in the columns provided, the birthweight (rounded to the nearest pound) and blood type as coded in the Census.

Note: For a small percentage of the children, the blood type was not recorded (code 9 for Variable B: see page 320). For each such child in your samples, you will need to draw an additional child in order to obtain the specified numbers (20 and 15) of children with known blood type. Extra blood-type blanks are provided for this purpose. Use the original sample for birthweight.

STEP 2   For boys and girls separately, construct the ungrouped frequency table as indicated, and compute the means and standard deviations by the coding method.

STEP 3   Using the results of step 2, compute the difference $\bar{x}_1 - \bar{x}_2$ and its standard error, and record on page 3 of the worksheet.

STEP 4   For boys and girls separately, determine the proportion of infants having blood type O (combining Rh-negatives and Rh-positives).

STEP 5   Using the results of step 4, compute the difference $p_1 - p_2$ and its standard error, and record on page 3 of the worksheet.

STEP 6   Record in Databank, File G-1, the first 20 ID numbers for boys and the first 15 ID numbers for girls. (The columns for Raven scores will be filled in later.)

STEP 7   Complete the inquiry.

BAYESIAN INFERENCE

We have on several occasions pointed out how great an advantage accrues to the statistician who has some knowledge about the population from which he is going to draw a sample. For example, knowing that a population is highly skewed may lead him to select the median rather than the mean as his measure of central tendency (Unit 8 and Unit 15). Knowing that a population is normal will justify the use of Formula 15 for the standard error of s (Unit 15). Knowing the approximate value of σ will help the statistician to select the proper sample size (Unit 14). But a knowledgeable statistician may also have less formal information about a population. Can such information help a statistical analysis?

Suppose, for instance, that you were going to draw a sample of male undergraduates at your college to estimate the center of the distribution of their heights. Because of what you know about human heights, you would know that the distribution is probably nearly normal and would accordingly decide to specify the center by the population mean μ and to use $\bar{x}$ as the estimate for μ. In addition you would know that σ is probably around 3 in, and this fact tells you how large a sample is needed for the estimate $\bar{x}$ to have a given degree of accuracy.

But of course you have advance knowledge not only about the shape and σ—knowledge gathered through your study of statistics. You also have some definite ideas, based on ordinary observations of your fellow students, about the value of μ itself. From such observations, you know that relatively few male undergraduates are taller than about 74 in and that relatively few are shorter than about 67 in. Thinking along these lines, you might be able to *guess* a pretty accurate value for μ, even before drawing a single person at random. Suppose that you make such a guess.

Now you go to work to draw the random sample and measure the heights of the male undergraduates so selected. The evidence from the sample can provide the estimate $\bar{x}$ with its properties as we have developed them in Unit 13. But what about the information contained in your prior guess as to the value of μ? Should it just be thrown away? Or would it make sense to combine the information known before the sample was drawn with the information in the sample itself?

There is a vigorous school of statisticians, known as bayesians, who answer this question with a firm Yes. The eponymous Thomas Bayes (1702–1761) was an English clergyman and Fellow of the Royal Society who wrote the first mathematical study suggesting how it is possible to combine prior ideas with sampling results. The mathematics involved in making the combination tends to get rather fierce in most cases. Fortunately, however, there is one very important problem that can be dealt with quite simply: estimating a population proportion P. We shall with the work of this unit ask you to try out the bayesian method on two examples of estimating P. In this way you can learn to make the combination of prior ideas and sampling results; and you can also, by combining your efforts with those of others in the class, see how well the bayesian inference actually works.

THE BAYESIAN METHOD FOR A PROPORTION

We shall begin by seeing how your prior ideas about a population proportion can be expressed in a form analogous to the results based on sample frequencies as developed in Unit 14; this will enable us easily to combine the two sources of information.

In Unit 14 we found that the sample proportion p of Yes answers is the natural estimate for the corresponding proportion P of Yes answers in the population. The results of the central limit theorem tell us, for example, that p is about as likely to fall above P as it is to fall below it. Analogously to this, the bayesian statistician attempts before the sample has been drawn to decide what he thinks is the best value of P. He recalls all that he knows or believes about the population in question and after *careful thought* comes up with his best guess. Let us denote this guess by p_B (we shall attach the subscript B for "Bayes" to the bayesian analogs of the formulas of Unit 14). This value p_B should be so chosen that the statistician thinks P is as likely to be above p_B as below it—just as, with actual samples, P is as likely to be above as below the estimate p.

Let us give an illustration. Suppose that the mayor of your city is running for reelection and you are interested in knowing the proportion P of voters who favor him. Think of all you know about the matter: What percentage of the votes did he get in the previous election? Have his policies been well received? Have there been scandals in his administration? Is the press favorable? Suppose that, after reflecting on all such considerations, your best guess is that the mayor will get 56 percent of the votes, that is, $p_B = 0.56$.

The bayesians have devised a sort of thought-experiment to help you check your value of p_B. They ask you to imagine that you are going to have to bet a large sum of money (say $100) on whether P will turn out to be above your chosen value p_B or below it. *Would you be equally willing to bet either way*? If not, the value of p_B needs adjustment. If you would, for example, rather bet that P is above p_B than to bet that P is below p_B, you need to think some more and increase p_B to bring it to a proper *central* value. (As Dr. Johnson might well have said: "Depend upon it, Sir; when a man knows he may lose a large sum of money, it concentrates his mind wonderfully.")

The analysis of Unit 14 gave us not only the estimate p but also its standard error SE_p, with the interpretation that the interval $p \pm SE_p$ has a 68 percent chance to cover P. What is the bayesian analog of SE_p? It would be a number—let us denote it by SE_B—such that (prior to drawing a sample) you think there is a 68 percent chance that P lies in the interval $p_B \pm SE_B$. In the reelection illustration, let us suppose that you decide to put $SE_B = 0.07$. This means that you believe there is a 68 percent chance that P falls somewhere in the interval 0.56 ± 0.07, which is to say somewhere between 0.49 and 0.63. Or, putting it the other way, there is a 32 percent chance that P fails to fall in this interval, according to your prior ideas.

This too can be stated as a wager. If you think the odds in favor of covering P are 68 to 32, then you should be willing to bet $68 that $p_B \pm SE_B$ covers P against $32 that it fails to cover. Would you indeed regard this as a fair bet? *Would you be equally willing to bet either way*? If not, an adjustment of SE_B must be made so that you *would* be equally willing to take either end.

The labor of contemplation, to translate your prior ideas into values of p_B and SE_B in accordance with the definitions of the statistics, is the hard part; but with the help provided by the suggested bayesian thought-experiment of wagering, it can be done. Once the values of p_B and SE_B are in hand, it is easy to convert them into a format suitable for direct combination with actual sample results. Let us see how this is done.

To begin with we need the bayesian analog of sample size n. Recall from Formula 13 of Unit 14 that the formula for SE_p has the term n in it, that is, $SE_p = \sqrt{pq/n}$. This equation can easily be solved for n by squaring both sides, multiplying by n, and dividing by $(SE_p)^2$:

$$(SE_p)^2 = \frac{pq}{n} \qquad n(SE_p)^2 = pq \qquad n = \frac{pq}{(SE_p)^2}$$

The analogous bayesian sample size is, accordingly, $n_B = (p_B q_B)/(SE_B)^2$, where of course $q_B = 1 - p_B$. Once we have computed n_B, we can state the bayesian analog of the number np of Yes answers in the sample as $n_B p_B$.

Before proceeding, let us illustrate this with the reelection problem. There it was assumed that after taking thought you had arrived at $p_B = 0.56$ and $SE_B = 0.07$. From these guessed values it follows that $q_B = 0.44$, $n_B = (0.56 \times 0.44)/(0.07)^2 = 50.3$ (which we round off to the nearest integer 50 to keep things simple), and $n_B p_B = 50 \times 0.56 = 28$. These numbers $n_B = 50$ and $n_B p_B = 28$, which are analogous to n and np, have a simple interpretation: If you had in fact drawn a sample of $n_B = 50$ voters and had found that $n_B p_B = 28$ of these 50 voters were in favor of reelection, you would then have computed, by the methods of Unit 14, the 68 percent confidence interval 0.56 ± 0.07 for P. (Check that this is so.) Thus we have calculated from your prior ideas an imaginary sampling experiment that would lead to the same interval which the bayesian has found by contemplation. We may reasonably say that the sample of $n_B = 50$, with $n_B p_B = 28$ Yes answers, is *equivalent* to your prior ideas and represents them in a form analogous to the results based on sample frequencies as developed in Unit 14.

The rest is easy. If you now draw an *actual* sample, say of $n = 40$ voters, and find that $np = 19$ of them favor reelection, you can at once combine this evidence with that offered by your prior ideas as represented by the equivalent (imaginary) sample. The *combined* sample size is $n_C = n_B + n = 50 + 40 = 90$, and the combined number of Yes answers is $n_C p_C = n_B p_B + np = 28 + 19 = 47$. (We use the subscript C for "combined.") The data as a whole, real together with imaginary, give the estimate $p_C = \frac{47}{90} = 0.522$, to which we attach the standard error

$$SE_C = \sqrt{\frac{0.522 \times 0.478}{90}} = 0.053$$

as calculated by Formula 13. The estimating interval 0.522 ± 0.053 represents the total information available to you: your prior ideas as well as what you found by drawing an actual sample.

Let us contrast this with the estimate available if you were to ignore your prior ideas and use only the sampling evidence. The actual sample gives $p = \frac{19}{40} = 0.475$, and this estimate has the standard error

$$\mathrm{SE}_p = \sqrt{\frac{0.475 \times 0.525}{40}} = 0.079$$

This pure-sampling interval 0.475 ± 0.079 is substantially wider than the combined interval 0.522 ± 0.053, in fact about 50 percent wider. This is natural, since the actual sample size is only 40 whereas the combined sample contains 90 "observations."

Let us summarize the bayesian procedure for proportions. First, by contemplation, determine p_B so that you think P is *equally* likely to fall above or below it. Then determine SE_B so that you think there is a *68 percent chance* that $p_B \pm \mathrm{SE}_B$ covers P. Calculate the equivalent sample size $n_B = p_B q_B / (\mathrm{SE}_B)^2$ and the equivalent number of Yes answers $n_B p_B$. Now take an actual random sample and add its data to the (imaginary) data of the equivalent sample. Analyze the combined data as in Unit 14. We shall give you a chance to try this out, using your own prior ideas and your own samples.

CRITIQUE OF THE BAYESIAN METHOD

As our illustration shows, the bayesian method combines prior ideas and newly gathered evidence and thereby comes up with shorter estimating intervals than could be found using the new evidence alone. Many statisticians find this bayesian approach satisfactory, and several persuasive advocates have urged its general use (see Box 6). However, the majority of statisticians remain skeptical.

One reason for resistance to the bayesian method has already been alluded to in Unit 2: Evidence based solely on a random sample is attractive to many investigators just because of the impartiality of honest dice or of a table of random numbers. Such mechanisms have no axe to grind (although they can, as you have seen, sometimes give quite wrong answers). In contrast, the prior ideas of the statistician are inherently personal and subjective. They reflect his knowledge and experience, but also his biases and prejudices. This would be acceptable if he were the only person to use the results of the analysis, but all worthwhile investigations are of concern to many people. Many members of the public using his results may not share his preconceptions. They may prefer that the analyst

BOX 6

TO BUILD ON SAND—OR ON THE VOID...

Bayesian statistics is usually dated from 1763, two years after Bayes' death, when his mathematical papers were published posthumously by a friend. The revival of bayesian statistics dates from the 1950s—which makes modern bayesian statistics a very recent discipline indeed. Novel and imaginative approaches to old problems frequently attract to them enthusiastic proselytizers, and the modern descendants of the Reverend Thomas Bayes are no exception to this rule. They are most zealous in their efforts to convince and convert.

The following short excerpts are from a bayesian article designed to demonstrate to experimental psychologists how superior bayesian inference is to classic statistical inference in the treatment of scientific data. They are offered here with the hope that they will transmit to the reader some of the flavor of the neo-bayesian's enthusiasm. Those whose interest is whetted might well read the whole paper from which these passages have been excerpted.

Bayesian statistics as a coherent body of thought is still too new and incomplete [to write a definitive text].

[T]here must be at least as many Bayesian positions as there are Bayesians. Still, as philosophies go, the unanimity among Bayesians reared apart is remarkable and an encouraging symptom of the cogency of their ideas.

The Bayesian approach is a common-sense approach. It is simply a set of techniques for orderly expression and revision of your opinions with due regard for the internal consistency among their various aspects and the data.

People noticing difficulties in applying Bayes' theorem remarked, "We see that it is not secure to build on sand [vague prior opinion], take away the sand, we shall build on the void."

Reflection shows that any [statistical] policy that pretends to ignore prior opinion will be acceptable only insofar as it is actually justified by prior opinion.

Ward Edwards, Harold Lindman, and Leonard J. Savage, Bayesian Inference for Psychological Research, *Psychological Review* **70**: 193–242 (1963).

"just give us the news, please" and let each reader of the report judge it in the light of his own prior ideas.

Consider, for example, a medical researcher who has developed a new therapy in which he strongly believes. He runs a clinical trial, the results of which are inconclusive. Should he just report those findings—or should he combine them with his own favorable prior ideas and announce that the results are positive? Most people would prefer that he report his actual experimental findings as such. He is then of course free to add his own views, clearly labeled as such.

A second reason why many statisticians are reluctant to accept the bayesian approach relates to the desirability of attaching to any estimate a statement about its accuracy (Unit 13). When a statistician gives his estimate in the form $p \pm SE_p$, for example, the user knows that there is a 68 percent chance that this interval covers the true P. This theoretical prediction is borne out in actual experience; such intervals do cover about 68 percent of the time, as you have seen. No such statement can be made about the prior intervals $p_B \pm SE_B$ or about the interval $p_C \pm SE_C$ derived in part from the prior ideas. The fact that you are willing to bet \$68 to \$32 that $p_B \pm SE_B$ covers P gives us no assurance that this event will in fact occur with any specified frequency. What do we know about the frequency with which you win your bets?

Many statisticians could live with these two disadvantages if they were indeed convinced that the bayesian method could be relied on to give estimates that were more accurate. But is this the case? The estimate p_C can be understood as a kind of average of the prior estimate p_B and the sampling estimate p. It follows that if your prior guess p_B is right, so that p_B is at or very close to P, averaging in p_B in this fashion will tend to improve the estimate p. Should your prior guess p_B be far off, however, averaging it in will only make things worse.

To return to our reelection illustration, suppose that in reality $P = 0.55$, which supposition would be quite consistent with the sampling results (it falls within the 68 percent confidence interval based on the sample, 0.475 ± 0.079). In that case, the combined estimate $p_C = 0.522$ would be more accurate than the sampling estimate $p = 0.475$—an error of $0.522 - 0.55 = -0.028$ as compared with $0.475 - 0.55 = -0.075$. Now, however, suppose that P is really 0.4, which supposition is equally consistent with the sampling evidence. In this case the combined estimate has the larger error: $0.522 - 0.4 = 0.122$ as compared with $0.475 - 0.4 = 0.075$. If your prior ideas are right, it helps to average them in with the experimental evidence; but they may be wrong, in which case they are harmful.

But are they helpful in *most* cases? This is not a question to be answered by mathematics or by throwing dice, since it concerns the accuracy of peoples' ideas and their ability to express them in numerical terms. We guess that the answer depends very much on various circumstances. For example: How much does *this* statistician know about *this* variable? What are the personality characteristics that determine guessing behavior, and how is *this* statistician endowed in this respect? Very little careful research has been done on this matter. We shall give you the chance to participate, with other members of your class, in two small investigations of this kind.

☐ You are now ready to try your own hand at bayesian inference about two proportions relating to the STATLAB population. To make the work as interesting and instructive as possible, we ask you to observe two precautions. First, please do not glance at the Census until you have recorded your prior ideas in writing—to take even a peek now would tend to confuse the distinction between sampling evidence and genuinely prior knowledge. Second, please spend some time reflecting on your prior ideas, so that the values p_B and SE_B which you record properly represent what you really think. Try the trick of imagining that you have to bet a lot of money on the outcome.

We have selected two STATLAB proportions P that have not been studied earlier in the book, but about which you may well have some prior ideas. The first is the proportion of STATLAB children who were born on a weekend (that is, on Saturday or Sunday). What do you know about the distribution of childbirths over time? Do obstetricians like to play golf on Saturdays, and can they manage to do so? On what day of the week were you born? Consider whatever you think may be relevant. No holds barred—except a peek at the Census.

The second exercise is to think about the proportion of STATLAB fathers who were reported at test as having quit smoking—i.e., for whom code Q is recorded. Think of your father and of the men you know who have 10-year-old children. Remember that the time of test in STATLAB was 1971–1972. Was that around the time of the Surgeon General's report on smoking and lung cancer, which frightened many men into quitting cigarettes? Recall that the STATLAB families had been enrolled for 10 years in a prepaid health plan, that the median income was around $15,000, and anything else that might be pertinent—but do not look in the back of the book.

WORK TO BE DONE

STEP 1 After careful consideration, record on the worksheet, page 2, your prior ideas about the birthday and smoking variables in the form of values of p_B and SE_B.

STEP 2 Convert these values into equivalent samples n_B and $n_B p_B$.

STEP 3 Now draw a fresh random sample of $n = 35$ families, and record on the worksheet, page 1, the indicator $x = 1$ if the child was born on the weekend (otherwise $x = 0$) and the indicator $x = 1$ if the father was, at time of test, a quitter of cigarette smoking (otherwise $x = 0$).

STEP 4 For each problem determine np; combine it with the equivalent prior samples; and compute the combined estimate. Use Formula 13 to attach a standard error to each estimate.

STEP 5 Record in Databank, File H-1, the ID numbers of your new sample of $n = 35$ (from page 1 of your worksheet) and the various statistics calculated on page 2 of your worksheet.

STEP 6 Complete the inquiry.

In this second self-test review you have an opportunity to check out your grasp of the basic principles of inferential statistics. Since the next major area of study—testing of statistical hypotheses—depends heavily upon inferential statistics, we strongly advise a conscientious use of this test. Whenever this test reveals a soft spot in your comprehension, we urge you to go back and review the appropriate unit.

☐ The *central limit theorem* describes what happens to the mean, the standard deviation, and the shape of an $\bar{x}$ *distribution* as the sample size increases without limit. The results of the central limit theorem permit the derivation of the *standard error* formula: $\text{SE}_{\bar{x}} = s/\sqrt{n}$. This in turn enables one to specify, in quantitative terms, the *accuracy* of a sample $\bar{x}$ as an estimate of μ and, as a further step, to determine any desired *confidence interval* for μ by recourse to the *normal table*. (UNIT 13)

☐ The central limit theorem and the concepts of standard errors and confidence intervals can be extended to qualitative or any *dichotomized* data by encoding such data into numerical values of an *indicator variable x.* This leads to the following conclusions for the indicator variable: $\bar{x} = p$, $\mu = P$, $SD_p = \sqrt{PQ/n}$, $SE_p = \sqrt{pq/n}$. For inferences which are based on the central limit theorem, several solutions are available for determining the sample n that will yield any desired confidence interval for μ. These solutions depend on the formula $SD_{\bar{x}} = \sigma/\sqrt{n}$. For estimating P from p, a simpler and more dependable solution derives from the use of the formula $SD_p = 0.5/\sqrt{n}$. (UNIT 14)

☐ For normal samples $s/\sqrt{2n}$ yields a good estimate of the *standard deviation* of s. Because the calculation of s involves squaring each deviation from the mean, however, $s/\sqrt{2n}$ is not a satisfactory estimate for samples drawn from populations with a heavy tail or tails. This lack of precision of s as an estimate of σ also limits the applicability of $\text{SE}_{\bar{x}}$, but it does not affect $\text{SE}_{\tilde{x}}$. Since the value of $\tilde{x}$ depends primarily on the center of the population, a useful $\text{SE}_{\tilde{x}}$ for nonnormal populations (the only ones where the occasion arises for computing $\text{SE}_{\tilde{x}}$) is given by $1.25 s_h/\sqrt{n}$. Because of this, if one or both tails of the population are heavier than the normal, the accuracy of the estimate of $\tilde{x}$ may be equal to or even superior to that of $\bar{x}$. (UNIT 15)

☐ From the probability theorem—*if two statistics arise from two independent samples, the variance of their difference is equal to the sum of the separate variances*—we can obtain the following statement: The *standard error of the difference* between two estimates is equal to the square root of the sum of the squares of the standard errors of the two estimates. This general statement can be specified into the following formulas:

$$SE_{\bar{x}_1 - \bar{x}_2} = \sqrt{\frac{s_1^{\,2}}{n_1} + \frac{s_2^{\,2}}{n_2}}$$

$$SE_{s_1 - s_2} = \sqrt{\frac{s_1^{\,2}}{2n_1} + \frac{s_2^{\,2}}{2n_2}} \qquad \text{(normal populations)}$$

$$SE_{p_1 - p_2} = \sqrt{\frac{p_1 q_1}{n_1} + \frac{p_2 q_2}{n_2}}$$

$$SE_{\tilde{x}_1 - \tilde{x}_2} = \sqrt{\frac{(1.25 s_{h_1})^2}{n_1} + \frac{(1.25 s_{h_2})^2}{n_2}} \qquad \text{(UNIT 16)}$$

☐ Departing from the sample frequency inferential statistics is the approach referred to as *bayesian inference*. This approach combines *prior ideas* with sampling results. To make this combination mathematically requires that prior ideas about a population be expressed in *forms analogous to sample statistics*. Thus in estimating population proportions à la Bayes, we must first determine, by reflection, p_B, q_B, SE_B, and n_B; we then determine, by sampling, p, q, and SE_p; and finally, by computation, n_C, p_C, q_C, and SE_C. A number of arguments may be cited in support of and in criticism of bayesian inference. (UNIT 17)

THE TESTING OF STATISTICAL HYPOTHESES

In Units 13 to 17 we considered the statistical problems of estimating a parameter from a statistic. We now turn to another set of problems, the *testing of statistical hypotheses*.

Estimation and hypothesis testing are related, but there are important and interesting differences between them. Let us indicate one. In most estimations of parameters we make no predictions about the value of the parameter; presumably we shall be equally accepting of any estimate we obtain, even though *after* the estimate has been calculated, we adopt a critical attitude and seek to specify its accuracy (and therefore its inaccuracy) by determining standard errors and confidence intervals. On the other hand, in hypothesis testing we *start* with predictions (and even with preferences) about parameter values before we have even drawn our sample. Since, as we shall see, the various tests that are available to test a hypothesis yield different results when applied to the same sample, and since we do have predilections for one or another outcome, we may be tempted to look at the sample data first and then choose that test which will best support our predictions. Because some statisticians can, like Oscar Wilde, resist anything but temptation, a set of "Thou Shalt Nots" has been promulgated. For example, the ethical statistician must decide *before* he sees the data what test he will use. Once the sample data are in and he has scanned them, he is prohibited from switching to a different test, even though it now appears that a different test would better support his predictions.

There are various kinds of statistical hypotheses we can put to test. In StatLab we shall concern ourselves with two major types. In the first of these the hypothesis specifies the value of a parameter from a single population. For example, an electrical goods manufacturer may assert that his fluorescent tubes will, on the average, burn for 1000 hours. We may think of each individual tube as having a burning life of x hours. The value of x will, of course, vary from one tube to another. Considering the output of all the tubes manufactured as a population, the various values of x will have a mean value that may be denoted by μ. In this notation, the manufacturer's claim or hypothesis may be written: $\mu = 1000$.

To test his claim we obtain a sample statistic $\bar{x}$ from the population and then determine whether the value of $\bar{x}$ is consistent with $\mu = 1000$. To help make this determination, there are a number of *one-sample tests* of statistical hypotheses. These tests are so called because the data of only one sample are required.

Hypotheses using one-sample tests are to be distinguished from those which must use *two-sample tests*. These latter hypotheses make assertions about parameters from two populations. For example, on the basis of biochemical and physiological theory, a medical researcher may propose the hypothesis that people who have lived on a diet rich in animal fat during their first 18 years do not live as long, on the average, as those who have not partaken so generously of the flesh pots. This hypothesis asserts that the mean age at death of the first population, μ_1, is less than that of the second population, μ_2, or $\mu_1 < \mu_2$. The testing of the hypothesis will require the calculation of *two x's*—one from each population—and then a determination of whether the values of $\bar{x}_1$ and $\bar{x}_2$ are congruent with the hypothesized relation between μ_1 and μ_2. The tests which help in this determination are two-sample tests.

As our examples have indicated, a statistical hypothesis may derive from a claim which may or may not be justified; or from a scientific theory which may or may not be correct; or from past experience which may or may not still be applicable.

In this unit we shall present a simple approximate test for statistical hypotheses and illustrate it for hypotheses involving means of one or two populations. In the course of this presentation we shall also consider several basic concepts involved in hypothesis testing. You will then be given an opportunity to apply all this to your own data and to test several hypotheses about the mental capacity of the STATLAB children. In Unit 19 we shall extend the application of the simple test to cover hypotheses involving proportions and medians and shall present some additional concepts of testing. In Unit 20 we shall examine several special one-sample tests—tests that, in certain circumstances, are superior to the simple test. In Unit 21 we shall deal with analogous special tests for the two-sample case. Finally, in Unit 22 we shall present a test that is useful for categorical data.

THE z-SCORE TEST

We begin, then, by considering a simple test for judging the validity of a statistical hypothesis. Let us return to our example of the manufacturer of fluorescent tubes and suppose that a skeptical consumer testing organization (CTO) sets out to test the manufacturer's claim that $\mu = 1000$. As a first step, CTO will buy (say) 10 tubes, will assume that these 10 tubes constitute a random sample of all tubes manufactured, and will then burn each tube under standard conditions to determine how long it lasts. In this way CTO will have obtained a random sample of 10 x values and thus be enabled to calculate $\bar{x}$. Suppose that it discovers the burning life (in hours) of the 10 tubes to be

974, 796, 1093, 839, 824, 1157, 879, 747, 1044, 897

and, therefore, $\bar{x} = 925$.

This arithmetic mean provides, of course, an estimate for μ, the mean of the population from which the 10 tubes were randomly drawn. If it is indeed true that $\mu = 1000$, we would then expect $\bar{x}$ to be "reasonably" close to 1000. Since $\bar{x}$ is 75 hours short of 1000, the question at issue becomes this: Is $\bar{x}$ farther below 1000 than might reasonably be expected to arise merely as a result of random sampling from a population where $\mu = 1000$?

To judge the significance of this departure, we can take advantage of what we already know about deviations around a mean. Recall from Unit 10 that in a *normal distribution* a constant percentage of the cases (labeled B in our normal table) will fall to the left of $\mu - z\sigma$, where z is the number of σ units away from μ. Indeed, as we have seen, it is this fact that makes it possible to construct the normal table. Thus suppose you were to be told that a given x value falls 0.5 σ units below μ (that is, $z = 0.5$). If you turn to the normal table, you will see that the probability B associated with $z = 0.5$ is $B = 0.309$. This means that there is a probability of 0.309 or about one chance in three that such a departure from μ this large or larger can occur as a result of random sampling alone.

Let us now see how this applies to our question. The implicit suggestion is that we can express the departure of $\bar{x}$ from μ in σ units, obtaining what we shall call a z score (by analogy with Formula 7 in Unit 12). If we could assure ourselves that we are dealing with a normal distribution, then by referring z to the normal table we could determine the chance that the observed departure of $\bar{x}$ from 1000 is a result of random sampling alone. Once we have that probability, we would be in a good position to decide whether $\bar{x} = 925$ is "reasonably" close to 1000.

Now recall from the central limit theorem two of its conclusions: that $\bar{x}$ is (nearly) normally distributed and that the standard deviation of $\bar{x}$ is $\sigma/\sqrt{n}$. The first of these conclusions gives us assurance that we are dealing with a normal distribution and therefore the use of z scores and the normal table is legitimate in this instance. The second conclusion, by providing us with a means of calculating the standard deviation of $\bar{x}$, helps us to express the departure of $\bar{x}$ from the hypothetical value of μ (call it μ_H) in terms of a z score:

$$z = \frac{\bar{x} - \mu_H}{\mathrm{SD}_{\bar{x}}} = \frac{\bar{x} - \mu_H}{\sigma/\sqrt{n}} \qquad \text{FORMULA } 25$$

This second conclusion, however, does not give us all the information we need. In the formula $\mathrm{SD}_{\bar{x}} = \sigma/\sqrt{n}$, n is the sample size and we know *that* value ($n = 10$); σ is the standard deviation of the population of x values and we do *not* know that. In statistical hypothesis testing, there are traditionally two methods which have been used to approximate the value of σ when using the z-score test. Let us illustrate each.

We have already seen, in several instances, that statisticians make whatever use they can of prior experience. As we pointed out in Unit 14, the biometrician who knows through experience that the σ for people's heights is approximately 3 in will use that information in planning the sample size for a new investigation of heights. Psychometricians (statisticians who work primarily with psychological measurement) have built up a store of knowledge about the behavior of certain kinds of mental tests and know what to expect about their variability. (For example, the Stanford-Binet test almost always shows a standard deviation of about 15 IQ points.) In the same way, testing laboratories (bioassay, engineering, metallurgical, etc.) also consult past experience in estimating population parameters.

To return to our present illustration: Suppose that CTO has tested many populations of fluorescent tubes and its general experience is that the burning times of such tubes have a standard deviation of about $\sigma = 110$ hours. Using this "historical" value of σ, CTO would find that $\mathrm{SD}_{\bar{x}} = 110/\sqrt{10} = 34.8$ hours. Accordingly,

$$z = \frac{925 - 1000}{34.8} = \frac{-75}{34.8} = -2.16$$

Following the reasoning we reviewed a moment ago, we next re-fer to the normal table and discover that there is only a chance of 0.0154 that our observed $\bar{x}$ would be this far out (-2.16 units) on the lower tail of the distribution. Another way of putting it is this: The odds are 1 in 65 that an $\bar{x}$ as low as 925 is merely a chance de-parture from 1000. Undoubtedly this result would greatly reinforce the skepticism of CTO about the manufacturer's claim, although it is of course *possible* that the manufacturer's claim is correct and that only bad luck in the sampling led to 10 such poor tubes. Query: Should CTO publish a warning to its subscribers that the claimed 1000-hour average life for the fluorescent tubes is false? Before we answer, let us examine again what the quantity 0.0154 means.

SIGNIFICANCE OF A TEST: THE P VALUE

The probability 0.0154 will be called the P value of the test. (In technical publications it is usually just denoted by P, but we shall write "P value" to avoid confusion with the population proportion P.) It measures, in probability terms, the strength of the evidence *against* the hypothesis in the following way: If the observed dif-ference between $\bar{x}$ and 1000 could arise by chance with a *large* probability, then, of course, that would strengthen our faith in the hypothesis $\mu = 1000$ (see Fig. 25). On the other hand, if the proba-bility is *small* that the difference between $\bar{x}$ and 1000 is due only to chance, then we lose our faith in $\mu = 1000$. In other words, the

FIGURE 25 CTO tests the manufacturer's claim. The manufacturer claims that the mean life of his fluorescent tubes is $\mu_H = 1000$ hours. CTO measures the life of $n = 10$ randomly chosen tubes. The observed times to failure are depicted by the 10 dots, with mean $\bar{x} = 925$ hours. If the manufacturer's claim is correct (and if the standard devia-tion of tube life continues to be $\sigma = 110$ hours), then $\bar{x}$ will have (approximately) the normal frequency curve shown. The shaded area represents the P value, .0154, associated with the observed $\bar{x} = 925$. That is, .0154 is the chance of getting $\bar{x}$ as low as 925 if the claim is correct.

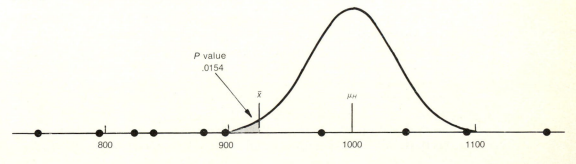

smaller the P value, the less reasonable is the assumption that our observed data are congruent with the hypothesis under test. It behooves you to read the preceding four sentences again until you are certain that you understand the argument here. Many students find it difficult to see that the *smaller the P value, the less tenable is the hypothesis under test*. Put still another way, the P value is a measure of the degree of surprise that a believer in the manufacturer's claim should feel when he finds an $\bar{x}$ value as low as 925: The smaller the P value, the greater the surprise, because a true believer would expect that any random $\bar{x}$ value would depart from 1000 only to such a degree as to make it obvious that this was a pure chance matter, i.e., the probability that the difference was a chance matter would be high.

But how much surprise should you feel before you are ready to abandon the hypothesis? The CTO, not knowing how much it takes to surprise any of its subscribers right out of their faith in the hypothesis, is best advised to publish not merely a conclusion but also the P value itself. In this way each reader of the report can determine for himself whether a P value of 0.0154 is damning evidence against the manufacturer's claim. The decision to accept or reject a hypothesis should not be dictated by the statistician. The most a statistician can do is to analyze the data (in this instance, provide P values) to help in the decision process. (We shall have more to say about this important question in the next unit.)

Most people, however, cannot tolerate ambiguity easily, nor do they thrive on making difficult decisions. For such people some statisticians have suggested that they make one guiding decision rule for a given investigation and then let all else follow unambiguously. (Indeed this is a recommended procedure for all people.) For example, we may decide to reject the hypothesis if the P value is 0.05 or less. This chosen probability of 0.05 is then called the *significance level*. (We could, of course, have chosen another probability.) Once we have chosen the significance level of 0.05, then, if in fact the P value for the hypothesis under test turns out to be 0.06 or even 0.051, we would without further travail accept the hypothesis; if it turns out to be 0.0154 or 0.04 or 0.05, we would reject it. A conventional way of describing such *rejection* would be to say: "The data are significant at the 5 percent level of significance." The term *significant* in that sentence means that the data constitute *significant evidence against the hypothesis under test*. We wish to emphasize that the level of significance (in this instance, 5 percent) must always be specified. Merely to say that "the data are statistically significant" (a statement frequently heard and read) is not enough; it almost suggests that there exists some generally recognized P value which, for all statisticians, differentiates between "statistically significant" and "statistically nonsignificant" P values. This is not so. It should be clearly understood that the significance level is idiosyncratic and purely arbitrary. It depends on the risks you are willing to take of rejecting a true hypothesis. The rule is this: Let each man decide the significance level for himself. To permit that, it is incumbent upon the statistician (and the CTO), when reporting the results of hypothesis testing, to report P values and not merely whether the data are, or are not, significant at a *prechosen* level.

In the preceding illustration of the z-score test we used a value, $\sigma = 110$, based on past experience. But historical values of σ for certain measures are rarely available, and sometimes, even when they are available, they are not trustworthy. A study of history not only tells us what things *were* but may also tell us that they no longer are. The actual value of σ for the population now under study may be quite different from the historical value. It would not be surprising, for example, if a new type of fluorescent tube being manufactured for the first time is more variable in its burning life than tubes of conventional types for which quality control is well established. If indeed the value of σ for the new population of tubes is substantially changed, then z, computed using the historical σ, will be substantially wrong. Quite generally, the use of an incorrect value for σ can seriously invalidate the results of the z-score test. This difficulty is avoided by the second method used to approximate the value of σ when using the z-score test.

If you do not have a historical σ, or if you think it may have changed, you can estimate a value for σ from your sample data by calculating s as given in Formula 5:

$$s = \sqrt{\frac{\Sigma fd^2}{n}}$$

The formula for z becomes

$$z = \frac{\bar{x} - \mu_H}{\text{SE}_{\bar{x}}} = \frac{\bar{x} - \mu_H}{s/\sqrt{n}}$$

FORMULA 26

Take again, for example, the fluorescent tube data. When we calculate the standard deviation of the 10 sample observations, we find $s = 129.4$. Using this value in place of $\sigma = 110$ in the formula of page 164, we get

$$z = \frac{925 - 1000}{129.4/\sqrt{10}} = -1.83$$

The P value corresponding to $z = -1.83$ is 0.034—much less significant than the P value of 0.0154 obtained when the historical σ was used, indicating the sensitivity of the z-score test to the value used for σ.

Although use of s to estimate σ when applying the z-score test is straightforward, a word of warning is in order. As we have emphasized several times in the preceding units, σ is not an easy parameter to estimate. Unless the sample is quite large one must anticipate that s will be rather inaccurate, and in consequence the computed z may also be somewhat unreliable. Indeed, $n = 10$ (as in our present example) is too small for s to be reliable. We shall return to this issue in Unit 20.

THE NULL HYPOTHESIS

In our example the manufacturer's claim was specific enough to permit direct formulation into a statistical hypothesis, $\mu = 1000$. But not all claims or scientific hypotheses can be that specific. In many instances, the state of the art is such that hypotheses can be phrased, at best, only in *relative* terms—in "more than" or "less than" statements but not in specific quantitative values. Thus a chemist may claim that his battery additive will prolong the life of a battery—just how long, he cannot tell. Or a pharmacologist might suggest the hypothesis that such-and-such a drug will *lower* blood pressure in hypertensive patients, but he simply does not know enough to say whether his hypothesis demands that the blood pressure drop by 5, or 8, or 17, or any other specific number of millimeters of mercury on the sphygmomanometer.

We have already seen that to test a hypothesis we must have a specified μ_H value in order to determine the P value of the distance of $\bar{x}$ from μ_H. Yet here are hypotheses we would like to test that do not provide μ_H values. This is the question to which we now turn.

Consider the following example: A psychologist proposes that a certain type of education will increase the IQ of mental retardates, but he cannot specify the number of IQ points expected to be gained. Admittedly this is not as sophisticated nor as revealing a hypothesis as it would be if it were cast in quantitative terms, but in this important and difficult field of mental retardation the psychologist has learned to be grateful for small favors. If he can demonstrate *any* beneficial effects of his training program, he will feel encouraged that he is on the right track. He therefore believes it important to test his hypothesis that merely asserts that $\mu_1 > \mu_2$, where μ_2 is the mean IQ score of the population of mental retardates as cared for at present and μ_1 is the mean IQ score that this population would have if given the special education.

To convert that hypothesis into a statistically testable one, we use a simple trick akin to that used to "prove" certain euclidean theorems. We use the "indirect proof": We first assert that our hypothesis $\mu_1 > \mu_2$ is *not* true (and this, as you will very soon see, can be easily stated in a quantitative form); then, if we can demonstrate that we may reasonably *reject* this negation of the original hypothesis, we would feel justified in holding fast to our original hypothesis $\mu_1 > \mu_2$. A concrete application of this to our present example will clarify the procedure.

We first reformulate our psychologist's hypothesis $\mu_1 > \mu_2$ into a negation of it: $\mu_1 = \mu_2$. (Note that negation does not necessarily mean opposite.) Thus where the psychologist's original hypothesis said that μ_1 is *greater than* μ_2, our reformulated hypothesis asserts that our psychologist is wrong and that μ_1 is *equal to* μ_2. This *null hypothesis*—so called because it asserts that there is *no* effect of the special treatment or that there is *no* difference between the μ values of the two populations—can be rewritten as $\mu_1 - \mu_2 = 0$. We have now expressed the hypothesis in terms of a *specified quantitative* (i.e., zero) difference between means. Now to test this quantitative null hypothesis.

First recall from Unit 16 that $\bar{x}_1 - \bar{x}_2$ is the usual estimate for $\mu_1 - \mu_2$. If it is indeed true that $\mu_1 - \mu_2 = 0$, then we would expect $\bar{x}_1 - \bar{x}_2$ to be reasonably close to zero. In other words, if our psychologist were to take two independently chosen random samples (say of size $n = 30$) of retarded children and provide the first sample with his special training while withholding his training from the second sample, and at the conclusion of his experiment measure the IQ scores for both samples, he could then calculate $\bar{x}_1 - \bar{x}_2$ to be used in testing the null hypothesis that $\mu_1 - \mu_2 = 0$.

Let us assume that he does so and finds that $\bar{x}_1 = 66$ and $\bar{x}_2 = 60$ and that therefore $\bar{x}_1 - \bar{x}_2 = 6$. He is encouraged to find that the children who received the special training have the higher mean score. But the question remains: Is this departure of $\bar{x}_1 - \bar{x}_2$ from zero larger than might reasonably be expected by chance? To answer this we can use precisely the same reasoning we used with our fluorescent tube example. We need only put $\bar{x}_1 - \bar{x}_2$ into standard units to obtain a z score; then by use of the normal table we can determine the P value of an $\bar{x}_1 - \bar{x}_2 = 6$. Since we know by Formula 19 from Unit 16 that the standard deviation of $\bar{x}_1 - \bar{x}_2$ is

$$\text{SD}_{\bar{x}_1 - \bar{x}_2} = \sqrt{\frac{\sigma_1^{\,2}}{n_1} + \frac{\sigma_2^{\,2}}{n_2}}$$

the z score for $\bar{x}_1 - \bar{x}_2$ becomes

$$z = \frac{\bar{x}_1 - \bar{x}_2}{\sqrt{\sigma_1^{\,2}/n_1 + \sigma_2^{\,2}/n_2}} \qquad \textbf{FORMULA 27}$$

If σ_1 and σ_2 are not available, they would have to be replaced by estimates in the familiar way, leading to

$$z = \frac{\bar{x}_1 - \bar{x}_2}{\sqrt{s_1^2/n_1 + s_2^2/n_2}}$$

FORMULA 28

Suppose that the psychologist computes from his samples the estimates $s_1 = 9.2$ and $s_2 = 10.1$. Then his z score will be

$$z = \frac{\bar{x}_1 - \bar{x}_2}{\sqrt{s_1^2/n_1 + s_2^2/n_2}} = \frac{6}{\sqrt{\frac{(9.2)^2}{30} + \frac{(10.1)^2}{30}}} = 2.41$$

Reference to the normal table shows that a z score as large as 2.41 has a P value of 0.0080. That is, there is only about 1 chance in 125 that a difference of 6 IQ points could occur by chance alone. Since the smaller the P value, the less confidence we have in the hypothesis being tested (the null hypothesis), it might be decided that this P value is small enough to cast considerable doubt on the null hypothesis. If, for example, the psychologist had decided in advance to use the 1 percent level of significance, he would now reject the null hypothesis. Therefore, having rejected the null hypothesis of no difference between μ_1 and μ_2 our psychologist now has some statistical justification for maintaining his *original* hypothesis that $\mu_1 > \mu_2$. Since the data are significant evidence against the null hypothesis (at the 1 percent level), our psychologist may in good conscience maintain his original hypothesis that his training has benefited the retarded children.

ONE-SIDED AND TWO-SIDED TESTS

Our psychologist is, however, not quite home free. The preceding analysis would be quite acceptable on the assumption that if μ_1 were not equal to μ_2, then it would necessarily follow that μ_1 has to be greater than μ_2. That is, if $\mu_1 - \mu_2 = 0$ is *not* true, then $\mu_1 > \mu_2$ *must* be true. But recall that the null hypothesis is not necessarily the opposite of $\mu_1 > \mu_2$. It may be quite true that the null hypothesis is wrong, but μ_1 may be *less than* μ_2. It is possible, for example, that the special training induces strain and tension and has a harmful effect on the mental capacity of retarded children. Were we to find that $\bar{x}_1 - \bar{x}_2 = -6$, we would be equally ready to drop the null hypothesis as when finding $\bar{x}_1 - \bar{x}_2 = 6$. In either case, the probability of getting that large a difference (whether plus 6 *or* minus 6 between $\bar{x}_1$ and $\bar{x}_2$) would be about 0.0080, or 1 chance in 125.

Thus the *combined* chances of getting a deviation in either direction would be doubled, that is, about 2 chances in 125 (a P value of $2 \times 0.0080 = 0.0160$), since both tails of the normal distribution must be added (i.e., *both* areas labeled B in the sketch at the head of the normal table). This is called a *two-sided test* (two-tailed test). By this two-tailed test the P value for the null hypothesis is 0.0160. Now it appears that the null hypothesis of our example is a bit more tenable, and therefore the original hypothesis is a bit less so. The psychologist would not reject the null hypothesis at the 1 percent level if he used a two-sided test.

When we are interested in deviations on one side only, as was true in our example of fluorescent tubes where we wished to evaluate the hypothesis that the average tube life was *no less* than 1000 hours, we use a *one-sided test* (one-tailed test).

The choice between a one-sided and a two-sided test is often not obvious, and the question has been much debated by statisticians. If you use a one-sided test, find a P value of 0.03, and then quote this result as a measure of the significance of the evidence against the hypothesis under test, you may be challenged by a believer in the hypothesis who asserts that you should have used a two-sided test and that therefore the P value is really 0.06. On the other hand, if you publish the P value of 0.06, someone skeptical of the hypothesis may come forward to assert that you should have used the one-sided test.

Much of the heat of the debate probably reflects the fact that the proper choice depends upon one's point of view. If the hypothesis under test is false, what alternative explanation is reasonable? Are there any *reasonable alternatives* on the other side of the parameter? If there are, then perhaps we should use two sides. Certainly in our two examples there are reasonable alternatives on both sides. That is, it is possible that the true burning life of our fluorescent tubes may be less than 1000 or more than 1000; it is also possible (given the state of psychological theory in this difficult area) that any kind of special education may help or may injure the mental capacity of the retardates. Therefore it would seem prudent to use a two-tailed test for those hypotheses.

But we might ask another question: Are you *interested* in alternatives (no matter how reasonable) on only one side or on both? For the fluorescent instance, CTO is interested in only one side—for if the average burning life of a tube should exceed the manufacturer's claim, it is not incumbent upon CTO to assess the P value of this largesse; a one-tailed test will do. From the point of view of CTO, the only values of $\bar{x}$ relevant to their interest are values less than 1000. (The manufacturer, of course, might have a different point of view.) For the IQ instance, it may be thought to be important to know whether the special training can be harmful; if so, both positive and negative values of $\bar{x}_1 - \bar{x}_2$ are relevant and therefore a two-tailed test is indicated. It is clear that the same data may properly be interpreted differently by different analysts if their interests and presumptions are different.

But now we must remind you of what we pointed out in the introduction to this unit. The ethical statistician cannot wait until *after* the data have been collected and the findings analyzed to decide whether it is expedient to test his hypothesis in a one-sided or a two-sided way. He does not have an option of *arbitrarily* selecting between a one-sided and a two-sided test so that it will support the case he wishes to make for or against the hypothesis. To avoid bias, this decision (along with others we shall discuss in later units) must be made in advance.

One final caution: The results of the z-score test are no more valid than the assumptions on which it is based. Is it reasonable to regard the sample as a random one? Is the historical σ still valid? Is the standard error of the estimate reliable? Is the statistic nearly normal in its distribution? In practice these are not always easy questions to answer.

☐ You are now ready to put into practice what you have learned about the testing of statistical hypotheses. To try out all these notions you will be asked to use the z-score test on the three following statistical hypotheses. Professor Read D. Tuddenham investigated the Raven scores of California children living in the San Francisco Bay Area. From his findings it appears that boys of the age of our STATLAB boys had a mean Raven test score of 26.2. Because the STATLAB population comes from the same area, we might suggest *Hypothesis 1: The mean Raven score for the* STATLAB *boys is 26.2.* Now consider the following facts: Girls mature more quickly in verbal ability than do boys; most tests of mental ability emphasize verbal facility; girls usually score as well as boys on most mental ability tests; the Raven test does not require verbal facility but emphasizes perception of design and spatial organization. These facts suggest *Hypothesis 2: The* STATLAB *boys exceed the* STATLAB *girls in mean Raven scores.* Finally, to round out the possibilities, we have *Hypothesis 3: There is a difference between the mean Raven scores of* STATLAB *boys and* STATLAB *girls.* To complete your work in testing the first hypothesis, you will need the fact that Tuddenham found boys' Raven scores have a standard deviation of 8.3 points.

WORK TO BE DONE

STEP 1   *Note: Before* proceeding with the succeeding steps of the work, turn to the inquiry, and answer items 1, 2, 3, 10, 11, 15, 16. *This is important.*

STEP 2   Retrieve from Databank, File G-1, the ID numbers of your STATLAB boys' ($n_1 = 20$) and STATLAB girls' ($n_2 = 15$) samples, and copy these ID numbers on the appropriate blanks, page 1 of the worksheet.

STEP 3   By reference to the Census, record the Raven scores for each of the 35 STATLAB boys and girls. Determine $\bar{x}_1$ (boys) and $\bar{x}_2$ (girls) by the direct method.

STEP 4   Construct the frequency tables on the worksheet, page 3.

STEP 5   On pages 2 and 3 of the worksheet, convert the ordinary graph paper into normal probability paper (using the inside back cover of this book), and determine, graphically, the values of s_1 and s_2.

(*Note:*   Do all your work for steps 6, 7, and 8 on page 4 of the worksheet. Write out *all* the formulas you use.)

STEP 6   Using the z-score test, obtain the value of z necessary to test Hypothesis 1, above, using the historical value of σ_1.

STEP 7   Repeat the test for Hypothesis 1, using s_1 as an estimate for σ_1.

STEP 8   Obtain the values of z necessary to test Hypotheses 2 and 3, using the appropriate estimates s_1 and s_2.

STEP 9   Record in Databank, File G-1, the Raven scores for the 20 boys and 15 girls; $\bar{x}_1$ and $\bar{x}_2$; s_1 and s_2; and the two frequency tables ($w = 5$).

STEP 10   Complete the inquiry.

FURTHER EXAMPLES AND CONCEPTS OF TESTING

This unit develops the ideas of Unit 18 in two ways. First, we shall present some further examples of the z-score test, showing you how to carry out tests for proportions and for medians, and then see generally for what sort of problem the test can be used. Second, we shall round out the conceptual framework of testing by presenting some additional basic ideas. With these ideas, you will understand more clearly how a test works and will also be able to plan an investigation so that the resulting test will accomplish its purposes.

TESTING A PROPORTION

A sample proportion p may be used to test the hypothesis $P = P_H$ that the corresponding population proportion P has a specified value P_H. This is done in much the same way that a sample mean $\bar{x}$ was used in Unit 18 to test the hypothesis $\mu = \mu_H$.

Recall from Unit 14 the standard deviation of p as given by Formula 12: $\text{SD}_p = \sqrt{PQ/n}$. If the hypothesis under test is correct, then P has the value P_H and in consequence Q has the value $1 - P_H = Q_H$. Using these values in Formula 12, we see that the sample proportion p will have the standard deviation $\text{SD}_p = \sqrt{P_H Q_H / n}$ if the hypothesis $P = P_H$ is correct. If we divide the deviation $p - P_H$ by SD_p, we convert $p - P_H$ to standard units and thus obtain the z score:

$$z = \frac{p - P_H}{\sqrt{P_H Q_H / n}}$$

FORMULA 29

This value of the z score can be referred to the normal table to obtain P values, provided that n is large enough that the sample proportion p has a distribution of nearly normal shape (see Unit 14).

As an illustration, consider the problem of Councilman Jones, who is trying to decide whether to challenge Mayor Smith at the next election. Smith has always been popular—he won 60 percent of the votes at the previous election and most politicos think he is holding his favor with the voters. Jones suspects that Smith has recently begun to slip and may be vulnerable, but Jones would not wish to give up his safe council seat to run against Smith unless he can find convincing evidence that Smith's popularity has indeed dropped. Being a modern politician, Jones will commission a sample survey of the voters to help him make his decision.

We can put Jones's problem in the form of a statistical hypothesis. Let P denote the proportion of all the voters in the city who now favor Smith. If things have not changed since the last election, then $P = 0.6$. This is the hypothesis under test, and if it is correct Jones would not want to make the race since he would presumably be defeated.

Jones figures that he can afford to hire a pollster to query a random sample of $n = 100$ voters and find the proportion p of the sample who now favor Smith. Clearly Jones is not interested in the possibility that P has risen *above* 0.6; he only wants to know whether it has fallen below 0.6. Therefore Jones would use a one-sided test and would reject the hypothesis $P = 0.6$ if his sample p is sufficiently low.

But how low must the sample proportion be to be "sufficiently" low? Jones is not interested merely in collecting the data, calculating a P value, contemplating the result, and drawing scholarly implications from it. He must actually make a hard decision: To run or not to run.

Jones therefore needs a decision rule; that is, he needs to select a significance level which will guide his decision unequivocally (see Unit 18). Suppose he decides to use the 5 percent level. This means that Jones will reject the hypothesis $P = 0.6$ and enter the race, provided that his test gives a P value of 0.05 or less.

Formula 29 permits us to express Jones's decision rule—to

run or not to run—in terms of the sample proportion p, by telling us how low the sample proportion must be for Jones to reject the hypothesis that the true proportion for the entire population is $P = 0.6$. In our example, $n = 100$, $P_H = 0.6$, and $Q_H = 0.4$. Putting these values into Formula 29 gives

$$z = \frac{p - 0.6}{0.0490}$$

This equation permits us to find the value of p corresponding to any selected value of z. It is clear that small values of p correspond to small values of z. If we look at the normal table, we find that 5 percent of the area lies below $z = -1.645$. And since Jones has decided to use the 5 percent significance level, we now want to know how low p must be to yield $z = -1.645$. Substituting $z = -1.645$ in the preceding formula gives

$$p - 0.6 = (0.0490) \times (-1.645) = -0.081$$

and hence $p = 0.519$. Thus z will be below -1.645 provided that p is below 0.519. In other words, if the hypothesis $P = 0.6$ is true, there is only about a 5 percent chance that the sample proportion p will fall below 0.519. Thus Jones, having first decided on a significance level of 5 percent, has now determined, *prior* to doing the survey, how low the sample proportion p must be to be "sufficiently" low for him to reject the hypothesis that $P = 0.6$.

To summarize, Jones's test may be stated this way:

Interview $n = 100$ voters chosen at random, and obtain the proportion p of voters who prefer Smith. If $p < 0.519$, reject the hypothesis that $P = 0.6$ (and conclude that Smith is slipping and enter the race for mayor); if $p > 0.519$, accept the hypothesis that $P = 0.6$ (and remain safely on the council).

The statistical value 0.519 of p, which in this test separates rejection from acceptance, is known as the *critical value* of the sample statistic p.

THE CONTINUITY CORRECTION

There is, however, a problem. Because p is a discrete variable, we shall see that it is impossible to obtain a p of exactly 0.519 (see Unit 3), and therefore Jones may not be able to use a significance level of exactly 5 percent. When the pollster asks the 100 voters if they favor Smith, the number of Yes answers must be a whole number; it might for example be 51 or 52. The corresponding values of p, found after division by $n = 100$, are 0.51 or 0.52. No value intermediate

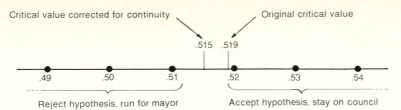

FIGURE 26 Councilman Jones's test of the hypothesis that $P = .6$. Corresponding to the 5 percent level of significance, he finds from the z-score formula the critical value .519. The continuity correction changes this to .515, midway between $p = .51$ and $p = .52$. The dots indicate possible values of p for $n = 100$.

between these can occur. Therefore, to reject the hypothesis if $p < 0.519$ really means to reject if $p \leq 0.51$; to accept if $p > 0.519$ really means to accept if $p \geq 0.52$. Any critical value between 0.51 and 0.52 would yield exactly the same test as the critical value 0.519 (see Fig. 26).

Perhaps the most natural choice for the critical value of Jones's test is 0.515, midway between the largest rejection value $p = 0.51$ and the smallest acceptance value $p = 0.52$. In any event, it is a very good idea to use such midway values when computing z by Formula 29. This is so because p is a discrete variable whereas the normal curve represents the distribution of a continuous variable. The attempt to make the continuous normal distribution give a good fit to the discrete distribution of the sample proportion will be most successful if we do the fitting at a point, such as 0.515, midway between consecutive values of the discrete variable. The interval from 0.51 to 0.52 is thereby split in two equal halves. One half is associated with the acceptance region and one half with the rejection region, in an evenhanded way. When this simple and useful trick is used, as it should be, we say that the normal curve has been "corrected for continuity."

As an illustration, let us use the *continuity correction* to find out how close Jones has come to the 5 percent significance level he wanted to use. Using Formula 29, the value of z corresponding to $p = 0.515$ is

$$z = \frac{0.515 - 0.6}{0.0490} = -1.73$$

The area under the normal curve to the left of -1.73 is 0.042 (see area B in the normal table). According to this more accurate reckoning, Jones's test has the significance level 4.2 percent, a little lower than the 5 percent level he at first decided to use.

Jones, however, is willing to take a risk somewhat greater

than 4.2 percent. Could he get closer to his desired 5 percent level by using a different critical value? Suppose he were willing to raise his critical value to 0.525. (This would mean that $p = 0.52$ is considered grounds for rejection, where formerly it led to acceptance.) Then

$$z = \frac{0.525 - 0.6}{0.0490} = -1.53$$

for which the significance level as read from the normal table is 6.3 percent, even farther from the 5 percent risk he is willing to run. As it happens, Jones cannot get closer to the 5 percent level than with a critical value 0.515. The discreteness of p prevents the exact achievement of a prechosen level of significance. Jones will therefore have to settle for a significance level of 4.2 percent.

THE TWO TYPES OF ERROR

Let us examine more closely the meaning of the significance level 4.2 percent associated with Councilman Jones's test. This value, 4.2 percent, is the chance that the sample p will be less than 0.515 when the true value of the population P is 0.6. Since 0.515 is the critical value for p, this means that there is a 4.2 percent chance that Jones will reject the hypothesis $P = 0.6$, *even if it is valid.* Or, to put it another way, 4.2 percent is the measure of the risk which Jones will run of entering the mayoralty race, and presumably being defeated, *if* Smith still has the allegiance of 60 percent of the voters. In still other words, it is Jones's risk of a *false rejection,* i.e., rejecting the hypothesis $P = 0.6$ when the hypothesis $P = 0.6$ is indeed true.

From Jones's point of view, it would be a serious error to enter the race if $P = 0.6$. His test gives him only a rather small chance (4.2 percent) of making this mistake, but would it not be wise for him to reduce even this small risk? He can easily do so. For example, suppose he were to modify his test and use 0.505, the next smaller critical value of p, rather than 0.515. With this modified test, he would reject the hypothesis $P = 0.6$ and enter the race only if 50 or fewer of the 100 voters favor Smith; he would accept the hypothesis if 51 or more favor him. With the new rule, Jones's chance of false rejection is reduced from 4.2 to 2.6 percent. [To be sure that you understand the calculation of significance levels using the continuity correction, we urge you to verify this value and find $z = (0.505 - 0.6)/0.0490 = -1.94$.] Surely it is better to have only a 2.6 percent chance of making a serious mistake instead of a 4.2 percent chance. And why not go still farther in this direction—would not 0.495, the next lower critical value, be still safer?

The answers to questions of this sort were clarified around 1930 in a series of papers published jointly by two statisticians, one (J. Neyman) a Pole, the other (E. S. Pearson) an Englishman. The Neyman-Pearson theory focused the attention of the statistical community on the fact that *two* kinds of error must be considered when making a test. We must be concerned not only about the possibility of falsely *rejecting* the hypothesis should it be correct but also about the possibility of falsely *accepting* it should it be wrong.

Let us use Jones's dilemma to illustrate the idea of two types of error. Hitherto we have only been considering what would happen if in fact $P = 0.6$. But Jones needs to consider also the possibility that (as he suspected in the first place) the mayor has lost favor, so that P is now well below 0.6. To be specific, let us consider the possibility that $P = 0.5$: that Smith is now favored by only half the voters. If this were true, Jones would want to run, believing that with vigorous campaigning he might well win. If indeed $P = 0.5$, Jones would regard it as a serious error falsely to accept the hypothesis $P = 0.6$ and stay on the council. Since Jones cannot know whether in fact $P = 0.6$ or $P = 0.5$, he must keep both possibilities in mind; and he must consider both the error of *false rejection* of the hypothesis under test ($P = 0.6$) and the error of *false acceptance* of that hypothesis. (Neyman and Pearson called these two errors, respectively, the "error of the first kind" or "Type I error" and the "error of the second kind" or "Type II error." The Type I error had, historically, been considered first. We shall use the more descriptive terms of *false rejection* and *false acceptance*.)

The chance of false acceptance can also be easily calculated with the aid of a z score. For example, if Jones were to use his original critical value of 0.515 (corresponding to level 4.2 percent) and his pollster found that $p > 0.515$, Jones would of course accept the hypothesis that $P = 0.60$. But suppose that, in fact, $P = 0.5$ and, of course, $Q = 0.5$ also. Then Formula 29 would become

$$z = \frac{p - 0.5}{\sqrt{(0.5 \times 0.5)/100}} = \frac{p - 0.5}{0.05}$$

Solving for z when $p = 0.515$, we obtain $z = 0.3$. Therefore if $p > 0.515$, z must be greater than 0.3. The chance of obtaining a $z > 0.3$ is 0.382 (see the normal table). Thus if in fact the population proportion P of Smith's support is 0.5 and Jones's pollster finds that $p > 0.515$, Jones's original decision rule (to reject the hypothesis that $P = 0.6$ in case $p < 0.515$) will mean that he risks a 38.2 percent chance of false acceptance of the hypothesis $P = 0.6$.

Table 20 shows what the chances of the two errors are, corresponding to various critical values Jones might consider using. As common sense would suggest, if Jones lowers the critical value,

TABLE 20 Councilman Jones's chances of error when guided by a sample of $n = 100$

If he rejects $P = 0.6$ when	His chance of false rejection will be	false acceptance (if $P = 0.5$) will be
$p < 0.545$	13.1%	18.4%
$p < 0.535$	9.2	24.2
$p < 0.525$	6.3	30.9
$p < 0.515$	4.2	38.2
$p < 0.505$	2.6	46.0

Each of the five critical values has been corrected for continuity—that is, each is just halfway between two consecutive possible values of p (0.50, 0.51, . . . , 0.55). The chances of error were then calculated by the z-score method.

his risk of false rejection goes down but his risk of false acceptance goes up. There is no such thing as a free lunch.

Jones is not going to like the values in Table 20. His original test (reject if $p < 0.515$) gives him an acceptable risk of false rejection (4.2 percent) but much too great a risk (38.2 percent) of missing out on a good chance to become mayor. He could get that risk down to a more tolerable 18.4 percent by using the critical value 0.545, but then the risk of false rejection becomes unacceptably high (13.1 percent). There is unfortunately no way (with his present polling plans) of changing the critical value to bring both risks simultaneously down to tolerable levels, as Table 20 shows.

The fact of the matter is that a sample of $n = 100$ just does not provide enough information to make a choice between $P = 0.6$ and $P = 0.5$ without a large risk of at least one of the false-rejection–false-acceptance errors. Jones must find the money to pay for a bigger survey. Table 21 shows what he could do with a sample of $n = 200$. The chances of error chosen in this table would be much more palatable to Jones than those in Table 20. For example, Jones might like to use the critical value of 0.5425 (which, following the continuity correction, is midway between the possible values $p = \frac{108}{200} = 0.540$ and $p = \frac{109}{200} = 0.545$). This would give him a 4.8 percent chance of false rejection and at the same time an 11.5 percent chance of false acceptance—much better risk control than was available when $n = 100$. He could, of course, force these risks down even farther by increasing n above 200. He must balance the costs of a larger survey against his desire to reduce his risks.

As a second illustration of the two types of error, let us recall the fluorescent tube illustration from Unit 18. The manufacturer claims that the mean burning life μ of the tubes he produces is 1000

TABLE 21 Councilman Jones's chances of error when guided by a sample of $n = 200$

If he rejects $P = 0.6$ when	false rejection will be	His chance of false acceptance (if $P = 0.5$) will be
$p < 0.5525$	8.5%	6.9%
$p < 0.5475$	6.5	9.0
$p < 0.5425$	4.8	11.5
$p < 0.5375$	3.5	14.5
$p < 0.5325$	2.6	17.9

The possible values of p are 0.530, 0.535, 0.540, 0.545, and so forth, and the (corrected) critical values are again half-way between them.

hours; the CTO suspects that $\mu < 1000$. They will purchase a sample of $n = 10$ tubes, find the mean life $\bar{x}$ of these n tubes, and reject the claim if $\bar{x}$ is too far below 1000. For simplicity, we shall here make the assumption that the standard deviation σ of burning times is known to be 110 hours, so that $\bar{x}$ has the standard deviation $110/\sqrt{10} = 34.8$ hours.

Let us suppose that CTO decides to use the 10 percent level of significance and express this decision in terms of the value of $\bar{x}$. We shall as usual refer

$$z = \frac{\bar{x} - 1000}{34.8}$$

to the normal table. Since 10 percent of the normal distribution lies below $z = -1.282$ and small values of $\bar{x}$ go with small values of z, there is a 10 percent chance that $\bar{x}$ will fall below

$$-1.282 \times 34.8 + 1000 = 955.4$$

(assuming that $\mu = 1000$). This is the critical value of $\bar{x}$. Since $\bar{x}$ has a continuous distribution, we are able to get just the significance level we set out to get—unlike the situation when testing a (discrete) proportion. Our test may be stated as follows:

Find the mean life $\bar{x}$ of $n = 10$ randomly chosen tubes. If $\bar{x} < 955.4$ hours, reject the hypothesis $\mu = 1000$ (and, since yours is an activist organization, warn your members not to trust the claim). If $\bar{x} > 955.4$, accept the hypothesis $\mu = 1000$ (and issue no warning).

The situation is illustrated in Fig. 27. The significance level of 10 percent means that CTO has 1 chance in 10 of false rejection, that is, of rejecting the hypothesis if the manufacturer's claim is correct. But CTO needs also to consider false acceptance. Suppose

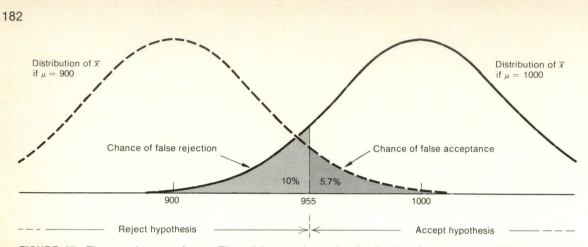

FIGURE 27 The two chances of error. The solid curve shows the distribution of $\bar{x}$ in case $\mu = 1000$, as the manufacturer claims. This claim or hypothesis will be rejected if $\bar{x}$ falls below the critical value 955, which corresponds to the 10 percent level of significance. The shaded area of 10 percent under the solid curve depicts the chance of falsely rejecting the claim if it is correct. The dashed curve shows the distribution of $\bar{x}$ in case $\mu = 900$, as CTO suspects. In that case, the chance of falsely accepting the claim is 5.7 percent, depicted by the shaded area under the dashed curve.

for example that the true value of μ is 900 hours. If that were so, CTO would consider that the product quality is seriously lower than claimed and that it is important to warn their members not to trust the manufacturer's claim. How well does their test enable them to do so? The z-score method again easily gives the answer. You should check that, when $\mu = 900$,

$$z = \frac{955.4 - 900}{34.8} = 1.59$$

so that CTO has only a 5.6 percent chance of falsely accepting the hypothesis $\mu = 1000$. As it happens, they are better protected against the error of remaining silent if the product is bad (a 5.6 percent chance of false acceptance) than against the error of speaking out if the product is good (a 10 percent chance of false rejection). This state of affairs will presumably please CTO better than it does the manufacturer, who may think that it is unfair to expose him to a 1-in-10 chance of being falsely accused of claiming too much.

We recommend that you spend some time studying Fig. 27 and then reread the related discussion above. The situation is a complex one because of the necessity of keeping in mind two possible distributions of $\bar{x}$.

THE POWER OF A TEST
To make the discussion as simple as possible, we have been considering just one specific value of P, namely $P = 0.5$, as an alter-

native to the value $P = 0.6$ under test or one specific value $\mu = 900$ as an alternative to the hypothetical $\mu = 1000$. A complete analysis should, however, take into account other possible values, such as $P = 0.55$ or $\mu = 950$.

Let us first consider the illustration related to the race for mayor. For every value of P less than 0.6 (Jones, as we have pointed out, would not be interested in values of P greater than 0.6), one can compute the chance of falsely accepting the hypothesis $P = 0.6$. We have done this for the test:

> *Interview 200 voters chosen at random. If the proportion p who favor Smith is below 0.5425, reject the hypothesis that P = 0.6. If p > 0.5425, accept the hypothesis.*

The results are shown as the solid curve in Fig. 28. Corresponding to each possible value of P, you can read the chance that the test will lead Jones to accept (falsely) the hypothesis that $P = 0.6$.

We also show in Fig. 28, as a dashed curve, the chance of rejecting (correctly) the hypothesis $P = 0.6$ for each possible P. This

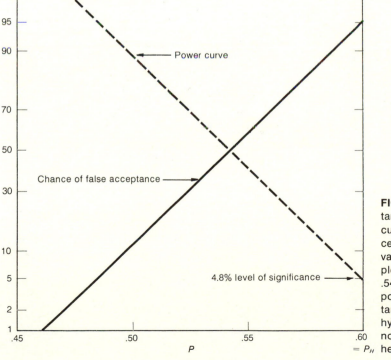

FIGURE 28 Chance of false acceptance and the power curve. The solid curve gives the chance of false acceptance, corresponding to various values of P, when Jones takes a sample of $n = 200$ and uses critical value .5425. The dashed curve, known as the power curve, shows the complementary chance of (correctly) rejecting the hypothesis $P = .6$. When plotted on normal probability paper, as done here, these curves are nearly straight.

is just the complement of the chance of accepting it. For example, at $P = 0.5$ the chance of (false) acceptance is 0.115, so the chance of (correct) rejection in that case is $1 - 0.115 = 0.885$. The dashed curve is known as the *power curve* of the test. It shows how good a chance the test has of noticing that $P = 0.6$ is false and rejecting it. The higher the power, the better the test is working. By definition, high power means the same thing as a low chance of false acceptance. If two tests have the same significance level (i.e., the same chance of false rejection), we would of course prefer the one with the higher power since it will have the lower chance of false acceptance.

One can always get higher power by raising the significance level or increasing the sample size. It sometimes happens that a test is not making full use of the available data; if so, use of a different kind of test may improve the power without the expense of a larger sample or raising the significance level. We shall give an illustration of this possibility in the next unit.

Figure 29 shows the power of various one-sided tests that CTO might use to test the hypothesis that $\mu = 1000$. Each of these tests is at the 10 percent level of significance, but they differ with regard

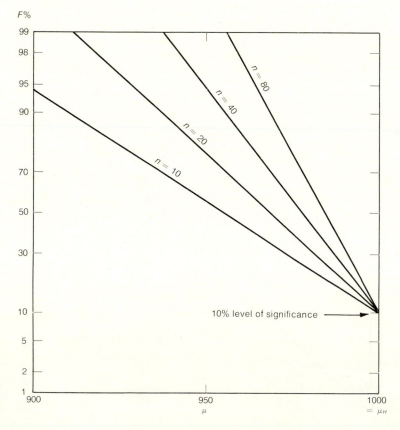

FIGURE 29 Power curves for four sample sizes: $n = 10, 20, 40, 80$. In each case the manufacturer's claim that $\mu = 1000$ hours is tested at the 10 percent level of significance (one-sided). The larger the sample, the more powerful the test.

to sample size. As you would expect, the power improves with increase in size of sample.

It is important to make sketches like Fig. 29 before you make a survey or conduct an experiment. When you do this, you may find that the sample size at first contemplated is much greater than needed, so that time and money would be wasted. A more frequent finding is that the sample size for the proposed experiment is inadequate for its purpose. For example, if CTO considered that $\mu = 950$ is far enough below the claimed $\mu = 1000$ to justify a warning to its members, then Fig. 29 would show them that $n = 10$ is inadequate; this sample size gives a 44 percent chance of false acceptance. To get the chance of false acceptance down to 5 percent when $\mu = 950$, CTO would need to burn $n = 42$ tubes.

It is common to see in scientific journals reports accepting a null hypothesis where the sample size used in the experiment was too small to give a reasonable chance of rejecting the hypothesis *even if it were substantially false.* [Thus an experimenter may find that a certain treatment yields results which appear to agree with his theory. When he tests for the statistical significance of these data, however, he discovers that he is not justified in rejecting the null hypothesis. Therefore he accepts the null hypothesis and reports that (unfortunately for his theory) the observed effect which seemed to be induced by his treatment is most probably a chance variation, and he concludes that his treatment has had no effect. The fault, however, lies not in his theory but in his experimental design. He was lost before he had even begun his experiment; he used such a small sample that the power of his statistical test was so *low* as to make it improbable that the test would reject the null hypothesis no matter what the true state of affairs might be.] Such experiments, such false acceptance of null hypotheses, and such reports in the journals can do great damage; they tend to close out lines of inquiry that ought to be pursued. An investigation of the power of the proposed statistical test of a hypothesis *before* the investigation is begun can avoid such grievous error.

OTHER USES OF THE z-SCORE TEST: DIFFERENCE OF PROPORTIONS AND MEDIAN

In Unit 16 we saw how the difference $p_1 - p_2$ of two sample proportions can be used to estimate the difference $P_1 - P_2$ of the corresponding population proportions. The statistic $p_1 - p_2$ has a distribution centered at $P_1 - P_2$, and this distribution will have a nearly normal shape if the two sample sizes n_1 and n_2 are not too small.

This statistic can also be used to test the null hypothesis that the two population proportions are the same: $P_1 = P_2$. If this hypothesis is correct, then $P_1 - P_2 = 0$, so that the distribution of $p_1 - p_2$

will be centered at zero. All that remains is to express $p_1 - p_2$ in standard units to get a z score for referral to the normal table. As you see, the reasoning is the same as that used in Unit 18, when the statistic $\bar{x}_1 - \bar{x}_2$ was employed to test the hypothesis that $\mu_1 = \mu_2$ in Formula 27.

To standardize $p_1 - p_2$, we need to recall from Formula 21 (Unit 16) that

$$SD_{p_1 - p_2} = \sqrt{\frac{P_1 Q_1}{n_1} + \frac{P_2 Q_2}{n_2}}$$

Since we do not know the value of P_1 and P_2, it might seem natural to replace P_1 by its estimate p_1 and replace P_2 by its estimate p_2, as we did in Unit 16. But there is a better estimate, one which takes advantage of the fact that P_1 and P_2 have the same value (according to the null hypothesis). We shall denote this common value by P. That is, $P_1 = P_2 = P$. It follows that Q_1 and Q_2 also have the same value, $Q = 1 - P$. The hypothesis does not tell us what the value of P is, but we can reasonably estimate it by the proportion p of Yes answers in the *two samples combined*. Again you must remember that the hypothesis asserts that P_1 and P_2 have the same value P; therefore the two sample proportions are estimates of a single P. We can substitute the estimate p for both P_1 and P_2 in the formula for $SD_{p_1 - p_2}$. Similarly, $q = 1 - p$ can be substituted for both Q_1 and Q_2. In this way we get a reasonable estimate for $SD_{p_1 - p_2}$ and can use it to standardize $p_1 - p_2$. This gives us the z score

$$z = \frac{p_1 - p_2}{\sqrt{pq/n_1 + pq/n_2}} = \frac{p_1 - p_2}{\sqrt{pq(1/n_1 + 1/n_2)}} \qquad \text{FORMULA 30}$$

As an illustration of this method, suppose you wanted to find out whether you are right in thinking that the men students at your college were more frequently cigarette smokers than were the women students. Let P_1 and P_2 denote respectively the true population proportions of men and women who smoke. The null hypothesis is that $P_1 = P_2$, and a one-sided test is reasonable since you are interested in distinguishing the possibility that the men smoke more than the women ($P_1 - P_2 > 0$) from the null hypothesis $P_1 - P_2 = 0$.

You draw a random sample of $n_1 = 150$ men and find by interview that 68 of them smoke; of $n_2 = 120$ randomly selected women, 41 are smokers. Since $p_1 = \frac{68}{150} = 0.453$ is bigger than $p_2 = \frac{41}{120} = 0.342$, it does appear that the men smoke in higher proportion than the women. How significant is the difference? Pooling the two samples, we find there are $68 + 41 = 109$ smokers among the $150 + 120 = 270$ students interviewed, which is a proportion of $p = \frac{109}{270} = 0.404$. This is a reasonable estimate for the assumed common value of P_1 and P_2. It follows that $q = 1 - 0.404 = 0.596$. Substituting all these values in Formula 30 gives

$$z = \frac{0.453 - 0.342}{\sqrt{0.404 \times 0.596 \times \left(\frac{1}{150} + \frac{1}{120}\right)}} = 1.85$$

which from the normal table corresponds to the P value 0.032. The difference between the proportions of men and women smokers (45.3 and 34.2 percent) is significant at the 5 percent level if the one-sided test is used.

Why have we not used a continuity correction here? For each combination of a value of p_1 with a value of p_2, the z-score statistic would have a different value. There are so many of these values, and they are so close together, that a continuity correction would make little difference in this problem.

TEST FOR A MEDIAN   As another example of the z-score testing method, recall from Unit 6 the use of a sample median $\tilde{x}$ to estimate the corresponding population median $\tilde{\mu}$. As mentioned in Unit 15, if the sample is not too small, $\tilde{x}$ will have a nearly normal distribution centered at $\tilde{\mu}$. The method given there for finding the standard error of $\tilde{x}$ at once permits us to compute a z score for testing a hypothetical value of $\tilde{\mu}$ by the following formula:

$$z = \frac{\tilde{x} - \tilde{\mu}_H}{1.25 s_h / \sqrt{n}} \qquad \text{FORMULA 31}$$

To illustrate this, recall that our sample of 50 STATLAB mothers had median weight $\tilde{x} = 129.5$ lb and we found for this estimate the standard error 3.5 lb (Unit 15). Suppose we wanted to test the hypothesis that our 1296 mothers had the same median weight of 132.7 lb found in another large population of women. We would use a two-sided test (why?), finding

$$z = \frac{129.5 - 132.7}{3.5} = -0.91$$

with P value $2 \times 0.181 = 0.362$.

For testing the null hypothesis that two medians are equal, the z-score test is

$$z = \frac{\tilde{x}_1 - \tilde{x}_2}{\sqrt{(1.25 s_{h_1})^2 / n_1 + (1.25 s_{h_2})^2 / n_2}} \qquad \text{FORMULA 32}$$

WHEN CAN A z-SCORE TEST BE USED?

We hope that the various examples presented in this and the preceding unit will have made it clear that the z-score method is of broad applicability. Whenever you wish to test that some parameter has a specified value, or that two parameters have a specified difference (such as zero), or indeed in many other similar formulations of hypotheses about parameters, the z-score test is useful if two conditions are met. First, you need an estimate for the parameter (or the difference of parameters, etc.) which has a nearly normal distribution, centered at the true value. Second, you need a reliable standard deviation or standard error to be able to express the deviation of estimate from hypothetical parameter value in standard units. The second requirement is likely to be harder to meet than the first.

☐ You are now ready to try out the concepts and methods of this unit in your own investigation of the churchgoing habits of the STATLAB families. In this investigation you will test two hypotheses.

We shall first ask you to examine the proportion P of the 1296 STATLAB mothers who went to church regularly, their report being taken when the child was about 10 years old. These mothers are the ones in families with either code 1 or code 2 for church attendance. In several studies of married women with young children, church attendance rates of over 50 percent have been found. Someone well acquainted with the STATLAB group speculated that in this group, however, only around 50 percent attend. Let us formulate the hypothesis

$$P = 0.5$$

and test it with a one-sided test where the relevant alternatives are $P > 0.5$.

It is often suggested that older women are more likely to attend church than younger ones. We may conjecture that this is true of STATLAB mothers. Is it? Let us divide them by age at childbirth into two nearly equal groups: those 27 or younger (at the time of the birth of their child) versus those 28 or older. Let P_1 denote the proportion of regular attenders (codes 1 or 2) among the younger mothers, and let P_2 denote the proportion among the older. To test our conjecture that $P_1 < P_2$, we first formulate the null hypothesis

$$P_1 = P_2$$

Because the alternatives $P_1 < P_2$ are the relevant ones (according to our conjecture), a one-sided test would again be in order.

WORK TO BE DONE

STEP 1 You are going to draw a sample of $n = 45$ StatLab families in order to test the first hypothesis ($P = 0.5$). Find, on page 1 of the worksheet, the critical value for this test that has a significance level as near to 10 percent as possible, and find the actual level associated with your critical value. (Be sure to use the continuity correction!) *Before going any further* with your work, answer item 1 of the inquiry.

STEP 2 Calculate on page 4 of the worksheet the power of your test, for the alternative values $P = 0.6$ and $P = 0.7$. Sketch the power curve on page 4 of the worksheet. *At this point* answer item 2 of the inquiry.

STEP 3 Draw a fresh sample of $n = 45$ StatLab families, and record on page 2 of the worksheet their ID numbers, mother's age, and church code. In listing these families, however, separate them according to mother's age as shown.

STEP 4 Count the number n_1 of mothers in your sample who are 27 or younger, and the number n_2 who are 28 or older. Also count the numbers $n_1 p_1$ and $n_2 p_2$ of regular church attenders in the two parts of your sample. Record these values on page 3 of the worksheet, and test the two hypotheses. (Test the hypothesis $P_1 = P_2$ at the 15 percent level, one-sided.)

STEP 5 Record in Databank, File I-1, only the ID numbers of your new sample, $n = 45$, drawn in step 3. Record also, from the work in step 4, n_1 and n_2; p_1 and p_2; and p.

STEP 6 Complete the inquiry.

THE t, SIGN, AND WILCOXON ONE-SAMPLE TESTS

The z-score test for a population mean, using the sample estimate s for σ (Formula 26, Unit 18), has been in use since the beginning of this century. It is simple to understand and apply, and it makes effective use of the data in most situations. It continues to be recommended provided that the sample size is reasonably large.

It was shortly after 1900 that statisticians began to wonder whether the P value of the z-score test was reliable with a small sample. As we have emphasized in Unit 15, σ is not an easy parameter to estimate. With a small sample, such as $n = 10$, the estimate s will frequently deviate substantially from σ. If the estimate s is a good deal smaller than the true σ, this will make the z score too large. If s is larger than σ, the z score will be too small. All this has the effect of giving extra "wobble" to z so that its distribution is more spread out than it would be if these errors in opposite directions did not exist. Since the size of the z score determines the P value, the net effect of a more dispersed z distribution will be more inaccurate P values. Can one allow for the extra wobble induced in the z distribution and adjust the P value accordingly?

THE t TEST

The first person to deal effectively with this question was W. S. Gosset, an employee of Guinness Breweries. In his study of the problem, published in 1908, Gosset assumed that he was dealing with a normal population—a natural assumption to make in that era (see Unit 10). He figured out how to allow for the effect on z of the variability of s so that correct P values could be calculated even when n is small. To do this he had to work out the correct P values *for each value of z, for each sample size.*

To make the charts and tables of P values usable for various problems (one of which will be taken up in Unit 21), it has been found convenient to replace n in Formula 26 by $n-1$. The resulting version of z is known as t. It may be written

$$t = \frac{\bar{x} - \mu_H}{s/\sqrt{n-1}}$$

FORMULA 33

In this formula for the one-sample t statistic, the quantity $n-1$ is known as the *degrees of freedom* of the test. This term may be explained as follows. There are n sample values, each of which may vary freely, but in computing s we use the deviations of the sample values from $\bar{x}$ (see Formula 5). As you know, the sum of these deviations always has to be zero (Unit 7). Therefore if you knew the value of $\bar{x}$ and the values of $n-1$ of the deviations, you could calculate what the remaining one must be. Thus although there are n deviations, once you have the value of $\bar{x}$ only $n-1$ of them are free to vary independently.

Guinness was perhaps the first industrial firm ever to hire a statistician. Gosset was so useful that the management wanted to keep their employment of a statistician a secret from competing breweries. Gosset was in consequence required to publish his mathematical papers under a pseudonym, and he chose "Student." For this reason, the test is often referred to as *Student's t test*.

The results of Gosset's discoveries are most conveniently presented in graphic form. In Fig. 30, each curve corresponds to a particular number of degrees of freedom, labeled "df" on the top of the chart. Using the appropriate curve, it is easy to read the P value that corresponds to a given t. The chart may be used with either positive or negative values of t, whichever is in the direction considered to be relevant. These P values are for the one-sided test and must be doubled if you are using a two-sided test.

As an illustration, recall the fluorescent tube example of Unit 18. There the manufacturer claimed that the mean burning life μ of his product is 1000 hours. The skeptical CTO found $\bar{x} = 925$ using a sample of $n = 10$; this deviates in the negative direction, which of course is the relevant direction for CTO because they sus-

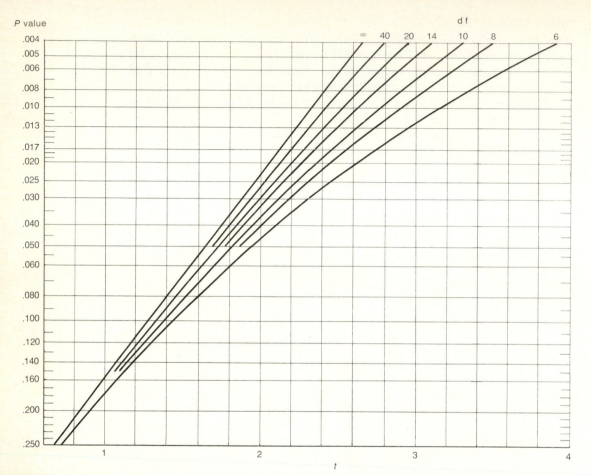

FIGURE 30 Relation between t and one-sided P values for the t test. The seven curves correspond to 7 degrees of freedom (df), as labeled: Interpolate for intermediate values. The vertical axis shows P values of the one-sided t test, plotted on the normal probability scale: The P value must be doubled if a two-sided test is used. The horizontal scale shows the corresponding value of t—positive or negative, whichever is relevant for the test. As explained in Unit 21, this chart may also be used for a two-sample t test. The chart also appears in Appendix B.

pect that the manufacturer is claiming too long a mean life. The sample estimate of σ is $s = 129.4$. Accordingly, $n - 1 = 9 = \text{df}$ and

$$t = \frac{925 - 1000}{129.4/\sqrt{9}} = -1.74$$

Figure 30 shows the curves for 8 and 10 df; 9 df would of course be about midway between them. Entering the chart at $t = 1.74$ on the bottom, reading up and then across, you will see that the P value for $t = 1.74$ and df $= 9$ is about 0.059. Thus as judged by the t test, these data are not quite significant at the 5 percent level (one-sided). That is, if CTO had decided to use the 5 percent significance level, they would accept the hypothesis $\mu = 1000$ and conclude that the 925-hour mean burning time is only a chance deviation. For the two-sided test (which the manufacturer would prefer), the P value is just twice as large: $2 \times 0.059 = 0.118$.

The z-score test (Unit 18) found these same data to be much more significant (P value 0.034 for the one-sided test). This shows that, with a sample size as small as 10, the refinement made by Gosset leads to a substantial change in the interpretation of the data. If you examine Fig. 30, you will see that the correction is less important when the sample size is large. For example, a t of 2 corresponds to a P value of 0.027 if based on $n = 40$; any larger sample size—all the way up to infinity (∞)—would make little change in the P value corresponding to $t = 2$.

Figure 30 can also be used to find how large t must be in order to achieve a preassigned significance level (see Unit 18). Suppose that you wanted to use the 2 percent significance level (one-sided) with $n = 15$ (df $= 14$). Enter the chart at P value 0.020, read across, and read down to $t = 2.26$; this is the critical value. Any value of t larger than this (or any t smaller than -2.26 if instead negative deviations are the relevant ones) would be significant at the prechosen level of 2 percent.

The values you can read from Fig. 30 are accurate enough for most practical work. In case more precise P values are needed, tables are available. The most precise is a six-decimal table of P values for all df up to 35, published by the Russian probabilist N. V. Smirnov. Table 22 illustrates the tabular presentation of P values. (We have chosen df $= 19$ for this illustration because that is the value of df arising in the work of this unit. Each df would, of course, have a different set of P values.) If, for example, you find $t = 2.96$ with a sample of $n = 20$, and if positive values of t are the relevant ones, the P value of 0.0041 can be obtained by interpolating in Table

TABLE 22 One-sided P values for the t test for 19 df

t	P value	t	P value	t	P value	t	P value
0	0.5000	1.0	0.1649	2.0	0.0300	3.0	0.0037
0.1	0.4607	1.1	0.1425	2.1	0.0247	3.1	0.0029
0.2	0.4218	1.2	0.1224	2.2	0.0202	3.2	0.0024
0.3	0.3837	1.3	0.1046	2.3	0.0165	3.3	0.0019
0.4	0.3468	1.4	0.0888	2.4	0.0134	3.4	0.0015
0.5	0.3114	1.5	0.0750	2.5	0.0109	3.5	0.0012
0.6	0.2778	1.6	0.0630	2.6	0.0088	3.6	0.0010
0.7	0.2462	1.7	0.0527	2.7	0.0071	3.7	0.0008
0.8	0.2168	1.8	0.0439	2.8	0.0057	3.8	0.0006
0.9	0.1897	1.9	0.0364	2.9	0.0046	3.9	0.0005

The P value is the chance that t, as computed from Formula 33 for a sample of $n = 20$, will have a value as large as, or larger than, the tabular entry. We assume that $\mu = \mu_H$ and that the population has a normal distribution.

22 between 0.0046 for $t = 2.9$ and 0.0037 for $t = 3.0$. How does this compare with the P value you can read from Fig. 30? Would the two-sided test be significant at the 1 percent level?

NONPARAMETRIC TESTS

The Student t test marked a milestone in the development of statistics when it was published in 1908. It is still very widely used—perhaps more widely than any other test. It has been found to possess several attractive features. For example, it has been proved that the t test is more powerful than any other for testing the center of a normal population.

It must be remembered, however, that all these findings, as well as the numerical results of Fig. 30, rest on the assumption that *the sample was randomly drawn from a normal population*. As time went on, statisticians became increasingly concerned with the departure from the normal shape of the distributions they worked with. For the reasons discussed in Unit 15, even a rather small departure from normality in the tails of the distribution (such as would arise from an occasional gross error) can substantially alter the distribution of s and hence that of t.

As statisticians have become more doubtful in recent decades about the correctness of the normality assumption, they have been trying to develop tests that are valid for small samples without assuming a precise shape (e.g., normal) for the population. Many tests have been devised that depend on much weaker and more generalized assumptions about the population shape than the assumption of normality. In this unit, we shall present two tests for the mean which require only that the population be *symmetric* (see Unit 5). This is a less stringent and more generalized assumption than normality because although the normal distributional shape is symmetric, so are an indefinite number of other quite different shapes, including curves with heavy tails. Statisticians have been more willing to rely on the generalized and weaker assumption of symmetry than on the highly specific assumption of normality.

The mathematical formula that expresses a specific frequency curve (such as the normal curve) in terms of its parameters (such as μ and σ) is known as a *parametric form*. Accordingly, tests that require only generalized assumptions (such as population symmetry) are called *nonparametric*. Additional examples of nonparametric tests, resting on other kinds of generalized assumptions, will be found in Units 21 and 24.

THE SIGN TEST

The sign test for the hypothesis $\mu = \mu_H$ is perhaps the simplest statistical test in general use. To apply it, all you have to do is to

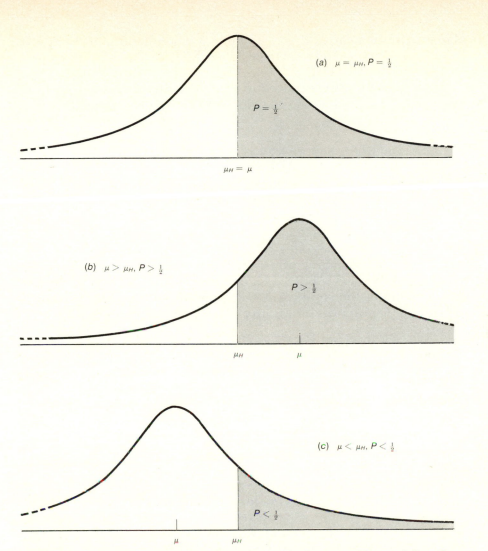

FIGURE 31 How P varies with μ. The proportion P of a symmetric population above μ_H is equal to $\frac{1}{2}$ when $\mu = \mu_H$, as shown in part *a*. If $\mu > \mu_H$, then $P > \frac{1}{2}$ (part *b*). If $\mu < \mu_H$, then $P < \frac{1}{2}$ (part *c*).

determine the proportion p of the sample that falls above μ_H and then see whether this proportion differs significantly from $\frac{1}{2}$. Remember: For this nonparametric test we are assuming that the population is symmetrically distributed about its mean μ. From this assumption it follows that the mean must also be the median and hence that half the population lies on each side of μ (see Unit 8). If the hypothesis $\mu = \mu_H$ is correct, then the proportion P of the population above μ_H will equal $\frac{1}{2}$—as illustrated in Fig. 31*a*. If, however, μ is greater than μ_H, as illustrated in Fig. 31*b*, then $P > \frac{1}{2}$. Finally, if $\mu < \mu_H$, then $P < \frac{1}{2}$ (Fig. 31*c*). Correspondingly, if $\mu = \mu_H$ is correct, we would expect the observed proportion p of the sample that falls above μ_H to be near $\frac{1}{2}$; if $\mu > \mu_H$, then p would tend to be above $\frac{1}{2}$; if $\mu < \mu_H$, then p would tend to be below $\frac{1}{2}$. These considerations provide the basis for a test.

Let us illustrate the procedure, using again the fluorescent tube example. We reproduce here, from Unit 18, the measured burning times of the 10 tubes:

$$974, 796, 1093, 839, 824, 1157, 879, 747, 1044, 897$$

You will see that three of them exceed $\mu_H = 1000$ hours while seven fall below this hypothetical mean. Thus the proportion p of the sample that has burning times greater than 1000 hours is $p = \frac{3}{10} = 0.3$. Since CTO suspects that $\mu < \mu_H$ and hence that $P < \frac{1}{2}$ (as in Fig. 31c), the sample p deviates from the hypothetical population $P = \frac{1}{2}$ in what CTO would regard as the relevant direction. But how significant is the value $p = 0.3$ as a departure from $P = \frac{1}{2}$? This is just the question of the significance of a proportion that we discussed in Unit 19. Making the continuity correction and calculating

$$\text{SD}_p = \sqrt{\frac{0.5 \times 0.5}{10}} = 0.158$$

we find from Formula 29 that

$$z = \frac{0.35 - 0.5}{0.158} = -0.95$$

From the normal table we see that the corresponding one-sided P value is 0.171—too large a P value to throw much doubt on the hypothesis. As judged by this test, the deviations of the observed burning times from the manufacturer's claim are not even significant at the 15 percent level. The two-sided test gives a P value of 0.342, still less significant.

This test is called the *sign test* because it is based on the proportion of the sample deviations from μ_H which have a positive sign. You may of course equally well use the proportion with deviations of negative sign—this just changes the sign of z without altering the P value.

It can happen that one of the sample values actually coincides with μ_H. If so, its deviation is zero and no sign can be attached to it. Such observations are customarily ignored when the sign test is used. If, for example, the ninth tube had burned for 1000 hours instead of 1044, we would just ignore it and use the remaining $n = 9$ values. The incidence of such lost values can be minimized by making precise measurements.

TABLE 23 Deviations of burning times from μ_H = 1000, ranked in order of increasing magnitude

x	$d =$ $x - 1000$	Rank	Signed rank
974	−26	1	−1
1044	+44	2	+2
1093	+93	3	+3
897	−103	4	−4
879	−121	5	−5
1157	+157	6	+6
839	−161	7	−7
824	−176	8	−8
796	−204	9	−9
747	−253	10	−10

The 10 deviations are ranked from smallest to largest without regard for sign, but then the sign of the deviation is attached to the rank. The sum of the ranks with positive signs is $V^+ = 2 + 3 + 6 = 11$.

THE WILCOXON TEST

Although the sign test is very easy to use and requires only the non-parametric assumption of symmetry, it is not really a satisfactory solution for the problem of testing $\mu = \mu_H$. The test does not have good power in most situations, because it does not make full use of the available evidence. Let us see why this is so.

Turn again to Fig. 31c. If it is true that $\mu < \mu_H$, then a sample is likely to have more negative deviations from μ_H than positive ones; and this is the feature of the data exploited by the sign test. However, it is also clear from the picture that when $\mu < \mu_H$ the negative deviations will tend to be larger in size than the positive ones; and this important aspect of the data is ignored by the sign test. A clever way of using *both* aspects was suggested in 1945 by a chemist named Frank Wilcoxon.

Table 23 shows, for the fluorescent tube example, the $n = 10$ burning times and also their deviations from the hypothetical mean $\mu_H = 1000$. We have here rearranged the observations in order of increasing magnitude of deviation (without regard to sign) from −26 to −253. Finally, the ranks of the deviations are given from smallest to largest, and to each rank is attached the sign of the deviation to which it corresponds. Not only are there fewer positive deviations (three) than negative ones (seven), but the positive deviations have ranks (2, 3, 6) which on the whole are smaller than the ranks of the negative deviations (1, 4, 5, 7, 8, 9, 10)—just as Fig. 31c would lead us to expect in case $\mu < \mu_H$.

Wilcoxon suggested that a good test could be made using the

TABLE 24 Distributions of V^- and V^+ for $n = 4$

Ranks					
1	2	3	4	V^-	V^+
+	+	+	+	0	10
−	+	+	+	1	9
+	−	+	+	2	8
+	+	−	+	3	7
−	−	+	+	3	7
+	+	+	−	4	6
−	+	−	+	4	6
+	−	−	+	5	5
−	+	+	−	5	5
+	−	+	−	6	4
−	−	−	+	6	4
+	+	−	−	7	3
−	−	+	−	7	3
−	+	−	−	8	2
+	−	−	−	9	1
−	−	−	−	10	0

Each of the four ranks may have either a + sign or a − sign attached to it. This makes the $2 \times 2 \times 2 \times 2 = 16$ possibilities listed in this table. Opposite each possible sign pattern is shown the resulting value of V^- and V^+. Thus the seventh pattern $(-+-+)$ gives $V^- = 1 + 3 = 4$ and $V^+ = 2 + 4 = 6$.

sum of the ranks with a positive sign (call it V^+). In our illustration, $V^+ = 2 + 3 + 6 = 11$. This value 11 is small for two reasons: There are only three positive deviations and these tend to be small in size, thus giving them low ranks. Conversely, the sum $V^- = 1 + 4 + 5 + 7 + 8 + 9 + 10 = 44$; this sum is large because there are many negative deviations and their ranks are on the whole large. Clearly, the small value of V^+ (or equivalently the large value of V^-) constitutes evidence against the hypothesis $\mu = 1000$ and in favor of the alternative $\mu < 1000$ as CTO suspects. Since Wilcoxon's statistic reflects more of the pertinent evidence than is used in the sign test, we may expect it to be more powerful.

But how significant is this evidence? What is its P value? This is where the nonparametric assumption of symmetry comes in. Let us leave our fluorescent tubes for the moment and consider the general case. If the hypothesis $\mu = \mu_H$ is correct (Fig. 31a), then the population being tested will contain just as many values of x that fall at a given distance to the right of μ_H as at the same distance to the left of μ_H. This means that any deviation from μ_H, having any specified magnitude $|x - \mu_H|$, is equally likely to be positive as negative. This is true of the smallest deviation in the sample, or the largest, or any in between. Therefore each *rank* of deviation is just as likely to have a plus sign or a minus sign attached to it. It is as if the plus or minus sign of the ranks are determined by tossing a fair coin n times where, say, heads means plus and tails means minus.

From this fact it is not hard to work out the distribution of V^- and V^+ (assuming $\mu = \mu_H$) and hence to get the P values of Wilcoxon's test for any given sample size. Table 24 shows how this works out for the case $n = 4$. There are four ranks each of which may be given either a plus or a minus sign. Therefore there are $2 \times 2 \times 2 \times 2 = 16$ ways to attach signs to them. These 16 possibilities are equally likely, and each has its value of V^- and V^+ as shown. We see that just one of the 16 cases gives $V^- = 0$, so the chance of finding $V^- = 0$ is $\frac{1}{16} = 0.0625$. This is the P value of $V^- = 0$, if small values of V^- are the relevant ones. Another case gives $V^- = 1$, so the chance of finding V^- to be 1 or smaller is $\frac{2}{16} = 0.125$. This would be the P value of $V^- = 1$. The distributions of V^- and V^+ are the same; and this common distribution is symmetric about its mean value 5.

Table 25 shows, for the Wilcoxon test for samples of 10 or fewer, the P values that are smaller than 0.2. This table is to be

TABLE 25   One-sided P values for the Wilcoxon one-sample test

				n				
V	3	4	5	6	7	8	9	10
0	1250	0625	0313	0156	0078	0039	0020	0010
1		1250	0625	0313	0156	0078	0039	0020
2		1875	0938	0469	0234	0117	0059	0029
3			1563	0781	0391	0195	0098	0049
4				1094	0547	0273	0137	0068
5				1563	0781	0391	0195	0098
6					1094	0547	0273	0137
7					1484	0742	0371	0186
8					1875	0977	0488	0244
9						1250	0645	0322
10						1563	0820	0420
11						1914	1016	0527
12							1250	0654
13							1504	0801
14							1797	0967
15								1162
16								1377
17								1611
18								1875

The P value is the chance that V^- or V^+, whichever is relevant when it is small, will have a value as small as, or smaller than, the tabular entry. For example, if $n = 9$, $V^+ = 8$, and small values of V^+ are relevant to your interest, then the P value is 0.0488. We assume that $\mu = \mu_H$ and that the population is symmetric.

used with either V^- or V^+, whichever is significant when its values are small. These are one-sided P values, but when doubled they give the two-sided values. Recall the fluorescent tube data of Table 23, where small values of V^+ are the relevant ones (from the point of view of CTO) and where $V^+ = 2 + 3 + 6 = 11$. Table 25 shows the P value for $V = 11$ and $n = 10$ to be 0.053 (rounded from 0.0527). That is, there is about 1 chance in 19 of getting $V^+ \leq 11$ if the manufacturer's claim is correct. (This is the one-sided P value; the two-sided value is just twice as large, $2 \times 0.053 = 0.106$.) This may be compared with the P value of 0.059 given for these same data by the t test. As this illustration suggests, the t test and the Wilcoxon test often give rather similar P values.

What if n is larger than 10? A table like Table 25 has been published covering all sample sizes up to 50. However, practical needs are often met by a simple *approximation* to the exact values given in such a table. It can be shown that

$$\text{Mean of } V \text{ distribution} = \tfrac{1}{4}n(n + 1) \qquad \text{FORMULA 34}$$

and the standard deviation of V is

$$SD_V = \sqrt{\tfrac{1}{24}n(n + 1)(2n + 1)} \qquad \text{FORMULA 35}$$

You can if you wish check these values for the case $n = 4$ against Table 24. Once we know how to determine the mean value and the standard deviation of V for any given n, we can calculate the z score for any obtained V:

$$z = \frac{V - \tfrac{1}{4}n(n + 1)}{SD_V} \qquad \text{FORMULA 36}$$

This z score can be referred to the normal table.

In computing z, it is best to apply the continuity correction to the discrete variable V, for the reasons discussed in Unit 19. For example, suppose you have drawn a sample of $n = 20$ and are testing $\mu = \mu_H$ against the alternative that $\mu > \mu_H$. According to this alternative (see Fig. 31b), there are likely to be few negative deviations and they will tend to have low ranks. This means that small values of V^- (or large values of V^+) are the relevant ones. Suppose you calculate the 20 deviations from μ_H, rank them in order of increasing magnitude, and find that there are seven negative deviations with ranks 2, 5, 6, 10, 13, 14, 17. Adding these gives $V^- = 67$. Since we are distinguishing $V^- \leq 67$ from $V^- \geq 68$, the value of V^- becomes 67.5 when corrected for continuity. Hence, from Formula 36,

$$z = \frac{67.5 - \tfrac{1}{4} \times 20 \times 21}{\sqrt{\tfrac{1}{24} \times 20 \times 21 \times 41}} = \frac{67.5 - 105}{26.8} = -1.40$$

The P value (one-sided) is 0.081. This is of course an approximate P value; a more exact value could have been obtained from a table like Table 24 for $n = 20$. However, this approximate P value of 0.081 is close enough to the correct value for all practical purposes. (You could instead have computed $V^+ = 143$ and found $z = +1.40$, with positive deviations significant, leading to the same P value. Try it.) The two-sided P value is $2 \times 0.081 = 0.162$.

One final comment: We have throughout assumed that it is possible to arrange the deviations in an unambiguous rank order according to magnitude. In practice, however, there will sometimes be two or more deviations of equal magnitude. When this happens, these deviations will be tied for two or more ranks. It is then customary (and reasonable) to give each of them the mean of the ranks for which they are tied, just as we did at page 100. For example, suppose our fourth tube had burned for 843 hours instead of 839. Then its deviation -157 would have been tied with the deviation $+157$ of the sixth tube for the ranks 6 and 7 (see Table 23). Each would then be given the mean rank $(6 + 7)/2 = 6.5$, and V^+ would have been 11.5 instead of 11. We cannot use Table 25 when there are ties in the ranks, but Formula 36 can still be used. Check that it gives $z = -1.63$ in this case, leading to the (approximate) P value 0.052. It is undesirable to have many ties in the ranks. If there are too many, an adjustment is needed in the standard deviation used in Formula 36. When you are working with continuous data, ties can be avoided by making the measurements precisely (see Unit 3).

CHOICE OF A TEST

We have now subjected the fluorescent tube data to analysis by five different tests: the z-score with assumed σ, the z-score with estimated σ, the t, the sign, and the Wilcoxon. This is only a selection of the many different tests that have been suggested for the one-sample hypothesis $\mu = \mu_H$. For most other common statistical hypotheses there are also many tests available. One of the main tasks of the statistician is to choose a test appropriate for his problem. Let us review our five tests in a way which will indicate the sort of considerations that lead to a wise choice. As in the similar choice of an estimate (Unit 15), statistical wisdom rests in large part on familiarity with the sort of population that is to be investigated.

If you are satisfied that your population has a distributional shape that is near enough to the normal, then the t test will be the best one to use, since for normal populations it is more powerful than any other test of the hypothesis $\mu = \mu_H$. How near to normal is near enough? It is hard to give a specific answer, but we can say that it is especially important that the resemblance be close in the tails of the population. Just a few percent extra cases far out in the tails can seriously alter the distribution of s and hence of t. When that happens, the t test will no longer be most powerful and the P values read from Fig. 30 will not be correct.

If you are not sure of normality, perhaps you are willing to make the weaker assumption that the population has a distribution which is nearly symmetric. If so, the Wilcoxon test is likely to work quite well. Slight departures from symmetry in the tail will not disturb this test very much, since it does not require the squaring of deviations that makes s so sensitive to extreme values. For many distributions encountered in practice, the Wilcoxon test is more powerful than the t test. And even if the population is perfectly normal, the power advantage of t over Wilcoxon is very small.

But suppose you think the population is not nearly symmetric. Perhaps it is heavily skewed, like family income or mother's weight. Then you can fall back on the z-score test of Unit 18, at least if you can find a reasonable way to express $\bar{x} - \mu_H$ in standard units. In some cases there will be a value of σ from past experience on which you are willing to rely—then Formula 25 can be used. If there is no trustworthy historical σ, then the estimate s can be used in Formula 26, but this is reasonable only if n is large enough for the estimate to be reliable. As a rule of thumb, n needs to be at least 30 or 40 before you can put much faith in a P value derived from Formula 26.

Finally, what if the population is skewed *and* σ is unavailable *and* n is small? This is a tough combination. Skewness rules out both t and Wilcoxon, and with σ unknown and n small the z-score test is unreliable. With three strikes against you, it is time to change the rules of the game.

Here are three suggestions. First, perhaps you can formulate the hypothesis differently. Instead of specifying the location of the population by its mean, why not use the median? Then the sign test can be used regardless of the population shape, since by definition half the population exceeds its median. Thus if the population median $\tilde{\mu}$ is equal to its hypothetical value $\tilde{\mu}_H$, then the proportion P of the population above $\tilde{\mu}_H$ will be $\frac{1}{2}$ and the sign test is applicable. Or the median test of Unit 19 may work well enough—it is not sensitive to extreme values as is the t test. Second, perhaps you can use a different scaling of your measurement. Experience may show that, in a particular domain of application, the quantity x tends to have a skewed distribution while $\sqrt{x}$ or $\log x$ is usually nearly symmetric. If so, then first convert all your x values into $\sqrt{x}$ or $\log x$. Third, perhaps a different experimental design is needed. For example, if you take 30 observations instead of 15, the z-score test becomes more reliable and in addition you will of course have greater power. If the hypothesis is an important one, the extra expense may be justified.

It is obvious that all this needs to be thought through *before* the data are collected. There is another reason for planning the data analysis at the beginning rather than treating it as an afterthought. Recall how different were the P values (one-sided) of our five tests of the *same* fluorescent tube data. They ranged from 0.171 for the sign test down to 0.0154 for the z-score test with historical σ. This is typical. Any shrewd statistician can find tests by which the same data give wildly different P values, high or low according to his wish. How can anyone be sure that the wish was not father to the choice? The only safe procedure is to select the test (as well as the choice between a one-sided or a two-sided version of it—Unit 18) before anyone has seen the data. The experimental protocol, drawn up in advance to specify what measurements are to be made and how they are to be made, should also contain a section on what statistical hypotheses are to be tested and how the data will be analyzed. Without that, the planning is incomplete and almost encourages statistical malfeasance.

□ You are now ready to try out the t test, the sign test, and the Wilcoxon test on your own data. We shall ask you to reanalyze the same data on boys' Raven scores already examined in Unit 18 so that you can see for yourself how these new tests compare with the z-score test and also convince yourself that the choice of a test can have a major effect on the apparent significance of a set of data.

WORK TO BE DONE

STEP 1   *Before* proceeding with the succeeding steps of the work, turn to the inquiry and answer item 2.

STEP 2   Retrieve from Databank, File G-1, the frequency table of the boys' Raven scores, and enter the first two columns on page 1 of the worksheet. Calculate $\bar{x}$ and s by the coding method.

STEP 3   Calculate on page 1 of the worksheet the value of the t statistic for the hypothesis $\mu = 26.2$. Find its two-sided P value, both from Fig. 30 and by interpolation in Table 22.

STEP 4   Retrieve from Databank, File G-1, the Raven scores of your sample of 20 boys, and enter them on page 2 of the worksheet.

STEP 5   Compute the deviations from μ_H, and by inspecting them determine the proportion p that are positive. Carry out the sign test (two-sided).

STEP 6   Complete the table on page 2 of the worksheet, compute V^+, and carry out the Wilcoxon test (two-sided) using Formula 36.

STEP 7   Record in Databank, File G-1, the s for boys' Raven scores computed in step 2, and the number of sets of ties found among the ranks in step 6.

STEP 8   Complete the inquiry.

THE t AND WILCOXON TWO-SAMPLE TESTS

In Unit 18 we presented the z-score test for the hypothesis that the means μ_1 and μ_2 of two populations are equal (Formulas 27 and 28). This involves finding the significance of the departure of $\bar{x}_1 - \bar{x}_2$ from its hypothetical value of zero by expressing $\bar{x}_1 - \bar{x}_2$ as a z-score. To apply the z-score test you must either have reliable values of σ_1 and σ_2 from past experience or else have both sample sizes n_1 and n_2 large enough that the estimates s_1 and s_2 can be relied upon. If neither of these conditions is met, then the P value of the z-score test will be suspect, as we have warned.

There are several tests of the hypothesis $\mu_1 = \mu_2$ that neither require you to know σ_1 and σ_2 nor to have large samples. However, these tests require additional assumptions about the populations. The situation is parallel with that examined in Unit 20, and we shall present two-sample analogs for the one-sample t and Wilcoxon tests. This will give us (together with the two z-score tests of Unit 18) four tests of statistical hypotheses to choose from whenever we are confronted with the null hypothesis $\mu_1 = \mu_2$. As in Unit 20 we conclude this unit with a discussion of the problem of rational choice among the various tests.

THE TWO-SAMPLE t TEST

You will recall from the preceding unit that the one-sample t test is able to deal effectively with the hypothesis $\mu = \mu_H$, even when n is small, provided that the population has a normal distribution. There is an analogous two-sample t test. Unfortunately, this test requires not only that the two populations both be normal but also that they have *the same amount of dispersion:* The two-sample t test requires the assumption that $\sigma_1 = \sigma_2$. (This burdensome additional requirement is imposed by the mathematical argument that underlies the test—an argument we shall not explore.)

If we do assume that σ_1 and σ_2 are equal, then we can write $\sigma_1{}^2 = \sigma_2{}^2 = \sigma^2$. Substituting σ^2 for both $\sigma_1{}^2$ and $\sigma_2{}^2$ in Formula 19 for $\mathrm{SD}_{\bar{x}_1 - \bar{x}_2}$ and factoring the σ^2 out from the square-root sign, we find

$$\mathrm{SD}_{\bar{x}_1 - \bar{x}_2} = \sqrt{\frac{\sigma^2}{n_1} + \frac{\sigma^2}{n_2}} = \sqrt{\sigma^2\left(\frac{1}{n_1} + \frac{1}{n_2}\right)} = \sigma\sqrt{\frac{1}{n_1} + \frac{1}{n_2}}$$

We must now estimate the common σ by finding the common s value. Both samples can make a contribution to doing this. The first sample, with n_1 observations, provides an estimate s_1 with $n_1 - 1$ degrees of freedom (see Unit 20); the second sample provides s_2 with $n_2 - 1$ df. These two estimates of σ may be pooled, since by assumption both samples come from populations having the same σ value. The formula for estimating the common σ is

$$s = \sqrt{\frac{n_1 s_1{}^2 + n_2 s_2{}^2}{n_1 + n_2 - 2}} \qquad \text{FORMULA 37}$$

The degrees of freedom for this estimate s for σ are $(n_1 - 1) + (n_2 - 1) = n_1 + n_2 - 2$. When s is used for σ in the formula for $\mathrm{SD}_{\bar{x}_1 - \bar{x}_2}$, we arrive at the two-sample t statistic for testing the null hypothesis that $\mu_1 = \mu_2$:

$$t = \frac{\bar{x}_1 - \bar{x}_2}{s\sqrt{1/n_1 + 1/n_2}} \qquad \text{FORMULA 38}$$

The value of t may be referred to Fig. 30, as before, to obtain one-sided P values, and when doubled these give two-sided P values just as in the one-sample case. Now, of course, you must use the curve for $\mathrm{df} = n_1 + n_2 - 2$.

As an illustration of the use of the two-sample t test, let us consider a nutritionist who believes he has found a dietary supplement that will hasten the weight gain of calves. He has facilities for feeding 15 calves under controlled conditions, and he randomly chooses that many calves from a large cattle ranch. The 15 calves can be divided into two random samples, of sizes $n_1 = 8$ and $n_2 = 7$, drawn from the population of holstein calves at the ranch. The eight calves in the first sample, who receive the standard diet without supplement, show the following weight gains (in pounds) during the experimental period:

62, 84, 53, 71, 109, 57, 74, 68

The seven calves in the second sample receive the supplement in addition to a standard diet, and they show the following weight gains:

81, 73, 132, 77, 114, 90, 66

The nutritionist is encouraged to note that the mean gain of the calves receiving the supplement (90.43 lb) exceeds that of the calves who did not get it (72.25 lb). He must now determine the statistical significance of this difference. He decides to use the 5 percent level of significance.

Following the reasoning described in Unit 18, he formulates the null hypothesis (which asserts that there is no treatment effect) as the hypothesis to be tested: $\mu_1 = \mu_2$. Here μ_1 is the mean weight gain of the population of holstein calves who receive the standard diet *without supplement;* μ_2 is the mean weight gain that this population would have made *if given the supplement* with their standard diet. For these populations, the control sample, $n_1 = 8$, yields $\bar{x}_1 = 72.25$ as the estimate for μ_1; the treatment sample, $n_2 = 7$, yields $\bar{x}_2 = 90.43$ as the estimate for μ_2. The nutritionist's hope that the supplement will tend to increase growth rate would be supported, of course, to the degree that the null hypothesis can be shown to be untenable. Since that hope, to state it more specifically, is tantamount to saying that $\mu_1 < \mu_2$, the nutritionist would consider positive values of $\bar{x}_2 - \bar{x}_1$ the relevant ones and would use a one-sided test. (It is conceivable that the supplement might actually impede growth, $\mu_1 > \mu_2$, but this possibility is not the one of concern to the experimenter, so values $\bar{x}_2 - \bar{x}_1 < 0$ would be irrelevant to him.)

The nutritionist may first consider using the z-score test of Formula 27 for his null hypothesis. Suppose, however, that there are no reliable historical values available for the two population standard deviations σ_1 and σ_2. The sample sizes 8 and 7 are of course too small for him to trust the z-score test based on the estimates s_1 and s_2 as in Formula 28. Consequently neither z-score test should be used. If, however, the nutritionist is willing to assume that the two population distributions of weight gains are both normal and that they have the same standard deviation ($\sigma_1 = \sigma_2$), then he can use the t test.

You can calculate for yourself (by the methods of Unit 9) that $n_1 s_1^2 = 2220$ and that $n_2 s_2^2 = 3455$. Substituting these values in Formula 37 gives

$$s = \sqrt{\frac{2220 + 3455}{7 + 8 - 2}} = 20.89$$

Now Formula 38 gives

$$t = \frac{90.43 - 72.25}{20.89 \sqrt{\frac{1}{7} + \frac{1}{8}}} = 1.68$$

Positive t values are the relevant ones, and this t has df $= 7 + 8 - 2 = 13$. Reference to Fig. 30 (Unit 20) shows that the P value for $t = 1.68$ and df $= 13$ is about 0.058. [As judged by the one-sided t test, these differences in weight gain are not quite significant at the 5 percent level that he chose at the outset of his statistical investigation. Our nutritionist, having thus failed to reject the null hypothesis, has thereby failed to demonstrate that his feed supplement will speed up the growth rate of calves.]

A NONPARAMETRIC ASSUMPTION

The two-sample t test is very widely used. It is an excellent test, provided that the assumption $\sigma_1 = \sigma_2$ is correct and that the two populations do indeed have the normal distribution. Of course, minor departures from these assumptions would not do great damage to the test. But many interesting variables have distributional shapes quite unlike the normal, and in such cases the t test cannot be recommended.

As we have pointed out in Unit 20, when statisticians became increasingly concerned with the nonnormality of many variables, the search began for tests based on weaker assumptions. Several useful tests have been found for the hypothesis $\mu_1 = \mu_2$ that require, in addition to $\sigma_1 = \sigma_2$, only that the two populations have the *same shape*, without placing any restriction on what the parametric form of that common shape may be.

The assumption of common shape, like that of symmetry in Unit 20, is thus a nonparametric assumption. It is much more plausible than an assumption that the distributions have a specific shape, such as the normal. A hypothetical illustration will make this clear: Suppose you wish to test the hypothesis that medical practitioners in Chicago have the same mean income as those in Detroit. Knowing what you know about how incomes are distributed, you would surely not feel comfortable in assuming that either population has a normally distributed income. However, it seems reasonable to assume that, whatever shape of income distribution may obtain for Chicago doctors, this same shape will also describe the incomes of Detroit doctors reasonably well. This nonparametric assumption holds for many comparisons between two populations, such as populations of weight gains, blood pressures, farm profits, star magnitudes, etc.

Let us now make the nonparametric assumption that the two populations do have the same distributional shape and in addition (as with the *t* test) that $\sigma_1 = \sigma_2$. Against this background, if the null hypothesis $\mu_1 = \mu_2$ is correct, then the distributions of the two populations will necessarily coincide. For if the two distributions are centered at the same point ($\mu_1 = \mu_2$), are dispersed to the same extent ($\sigma_1 = \sigma_2$), and have the same shape, then the two distributions will be indistinguishable. Under these circumstances, one can say that the two samples used to test the hypothesis $\mu_1 = \mu_2$ are, for practical purposes, *drawn from the same population*. This means that the individual values from the two samples will be randomly intermingled with each other. This is the fact that provides the basis for nonparametric tests of the hypothesis $\mu_1 = \mu_2$, as we shall now see.

THE TWO-SAMPLE WILCOXON TEST

Among the many nonparametric tests that have been suggested for the hypothesis $\mu_1 = \mu_2$, perhaps the most attractive is the one proposed by Frank Wilcoxon in 1945. (In 1947, H. B. Mann and D. R. Whitney published additional *P* values for the test and derived a normal approximation for it. As a result the test is often called the *Mann-Whitney test*.) Like his test for the one-sample problem, Wilcoxon's two-sample test uses a sum of ranks. We shall use the nutritionist's weight-gain example to describe and illustrate the two-sample test.

Table 26 shows the weight-gain data we have just analyzed by the *t* test, but now arranged in order of increasing size. As you see, although we have kept the two samples separate, we have ordered and ranked them as if they were combined. Rank 1 is attached to the smallest of the 15 weight gains, 53 lb, which happens to be in the first (control) sample. Rank 15 is attached to the largest of the 15 gains, 132 lb, which happens to be in the second (treatment) sample. We next sum the ranks of the observations in the second sample. The sum of the combined ranks (71) for the data in the second sample is denoted by *W*.

Now how should one interpret this value $W = 71$? Figure 32 gives the answer. Figure 32*a* illustrates how the two population distributions might look in case the null hypothesis were correct; that is, the supplement does not change the growth rate, so that in truth $\mu_1 = \mu_2$. In this case the weight gains in the two samples would tend to be of about the same size. Figure 32*b* represents the situation that the nutritionist hopes may obtain. Here the second population (of weight gains with the supplement) is shifted to the right of the first population (of unsupplemented weight gains), so that $\mu_2 > \mu_1$. In this case, the gains in the second sample would tend to be larger than the gains in the first and would therefore tend to have the larger ranks. As a result, W would be big. Finally, Fig. 32c represents the possibility (in which the nutritionist is not interested) that the supplement will slow down the growth of the calves. Then W would tend to be small. In summary, large values of W are relevant for the alternative $\mu_2 > \mu_1$ while small values of W are relevant for $\mu_2 < \mu_1$. In view of his interests, the nutritionist, using a one-sided test, would reject $\mu_1 = \mu_2$ provided that W is "too large."

The data of Table 26 are also displayed graphically in Fig. 33. Each dot represents one calf, plotted at its weight gain and labeled by its rank. If you compare this figure with the three parts of Fig. 32, you will see that the actual data are suggestive of the situation portrayed in Fig. 32b: It looks as if $\mu_2 > \mu_1$. But how significant are these data as evidence against the null hypothesis $\mu_1 = \mu_2$? In terms of the Wilcoxon test, how large must W be to justify the rejection of the null hypothesis?

TABLE 26 Weight gains of the 15 calves

First sample (control, without supplement), $n_1 = 8$		Second sample (treated, with supplement), $n_2 = 7$	
Gain	Rank	Gain	Rank
53	1		
57	2		
62	3		
		66	4
68	5		
71	6		
		73	7
74	8		
		77	9
		81	10
84	11		
		90	12
109	13		
		114	14
		132	15
			$\overline{71 = W}$

The 15 weight gains of the two samples combined are arranged in order of increasing size and ranked from 1 to 15 while keeping the two samples separate. This permits us to find the sum of the ranks of the second (treated) sample: $W = 4 + 7 + \cdots + 15 = 71$. The same data are depicted in Fig. 33.

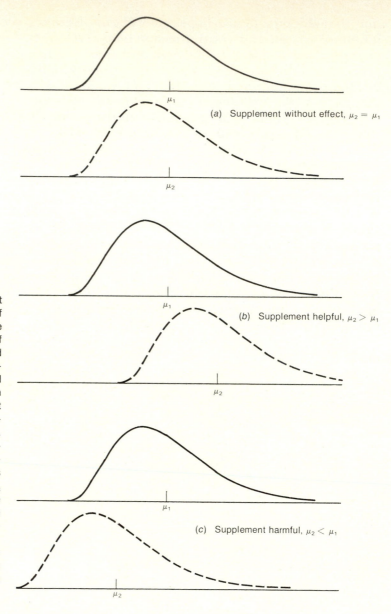

(a) Supplement without effect, $\mu_2 = \mu_1$

(b) Supplement helpful, $\mu_2 > \mu_1$

(c) Supplement harmful, $\mu_2 < \mu_1$

FIGURE 32 Three hypotheses about the dietary supplement. In each part of the figure, the solid curve depicts the distribution of the population of weight gains when all calves are fed the standard diet—the same distribution in each case. The three dashed curves depict the distribution when the supplement is fed to all calves, but according to three different hypotheses about the effect of this treatment. In part *a,* it is assumed that the supplement has no effect, so the distribution with treatment is the same as when no supplement is used and $\mu_2 = \mu_1$. In part *b,* it is assumed that the supplement tends to increase the weight gain: Thus the treated distribution is shifted to the right and $\mu_2 > \mu_1$. In part *c,* it is assumed that the supplement impedes weight gain: Thus the treated distribution is shifted to the left and $\mu_2 < \mu_1$.

Here is where the nonparametric assumption comes in. Recall that under that assumption, if the null hypothesis $\mu_1 = \mu_2$ is correct, then the two samples have in effect been drawn from the same population and will be randomly intermingled with each other. It is as if the ranks of the seven observations in the second sample were randomly chosen from the 15 available ranks. This fact permits one to work out the distribution of W and hence to associate a P value with any observed W.

To see how one could determine the *P* value for any observed *W*, let us leave our nutritionist and his calves for a moment and look at Table 27. This table illustrates the general procedure, using two very small sample sizes, $n_1 = 3$ and $n_2 = 4$. Combining these two samples there are seven ranks: 1, 2, 3, 4, 5, 6, 7. The four ranks of the second sample are randomly chosen from among these seven values. There are 35 possible samples of four, each listed in Table 27. For each of the samples, one can compute the resulting value of *W*, as shown. For example, if you happen to get ranks 2, 3, 5, and 7 for your sample, you will get $W = 2 + 3 + 5 + 7 = 17$. The 35 samples are equally likely (recall from Unit 2 what a random sample means), and that fact permits us to get the *P* value. Thus if small values of *W* are the relevant ones and if you observe $W = 1 + 2 + 4 + 6 = 13$, then Table 27 shows that there are just 7 of the 35 equally likely samples that will give a value of *W* less than or equal to 13. The chance of getting $W \leq 13$ is therefore 7 out of 35, so your *P* value is $\frac{7}{35} = 0.200$.

The 35 equally likely values of *W* form a distribution which may be displayed as a simple frequency table (Table 28) or as a histogram (Fig. 34). As you see, the distribution of *W* is symmetric about its mean value, which is 16. By the method of Unit 9 it is easy to calculate that the standard deviation of *W* is 2.83.

An inspection of Fig. 34 suggests that the distribution of *W* has something like a normal shape. If so, we can use the normal approximation to find the chance that $W \leq 13$ and then compare this approximation with the exact value (0.200) worked out above. Using the same procedure employed with the one-sample Wilcoxon statistic of Unit 20, we must express *W* in standard units for reference to the normal table. Since *W* is a discrete variable, the continuity correction calls for the use of 13.5, midway between 13 and 14, when distinguishing $W \leq 13$ from $W \geq 14$. We can convert this value of 13.5 into standard units by subtracting the mean of the *W* distribution (16) and dividing by the standard deviation (2.83). We get

FIGURE 33 Weight gains of 15 calves (data of Table 26). The gains (in pounds) of the $n_1 = 8$ untreated or control calves in the first sample are plotted on the upper line; the gains of the $n_2 = 7$ treated calves are shown on the lower line. Above the dot representing each calf is written the rank of its gain in the combined samples. The sum of the seven treated ranks is $W = 71$.

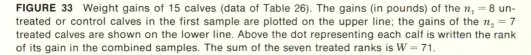
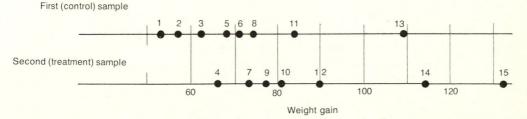

TABLE 27 The 35 ways to choose $n_2 = 4$ of the ranks from 1 to 7

Ranks in the second sample				W	Ranks in the second sample				W
1	2	3	4	10	2	3	4	7	16
1	2	3	5	11	2	3	5	6	16
1	2	3	6	12	1	3	6	7	17
1	2	4	5	12	1	4	5	7	17
1	2	3	7	13	2	3	5	7	17
1	2	4	6	13	2	4	5	6	17
1	3	4	5	13	1	4	6	7	18
1	2	4	7	14	2	3	6	7	18
1	2	5	6	14	2	4	5	7	18
1	3	4	6	14	3	4	5	6	18
2	3	4	5	14	1	5	6	7	19
1	2	5	7	15	2	4	6	7	19
1	3	4	7	15	3	4	5	7	19
1	3	5	6	15	2	5	6	7	20
2	3	4	6	15	3	4	6	7	20
1	2	6	7	16	3	5	6	7	21
1	3	5	7	16	4	5	6	7	22
1	4	5	6	16					

This table illustrates the calculation of the distribution of W, according to the null hypothesis, for samples of sizes $n_1 = 3$ and $n_2 = 4$. The four observations in the second sample may have any of the ranks from 1 to 7. There are 35 possible combinations of four ranks, as shown, and to each corresponds its sum of ranks W. If the null hypothesis $\mu_1 = \mu_2$ is correct (and if the two populations have the same shape and dispersion), then each of the 35 possibilities is equally likely.

$$z = \frac{13.5 - 16}{2.83} = -0.88$$

Now refer this standardized value $z = -0.88$ to the normal table. A glance shows area $B = 0.189$ to the left of -0.88. Therefore 0.189 is the normal approximation to the shaded area of Fig. 34. This approximation is reasonably close to the exact value (0.200) worked out above.

Although it is easy enough to work out an exact P value by the method of Table 27 when the sample sizes are as small as 4 and 3, the labor rapidly becomes prohibitive as the sample sizes are increased. A table like Table 27 for the nutritionist's example of sizes 8 and 7 would have 6435 rows! And even after the P values have been worked out by various mathematical shortcuts, it would scarcely be feasible to publish a table like Table 25 of Unit 20, giving all P values of practical interest. Such a table would need hundreds of columns, a separate column for each *pair* of sample sizes that an experimenter might use.

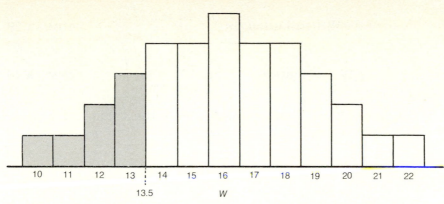

FIGURE 34 Distribution of W for $n_1 = 3$ and $n_2 = 4$. This histogram depicts the distribution of Table 28. The shaded area represents the chance (.200) that $W \le 13$, assuming that the two samples came from populations with the same distribution.

Fortunately, however, the normal approximation comes to the rescue. We have seen that it works reasonably well for the very small sample sizes 4 and 3, and it gets better as the sample sizes are increased. To use it, you need only know the mean and standard deviation of the W distribution, and they are given by two simple formulas:

TABLE 28 Distribution of W for $n_1 = 3$ and $n_2 = 4$

W	f
10	1
11	1
12	2
13	3
14	4
15	4
16	5
17	4
18	4
19	3
20	2
21	1
22	1
	35

The 35 equally likely values of W shown in Table 27 are here organized into a simple frequency table. This same distribution is depicted in Fig. 34.

$$\text{Mean of the } W \text{ distribution} = n_2 \frac{n_1 + n_2 + 1}{2} \qquad \text{FORMULA 39}$$

$$\text{SD of the } W \text{ distribution} = \sqrt{n_1 n_2 \frac{n_1 + n_2 + 1}{12}} \qquad \text{FORMULA 40}$$

You should check that these formulas give the values we worked out above from Table 27, in the case that $n_1 = 3$ and $n_2 = 4$:

$$4 \times \frac{3 + 4 + 1}{2} = 16$$

and

$$\sqrt{3 \times 4 \times \frac{3 + 4 + 1}{12}} = 2.83$$

Once we have the mean and SD_W, we can obtain the z score for any W by the formula

$$z = \frac{W - n_2 \dfrac{n_1 + n_2 + 1}{2}}{SD_W} \qquad \text{FORMULA 41}$$

Now let us solve the nutritionist's problem. He has $n_1 = 8$ and $n_2 = 7$, so that his W distribution has mean $7 \times (15 + 1)/2 = 56$ and standard deviation

$$\sqrt{8 \times 7 \times \frac{15 + 1}{12}} = 8.64$$

The value of W observed in his experiment was 71, and for him large values of W are the relevant ones. Using the continuity correction, we seek therefore the chance that $W \geq 70.5$. Applying Formula 41,

$$z = \frac{70.5 - 56}{8.64} = 1.76$$

Now look in the normal table to find that the area B to the right of $z = 1.76$ is 0.046. This is the one-sided P value of the nutritionist's data, according to the Wilcoxon test. (The exact P value, by the method of Table 27, turns out to be $\frac{302}{6435} = 0.047$.) If he decides instead to use a two-sided test, the nutritionist need only double the one-sided P value: $2 \times 0.046 = 0.092$.

The (one-sided) P values of the Wilcoxon test (0.047) and the t test (0.058) are reasonably close to one other. This is usually the case, both in the two-sample problem and in the one-sample problem (see Unit 20). Although both P values are close, the Wilcoxon test does permit the nutritionist to reject the null hypothesis at his prechosen 5 percent level, and therefore he can continue to maintain that $\mu_1 > \mu_2$—that his feed supplement does have a positive effect.

The steps required in using the two-sample Wilcoxon test may be summarized as follows:

1 You must be willing to assume that the two populations have nearly the same dispersion and nearly the same shape of distribution—whatever that common shape may be.
2 You must decide in advance of gathering the data whether large values of W, or small values of W, or both, are the ones relevant to your interest in testing the null hypothesis that $\mu_1 = \mu_2$.
3 Once the data are in hand, you make a table like Table 26, arranging the two samples in a common order of size but keeping them separate. Then label the observations with their ranks, from 1 to $n_1 + n_2$, and find the sum W of the ranks in the second sample.
4 Then you find the mean and standard deviation of W with the aid of Formulas 39 and 40.
5 Finally, you apply the continuity correction to the value of W computed from your data and then convert the corrected value to standard units, using the results of step 4, in Formula 41. Refer the z score to the normal table to get the one-sided P value. (Double this if a two-sided test is being used.)

One final comment: We have been assuming that the $n_1 + n_2$ observations can be ranked in an unambiguous order from smallest to largest, no two weight gains being equal. If you are dealing with continuous variables such as weight gain, this would be the case if the measurements are made and recorded with enough precision. In practice, however, ties are often found in a set of rounded data, and with discrete variables ties cannot be avoided by greater precision of measurement. Just as with the one-sample Wilcoxon test (Unit 20), the custom is to use mean ranks when there are ties. For example, if the calf in the second sample who gained 74 lb had instead gained 73, it would have been tied with the 73-lb gainer in the first sample (see Fig. 33). The two calves would be tied for ranks 7 and 8, so we would give each of them the mean rank $(7+8)/2 = 7.5$. As a result, W would have been 71.5 instead of 71. We are distinguishing between $W \geq 71.5$ and $W \leq 71.0$, so by the continuity correction we take the point midway between them, which is 71.25. In standard units this is

$$z = \frac{71.25 - 56}{8.64} = 1.76$$

giving a P value of 0.039. Ties are undesirable when using non-parametric tests, and continuous data should be recorded precisely enough to avoid them. If there are too many, Formula 40 needs a slight correction.

CHOICE OF A TEST

We have considered four tests for the hypothesis $\mu_1 = \mu_2$: two z-score tests (with historical σ_1 and σ_2 or using estimates s_1 and s_2), the t test, and the Wilcoxon test. As with tests for the one-sample mean (Unit 20), it is important to choose the test in advance of collecting the data. The considerations governing the choice are rather similar to those reviewed earlier.

If you are prepared to assume that both populations are normal and that $\sigma_1 = \sigma_2$, then the best test to use is the t test. In these circumstances it will have greater power than any other test. The t test is used more frequently than any other two-sample test, often unfortunately in circumstances not appropriate for it, in which case the P value is not reliable.

If you are not sure about normality, it would be better to use the Wilcoxon test, which requires (in addition to $\sigma_1 = \sigma_2$) only the weaker assumption that the two populations have the same distributional shape, whatever that may be. Thus a Wilcoxon P value will be valid in many situations where the t test's P value is not. Furthermore, even if the common shape should turn out to be normal and therefore the t test would be appropriate after all, the Wilcoxon test is only slightly less powerful than the t test. The Wilcoxon test can be considerably more powerful than the t test for shapes with one or two heavy tails of the kind so common in practice.

We have emphasized that the choice of test should be made before the data are gathered, so the question of normality needs to be settled on the basis of *general* experience with data of the kind to be collected rather than from an examination of the samples themselves. [Recall that had our nutritionist chosen the t test, he would have been led to doubt the virtues of his feed supplement; had he chosen the Wilcoxon, he would have been supported in his belief about the supplement. Being a nutritionist, he would have known that weight gains, like weights themselves, tend to be skewed. This would have been grounds enough to select the Wilcoxon test in advance. Looking at Fig. 33 (after the data were in) would certainly have reinforced his belief that this sort of population tends to be skewed.]

If you suspect that σ_1 and σ_2 are unequal, or that the two population shapes differ from each other, then neither the t test nor the Wilcoxon test is safe to use. You can fall back on the z-score test, which does not require that $\sigma_1 = \sigma_2$, provided that σ_1 and σ_2 are known or can be accurately estimated. As a rule of thumb, if each sample is of size 20 or more, the z-score test based on the estimates s_1 and s_2 is usually acceptable.

Finally, if σ_1 and σ_2 are unknown and presumed to be unequal and if either n_1 or n_2 is small, then the situation is difficult. There is a rather complicated method (due to Welch) for modifying the z-score test if both populations are normal but with different standard deviations. If each sample is, say, at least 15, you may be able to use the method of Unit 15 to find standard errors for the two sample *medians* and use $\tilde{x}_1 - \tilde{x}_2$ to test the hypothesis $\tilde{\mu}_1 = \tilde{\mu}_2$. Or perhaps you can arrange for larger samples.

☐ You are now ready to try out the two-sample t and Wilcoxon tests on your own data. We shall ask you to reanalyze the Raven samples to which in Unit 18 you have already applied the two-sample z-score test. Discussion in class will permit you to compare your results with those of other students and with the true facts about STATLAB boys' and girls' Raven scores as revealed by your instructor.

WORK TO BE DONE

STEP 1   Retrieve from Databank, File G-1, the frequency table of the girls' Raven scores, and enter the first two columns on page 1 of the worksheet. Calculate the mean $\bar{x}_2$ and standard deviation s_2 by the coding method.

STEP 2   Retrieve from Databank, File G-1, the mean $\bar{x}_1$ and standard deviation s_1 computed in Unit 20 for the boys' Raven scores. Enter these values on page 1 of the worksheet, and carry out the t test of the hypothesis $\mu_1 = \mu_2$ (two-sided).

STEP 3   Retrieve from Databank, File G-1, the Raven scores of your samples of 20 boys and 15 girls. Enter these scores on page 2 of the worksheet, but list the scores of each sample in order of increasing size.

STEP 4 Assign to each score the rank, from 1 to 35, which it has in the two samples when combined. Calculate the sum W of the girls' ranks, correct it for continuity, and convert it to standard units. Find the two-sided P value for the hypothesis $\mu_1 = \mu_2$ according to the Wilcoxon test.

STEP 5 Complete the inquiry.

THE CHI-SQUARE TEST

In Unit 14 we pointed out that data may be qualitative as well as quantitative. Let us review some of the characteristics of qualitative data. Qualitative data result from the answers to such questions as these: What are the children's blood types? What are the colors of the flowers? What are the fathers' occupations? Such qualitative or categorical data correspond to the classification of a population, or of a sample, into two or more categories. In the simplest case, in which there are only two categories, the question can be framed so as to call for a Yes or No answer: Is the child female? Has the mother ever smoked? In this two-category case, the data may be handled by comparing the proportions of Yes items in the population and in the sample.

We have studied techniques for estimating such proportions in Unit 14. In Unit 19, we saw how one could test the hypothesis that the proportion P of Yes items in a population has a specified value or that two proportions are equal. We shall now extend those methods to deal with data classified into more than two categories.

In this unit you will become familiar with a statistical test, known as the *chi-square* test, which examines simultaneously the proportions of items in all the categories into which a population has been divided. If a population has been divided into three categories, for example, the chi-square test will examine the hypothesis that the proportions in all three categories have the three specified values. The chi-square test is one of the oldest statistical tests and is still among the most widely used.

THE CHI-SQUARE TEST FOR A SIMPLE HYPOTHESIS ABOUT PROPORTIONS

As an illustration to help us understand the chi-square test, let us consider the classification of people according to the blood types O, A, B, and AB that are so important to recipients of blood transfusions. Each person has blood of just one of these types, and the proportions of the four types vary somewhat from one human population to another. According to a recent monograph, the population of the United States is divided among these four types about as follows:

Blood type	O	A	B	AB	Total
Proportion	0.454	0.395	0.111	0.040	1.000

We drew $n = 80$ STATLAB mothers at random and found for them this distribution of blood types:

	Blood type				
	O	A	B	AB	Total
Observed number	35	24	15	6	80
Proportion	0.437	0.300	0.188	0.075	1.000

As you see, the proportions of the four blood types among the mothers of our sample do resemble those of the national proportions in a general way. Yet there are differences. For example, our sample seems somewhat high in both types B and AB. Is the degree of agreement between our sample and the national population good enough that the discrepancies could reasonably be attributed to chance effects? Or are the departures of our sample from the national figures so great that we should conclude that STATLAB mothers are different from the national population in blood-type distribution? Obviously, this is the sort of question for which a statistical test can be helpful. We may state, as the hypothesis to be tested, this assertion: STATLAB mothers have the national distribution of blood types. Then we want to carry out a test of this hypothesis, leading to the calculation of a P value to help us make a rational judgment about the correctness of the hypothesis.

In presenting the chi-square test, let us formulate the problem generally while using our blood-type data as illustration. Suppose that the population has been classified into C categories, numbered in some arbitrary order from 1 to C. (In our blood-type illustration, $C = 4$.) Suppose that the hypothesis to be tested asserts that a stated proportion P_1 of the population falls into category 1, that a stated proportion P_2 falls into category 2, and so on up to P_C in category C. (In our illustration, it will be recalled, $P_1 = 0.454$, $P_2 = 0.395$, and so forth.)

Now we draw a random sample of size n from this population (in our illustration, $n = 80$) and sort the sample items into the C categories. We determine the sample proportion p_1 in category 1, p_2 in category 2, and so on. (In our illustration, $p_1 = 0.437$, $p_2 = 0.300$, and so forth.) The question is this: Are the sample p's in reasonable agreement with the hypothetical population P's?

In Unit 19 we saw how to deal with this problem when there are only $C = 2$ categories—by means of a z-score test for a proportion. Let us try to employ the ideas used there. We might look at the deviation of the sample p_1 from the population P_1 using Formula 29, express this deviation in standard units, getting the z score for category 1:

$$z_1 = \frac{p_1 - P_1}{\sqrt{(P_1 Q_1)/n}}$$

If z_1 is too large in either the positive or negative direction, we would be suspicious of the hypothesis under test—a hypothesis which asserts that our sample is drawn from a population in which the proportion of items of category 1 is indeed equal to P_1. Since large deviations of *either* sign are relevant, we can simplify things by squaring z_1. As you know, if z_1 is either large negative or large positive, then z_1^2 will be large positive—so it is large positive values of z_1^2 that are relevant to our interest.

At this point it will be helpful to make some algebraic simplifications. On squaring z_1 we find

$$z_1^2 = \frac{(p_1 - P_1)^2}{(P_1 Q_1)/n}$$

Now let us multiply both numerator and denominator by n^2, getting

$$z_1^2 = \frac{n^2(p_1 - P_1)^2}{nP_1 Q_1} = \frac{(np_1 - nP_1)^2}{nP_1 Q_1}$$

Here np_1 has a familiar meaning: It is the number of items in the sample that falls into the first category. Let us call this the *observed* number in that category. We can give a somewhat similar interpretation to nP_1: It is the number of items of category 1 that you might expect to find in the sample, on the average, if the hypothesis is correct. Let us call nP_1, therefore, the *expected* number in category 1. [We shall enclose (nP) in parentheses to emphasize its special meaning.]

Now let us turn to category 2. We should also be suspicious of the hypothesis under test if the second category showed a large deviation of p_2 from P_2, whether that deviation was positive or negative—that is, if

$$z_2{}^2 = \frac{(np_2 - nP_2)^2}{(nP_2)Q_2}$$

were large. Similar reasoning applies to all the other categories.

How should we combine all these separate possible suspicions into one omnibus measure of doubt? The obvious and simple way is to add up the z^2's, getting

$$z_1{}^2 + z_2{}^2 + \cdots + z_C{}^2$$

If the agreement between p's and P's is good in all the categories, each z^2 will be small. If the agreement is poor in one or more category, however, one or more of the z^2's will be large and the sum will be large. Alternatively, if many z^2's are just a little suspicious, the sum of all of them may be so big as to indicate serious doubts about the hypothesis; as the Scottish phrase has it—*mony a mickle makes a muckle*. In any case, big values of the sum of the z^2's are the relevant ones, i.e., the ones that would make us tend to reject the hypothesis under test.

So much is straightforward: Now comes the touch of genius. In the year 1900 the English statistician Karl Pearson (father of statistics and of the E. S. Pearson mentioned in Unit 19) saw that, by making a rather minor change in the z^2's, he would get accurate P values for the test in an easy way. The change consisted in dropping the inconsequential Q factors from the chain of

$$\frac{(np_1 - nP_1)^2}{(nP_1)Q_1} + \frac{(np_2 - nP_2)^2}{(nP_2)Q_2} + \cdots + \frac{(np_C - nP_C)^2}{(nP_C)Q_C}$$

Thus instead of ending with the sum of z^2's, we end with what Pearson called *chi square* (χ is the lowercase Greek letter chi):

$$\chi^2 = \frac{(np_1 - nP_1)^2}{nP_1} + \frac{(np_2 - nP_2)^2}{nP_2} + \cdots + \frac{(np_C - nP_C)^2}{nP_C}$$

FORMULA 42

Here too, big values of χ^2 would tend to make us suspicious of the hypothesis under test. Formula 42 is appropriate for testing a *simple* hypothesis, one that specifies a numerical value for each proportion.

To show how easy it is to calculate Pearson's χ^2, let us use our blood-type illustration. The data can be conveniently laid out as shown in Table 29. For each of the $C = 4$ blood-type categories, we first list the number of mothers of that type found in our sample. Below this we write the number that would have been expected according to the hypothesis. Thus if our STATLAB mothers do have the national blood-type distribution, the proportion $P_1 = 0.454$ of them would have blood type O, so that in a sample of $n = 80$ we would expect, on the average, to find $nP_1 = 80 \times 0.454 = 36.3$ mothers of type O. Similarly, $nP_2 = 80 \times 0.395 = 31.6$ would be expected to have blood type A and so forth.

Next we subtract the expected numbers from the observed numbers, getting the differences

$$\text{Observed} - \text{expected} = np_1 - nP_1 = 35 - 36.3 = -1.3$$

and so forth. These differences are then squared, getting $(np_1 - nP_2)^2 = (-1.3)^2 = 1.69$ and so on. The squared differences are divided by the expected numbers, getting $(np_1 - nP_1)^2/nP_1 = 1.69/36.3 = 0.05$ and so on. Finally, the sum of the terms in the bottom line is $\chi^2 = 8.51$. This is a global measure of the deviation of the observed numbers taken as a whole, from the values expected for them according to the hypothesis under test.

But how significant is this value of χ^2—that is, what P value is associated with $\chi^2 = 8.51$? Here is where Pearson's ingenuity pays off. By virtue of the way he defined his statistic, the P value of a given χ^2 value can be shown to *depend only on the number C of categories*—at least to a good approximation. (Had we used our original statistic $z_1^2 + z_2^2 + \cdots + z_C^2$, the P value of this sum would have depended on the values of the hypothetical population proportions $P_1, P_2, \ldots, P_C$.) Thanks to this great simplification, it is

TABLE 29 Calculation of χ^2 in the blood-type example

	Blood type				
	O	A	B	AB	Total
Observed number	35	24	15	6	80
Expected number	36.3	31.6	8.9	3.2	80.0
Difference	−1.3	−7.6	6.1	2.8	0.0
(Difference)^2	1.69	57.76	37.21	7.84	
(Difference)^2/expected	0.05	1.83	4.18	2.45	$8.51 = \chi^2$

This is a convenient layout for computing χ^2 according to Formula 42.

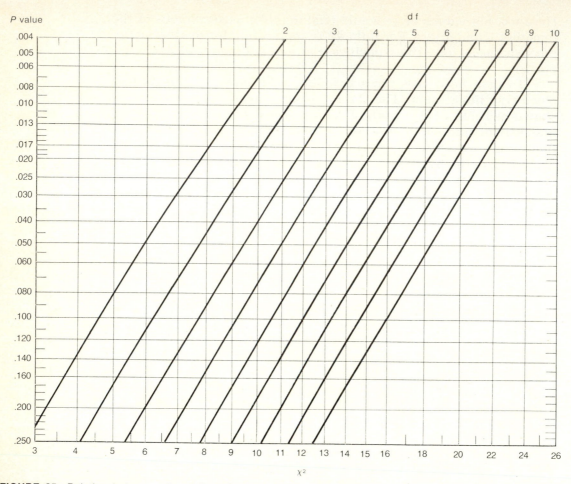

FIGURE 35 Relation between χ^2 and P values for the chi-square test. The nine curves correspond to 9 degrees of freedom (df), as labeled. The vertical axis shows P values for the chi-square test, plotted on the normal probability scale. The horizontal axis shows values of χ^2. The chart also appears in Appendix B.

possible to present, in Fig. 35, all the P values that will ordinarily be needed when using the chi-square test.

The chi-square test, like the t test of Unit 20, involves the concept of degrees of freedom. There are C observed numbers (np_1, $np_2, \ldots, np_C$) in the various categories, but they are subject to a constraint: They must add up to n. Once $C - 1$ of the observed numbers have been specified, the Cth number is determined by this constraint. Only $C - 1$ of the observed numbers may vary freely, so we may say that the data have $C - 1$ degrees of freedom.

As with the chart for the t test (Unit 20), Fig. 35 has a separate curve for each value of df. In our illustration, $\chi^2 = 8.51$ and df =

$C - 1 = 4 - 1 = 3$. An inspection of Fig. 35 shows that the P value for our data is about 0.037. As with all other tests, this measures the strength of the evidence against the hypothesis under test and thus tells us whether to accept or reject it. If we were using the 5 percent level, for example, we would now reject the hypothesis under test and decide that STATLAB mothers have a blood-type distribution different from that of the United States population.

As we have mentioned, the P values of the chi-square test are approximate ones, but the approximation is accurate provided that the expected numbers in the various categories are not too small. A rule of thumb is that all the expected numbers should be at least 5, but it is permissible to have one of them somewhat smaller than 5 if there are several others of a good size—as is true of our example.

If your χ^2 has more than 10 degrees of freedom, you may obtain a sufficiently accurate P value by referring

$$z = \sqrt{2 \times \chi^2} - \sqrt{2 \times df}$$

to the normal table. Suppose, for example, that you find $\chi^2 = 20.3$ with $df = 10$. Then

$$z = \sqrt{2 \times 20.3} - \sqrt{2 \times 10} = 6.37 - 4.47 = 1.90$$

for which the normal table gives the P value 0.029. Check this against Fig. 35.

WHEN TO USE CHI SQUARE

The chi-square test is appropriate for detecting any sort of departures from the P's specified by the hypothesis—departures in *either* direction in *any* category. Because of this, the chi-square test cannot be as powerful against a specific sort of departure as would be a test that concerned itself only with that specific sort. As we have mentioned, the data of Table 29 look as if STATLAB mothers have the blood types B and AB more frequently than the general United States population. These are the types caused by inheritance of a particular B gene, so perhaps that gene is unusually common in our population. If this conjecture had been raised *before* the data were in hand, we might have decided that the relevant departures from the hypothesis would consist of too many mothers of these two types. In that circumstance the chi-square test would not have been the right one to choose. Instead we could have used the test of Unit 19 for the hypothesis $P = P_H$ (Formula 29) to compare the observed proportion $(15 + 6)/80 = 0.262$ of such mothers with the expected proportion $0.111 + 0.040 = 0.151$. The z value (using the continuity correction) is

$$z = \frac{(20.5/80) - 0.151}{0.0400} = 2.63$$

with (one-sided) P value 0.0043.

Of course, what is not permitted is to settle on this specific test *after* seeing the data. As always, the test must be chosen in advance if it is to have any validity. Once you are able to examine the observed numbers in the various categories and can see what sorts of deviations from expected values have occurred, it is usually possible to devise some plausible reason for thinking that just those departures are the relevant ones. The test directed against those specific departures will then give highly significant (but highly erroneous) results. This is the statistician's brand of the post hoc fallacy: picking an extreme instance and pretending it was not chosen because it is extreme.

A CHI-SQUARE TEST FOR INDEPENDENCE IN A CONTINGENCY TABLE

The hypothesis about blood groups which we have just tested is a simple one: It clearly states for each category a specific numerical value for the proportion of the population that is expected in that category. (For a simple χ^2 hypothesis test that led to much wonderings and shakings-of-heads among historians of biological science, see Box 7.) Chi square can also be used to test more complex hypotheses: The most important of these *composite* hypotheses involve the classification of a population in two different ways *at the same time*. These hypotheses make certain assertions about the proportions to be found in each category, but they do not specify the actual values to be expected in these categories—indeed, the expected values must be estimated from the data.

For example, the students in your school might be classified into four groups according to color of hair (say: black, brown, red, blond) and at the same time classified into three groups according to color of eyes (say: dark, hazel, blue). Each student would thereby be placed into one of $4 \times 3 = 12$ categories, e.g., blond hair and blue eyes, or brown hair and dark eyes, etc. Would these two methods of classification turn out to be related to each other, so that a certain hair color would tend to go along with a certain eye color? That is, would eye color be dependent upon, or "contingent upon," color of hair? Or would eye color be independent of hair color, with the same proportion of blue-eyed people among the blonds as among the red, the brown, and the black-haired? Such a two-way classification is called a *contingency table*. Chi square is often used to test the hypothesis that the two ways of classifying the population are independent of each other. Note, in accordance with our characterization of a composite hypothesis, that the *independence hypothesis* does not specify the expected numbers in each of the categories; all it does is assert that the *same* proportion of blue-eyed people will be found among the blonds as among the brown-haired, etc. What this common proportion may be can only be determined after collecting the data.

Although we had not used the terms *independence* and *contingency table*, you have known since Unit 19 how to test for independence in the simplest version of a contingency table, where there is a 2×2 classification. For example, suppose that each student were classified merely as light-haired or dark-haired, and at the same time as light-eyed or dark-eyed. Then independence would mean that the proportion, say P_1, of light eyes among the light-haired is the same as the proportion, say P_2, of light eyes among the dark-haired. You could draw a random sample of n students, make the necessary classifications, and then, with Formula 30, use the z-score test of Unit 19 to decide whether $P_1 = P_2$.

To see how to deal with more general classifications, suppose that an investigator wishes to know whether attitude toward capital punishment depends upon political affiliation. The voters in a state may be classified according to whether they answer Yes or No to the question: Do you favor capital punishment? At the same time, they may also be separately classified as to whether they regard themselves to be Republicans, Democrats, or Independents. When the first two-way classification is cross-tabulated according to the second three-way classification, we get a 2×3 contingency table.

For illustration, suppose that a random sample of $n = 100$ voters are interviewed with the results shown in Table 30*a*. Thus of the 26 Republicans in the sample, 19 (or 73 percent) favor capital punishment while the other 7 oppose it. The 53 Democrats are almost evenly split, with 49 percent answering Yes. The Independents are in between, at 62 percent Yes.

The data in Table 30*a* do seem to support the idea that there is a relation between political affiliation and attitude. But one must always consider the possibility that such an appearance is merely the result of chance in the sampling. It is possible that in the entire population of voters, the ratio of Yes to No answers would be the *same* in each of the three party groups. In other words, we need a test of the statistical hypothesis that attitude toward capital punishment is *independent* of party affiliation. Let us work out, in some detail, how the chi-square test does this.

The sample is classified into $2 \times 3 = 6$ categories, in each of which there is an observed number (Table 30*a*). To compute χ^2 we must also have an expected number for each category. The inde-

BOX 7

TOO MANY SMALL χ^2'S OR HANKY-PANKY IN THE MONASTERY?

As every schoolboy knows, the Austrian monk Gregor Johann Mendel bred and raised many garden varieties of peas in his spare time at the Augustinian monastery of Brünn. From an analysis of their off-spring he promulgated (in about 1850) major laws of heredity and thereby laid the foundations for the science of genetics. His great achievement lay in this: For the offspring arising from the mating of two parents with specified unlike traits, and on the assumption of dominance of certain of these traits, Mendel could calculate precisely the theoretical proportions of the progeny who would have the traits of one parent or those of the other. His laws which enabled him to do this asserted that the hereditary "factors" (today we use the term *genes*) contributed by each parent did not *blend* in the offspring but remained *segregated* and in later generations were *recombined in a purely random order.*

Mendel's predictions can be stated as "simple" chi-square hypotheses. To illustrate with a mendelian prediction and an empirical test taken from his own work: All possible recombinations of hereditary factors resulting from the crossing of a male parent (pollen) with *smooth* and *yellow* seeds and a female parent (ovule) with *wrinkled* and *green* seeds will yield four categories of progeny: peas with smooth-yellow, smooth-green, wrinkled-yellow, and wrinkled-green seeds. It can easily be shown—assuming *random recombinations* of the original four parental factors and dominance for the smooth and yellow factors—that in the second generation, the expected valves shown in the table *must* hold.

Empirical data will occasionally yield as excellent a fit to valid theory as shown in the table, where the $\chi^2 = 0.470$ with a P value of 0.9256. However, we cannot expect that good a fit from many sets of empirical data—and Mendel reported many highly successful experiments. This is precisely what bothered the English statistician R. A. Fisher when (about 50 years after Mendel's death) he reexamined Mendel's experimental reports: Almost every set of empirical data seemed to fit Mendel's predictions beautifully—too beautifully? The suspicion having been raised, careful checking was called for:

> The possibility that the data from Mendel's experiments . . . do not represent objective counts, but are the products of some process

	Smooth-yellow	Smooth-green	Wrinkled-yellow	Wrinkled-green
Expected	0.5625	0.1875	0.1875	0.0625

Mendel cross-bred his peas, examined 556 seeds of the second generation, and this is what Mendel

Observed	0.5666	0.1942	0.1816	0.0556

The differences in proportions between theory and fact were

	−0.0041	−0.0067	0.0059	0.0069

of sophistication [the English understatement for "fudging"] is not incapable of being tested. Fictitious data can seldom survive a careful scrutiny, and since most men underestimate the frequency of large deviations arising by chance, such data may be expected generally to agree more closely with expectation than genuine data would.

The "careful scrutiny" Fisher proposed was this: Since χ^2 measures the degree to which the observed proportions in the various categories deviate from the expected proportions, and since we know the shapes of the χ^2 distributions, all we need do is calculate the χ^2's for Mendel's numerous experiments and see whether the resulting χ^2's have more *small* values than can be expected from the accidents of sampling. Fisher did just that and found just that. Mendel's data were too good to be true. Fisher then made the Charge Direct:

> Although no explanation can be expected to be satisfactory, it remains a possibility among others that Mendel was deceived by some assistant who knew too well what was expected. . . . The data of most, if not all of the experiments have been falsified so as to agree closely with Mendel's expectations.

Sewall Wright, an American geneticist, repeated the chi-square test from an independent tabulation 30 years after the publication of Fisher's paper and came out with substantially the same result as Fisher. But instead of assuming that deliberate falsification had taken place at the monastery, Wright says, "I am afraid that it must be concluded that [Mendel] made occasional subconscious errors in favor of expectations."

Another explanation, quite different from falsification, deliberate or unconscious, has more recently been proposed: It is that Mendel committed what we now recognize as a cardinal sin in experimental method, namely, that he *stopped counting* just at the points where his accumulating data happened very closely to fit his theoretical expectations.

Regardless of which explanation is the correct one, it is clear that *something* untoward happened in the monastery. All of which inspired the following irreverent account of the genesis of genetics:

> In the beginning there was Mendel, thinking his lonely thoughts alone. And he said "Let there be Peas," and there were peas and it was good. And he put peas in the garden saying unto them "Increase and multiply, segregate and assort yourselves independently," and they did and it was good. And now it came to pass that when Mendel gathered up his peas, he divided them into round [i.e., "smooth"] and wrinkled, and called the round dominant and the wrinkled recessive, and it was good. But now Mendel saw that there were four hundred and fifty round peas and a hundred and two wrinkled ones: this was not good. For the Law stateth that there should be only three round for every wrinkled. And Mendel said unto himself "Gott in Himmel, an enemy has done this, he has sown bad peas in my garden under the cover of night." And Mendel smote the table in righteous wrath, saying "Depart from me, you cursed and evil peas, into the outer darkness where thou shalt be devoured by the rats and mice," and lo it was done and there remained three hundred round peas and one hundred wrinkled peas, and it was good. It was very very good. And Mendel published.
>
> Gregory G. Doyle, 1968

Fisher's paper was first published in *Annals of Science,* 1936, pp. 115–137. This paper as well as Sewall Wright's analysis can be found in "The Origin of Genetics," edited by Curt Stern and Eva R. Sherwood (Freeman, San Francisco, 1966). Mr. Doyle's contribution was brought to our attention by Professor Curt Stern. Mendel's discovery, assessed by many as one of the great triumphs of the human mind, was robust enough to survive even overly good data.

TABLE 30 Testing the independence of party affiliation and attitude toward capital punishment

a Observed Numbers

	Republicans	Democrats	Independents	Total
Yes	19	26	13	58
No	7	27	8	42
Total	26	53	21	100

b Expected Numbers

	Republicans	Democrats	Independents	Total
Yes	15.1	30.7	12.2	58.0
No	10.9	22.3	8.8	42.0
Total	26.0	53.0	21.0	100.0

c Differences: Observed − Expected

	Republicans	Democrats	Independents	Total
Yes	3.9	−4.7	0.8	0
No	−3.9	4.7	−0.8	0
Total	0	0	0	0

A fictitious random sample of 100 voters is classified in two ways—according to party affiliation (Republican, Democrat, Independent) and according to answer (Yes or No) to the question "Do you favor capital punishment?" This gives the 2 × 3 table of observed numbers in part a. From the marginal totals in part a, and from the hypothesis of independence, we compute the expected numbers of part b. Finally, part c shows the differences between observed and expected numbers.

pendence hypothesis (being a composite hypothesis) does not give the expected numbers directly, but it does provide them indirectly through its assertion that the proportion of Yes answers is the same in each party. If this is true, it is easy to see that this common proportion, whatever it may be, must hold for the entire population. The hypothesis does not specify the value of the common proportion, but we can estimate it from our sample. Taking the sample as a whole, 58 percent answered Yes. If this (estimated) common value applies to the 26 Republicans, we would expect 58 percent of them to answer Yes, or $0.58 \times 26 = 15.1$. Similarly, $0.42 \times 26 = 10.9$ of the Republicans would be expected to answer No. Proceeding in this way, we can compute a 2×3 table of numbers to be expected in the six categories according to the hypothesis of independence (see Table 30b).

We are now ready to compute the terms of χ^2, just as in the earlier example. As a first step, we tabulate the six differences between the observed and the expected. They are shown in Table 30c. It is of course not accidental that all marginal totals in this table are zero. A little thought will show you how this follows from the way the marginal totals in Table 30a were used in computing the expected numbers in Table 30b. To find the value of χ^2 we need only square each difference and then divide by the corresponding expected number. The sum of the six ratios, one for each category, is the desired χ^2:

$$\chi^2 = \frac{(3.9)^2}{15.1} + \frac{(-4.7)^2}{30.7} + \frac{(0.8)^2}{12.2} + \frac{(-3.9)^2}{10.9} + \frac{(4.7)^2}{22.3} + \frac{(-0.8)^2}{8.8}$$
$$= 1.01 + 0.72 + 0.05 + 1.40 + 0.99 + 0.07$$
$$= 4.24$$

Although the calculation of χ^2 proceeds just as before, the degrees of freedom are different. Inspection of Table 30c shows that there are not $C - 1 = 6 - 1 = 5$ df in this case. Rather, once we have specified two of the differences (say 3.9 and -4.7), the other four can be calculated from the fact that all marginal totals are zero. There are accordingly only 2 degrees of freedom in the χ^2 in this case. [A similar argument shows that in general there are $(r - 1)(c - 1)$ df in a contingency table having r rows and c columns.]

If you refer to Fig. 35, you will see that for df $= 2$ the value $\chi^2 = 4.24$ corresponds to the P value of about 0.119. Thus, for example, the data of Table 30a fail to be significant at the 10 percent level and we cannot conclude that attitudes toward capital punishment are associated with party affiliation. But as we mentioned before, the chi-square test is not very powerful because it is an omnibus test against all sorts of deviations. This test would not have a good chance of detecting a moderate degree of dependence of attitude on affiliation.

It might have been better to use a test that pays attention to some logical ordering of the political parties. Thus we might test the hypothesis of independence specifically against the alternative that Yes answers *increase* as you go from Democrats to Independents to Republicans. Such tests are a little too complex for presentation here—and in any case, the decision to use such a test would have to be made *before* seeing the data.

□   You are now ready to try out the chi-square test in two investigations of your own data. The first investigation will seek to determine whether, in all your STATLAB sampling, you have indeed used as good a randomizing agent as advertised: In Cardano's words (Box 1), have your dice been "honest"? To this end you will test the fairness of one die—the red one. To date, each one of you has drawn approximately 285 STATLAB families for the various samples with which you have worked. (We say "approximately" because in Unit 14 you determined the sample size yourself—within limits.) In drawing any single family it was necessary, of course, to make two throws of your dice. This means that you have thrown your red die about 570 times. If the die is fair, then in a very large collection of throws each of the six faces would appear about one-sixth of the time. You may regard your data as a sample of about 570 drawn from a "population" of such throws and use the chi-square test to test the hypothesis that each face has the same expected number of appearances.

As a second investigation, we shall ask you to inquire into the relation, if any, between the educational level of StatLab fathers and whether or not they were cigarette smokers at time of test. Spend a few minutes speculating about such a relation. On the basis of general knowledge, would you think that fathers in their thirties and forties around 1971 would tend to smoke more, or smoke less, if they have attended college? Can you support your conjecture with evidence, information, or reasoning?

Important: Before proceeding with the work, turn to the inquiry and answer items 1, 5, and 7.

WORK TO BE DONE

STEP 1   Tabulate, on page 1 of the worksheet, the number of times each face of your red die is recorded in the ID numbers in the Databank. Note that each ID number records *two* throws of the red die. These ID numbers can be found in Databank, Files B-1, C-1, E-1, F-1, G-1, H-1, and I-1.

STEP 2   Apply the chi-square test to see whether your red die is fair, i.e., whether the data you have tabulated agree reasonably with the hypothesis that each face has the same expected number of appearances.

STEP 3   Draw a fresh sample of $n = 50$ StatLab fathers, and record for each on page 2 of the worksheet (1) whether the father is or is not a cigarette smoker at time of test and (2) the father's educational level.

STEP 4   Tabulate the results of step 3 to form a 2×3 contingency table, Table a on page 3 of the worksheet. Note that the fathers' educational levels are grouped into three categories: not a high school graduate (codes 0 and 1), a high school graduate with no college (code 2) and some college (codes 3 and 4).

STEP 5   Apply the chi-square test for the hypothesis that smoking and education are independent for StatLab fathers.

STEP 6   Complete the inquiry.

The section you have just concluded is a complex and intellectually challenging one (which statement will probably be greeted by a fervent "Amen!"). The many new concepts require the longer than usual summary paragraphs to cover each of the five units. There is another reason for these long paragraphs. The last five units are very closely interrelated, and this review has been written to stress this interdependence of concepts and to help you achieve a further understanding of what a statistician does when he tests statistical hypotheses. And this is important, since what he does influences the decisions of a significant section of society: It helps determine which claims, hunches, and generalizations of the medical researcher, the economist, the pollster, the politician, the scientist, the market surveyor, etc., will be rejected and which will be accepted.

☐   *One-sample* or *two-sample tests of statistical hypotheses* yield *P values* that express, in probability terms, the strength of the evidence against the *hypothesis under test:* The smaller the P value, the less tenable is the hypothesis. (This frequently means that the original assertion which led to the statistical testing of the hypothesis becomes more firmly held.) To understand what a P value is and how it relates to the hypothesis under test, as well as to the assertion which led to statistical testing, is to understand the central concept of statistical hypothesis testing. The P value makes possible the adoption of a *decision rule* whether to reject or accept the hypothesis. That is, prior to the inspection of the data, one can predetermine the decisive *significance level,* i.e., the minimum P value required for the rejection of the hypothesis under test. The *z-score test* provides us with a basic test of statistical hypotheses and, in one form or another, appears in almost every test we examine here. In its simplest form, the z-score test uses the following formula for the case of a mean of one sample:

$$z = \frac{\bar{x} - \mu_H}{\sigma/\sqrt{n}}$$

where σ is either a *historical* σ or is estimated by s. The value obtained for z is then referred to the normal table, from which we obtain the P value corresponding to that z. In determining the P value, either a *one-sided (one-tailed)* or *two-sided (two-tailed) test* can be chosen (prior to inspection of the data). The choice is determined by the interests and assumptions of the analyst. In two-sample problems the hypothesis under test is the *null hypothesis.* Conjectures that do not specify μ_H values are cast into the null hypothesis form to enable statistical testing. In the two-sample case the null hypothesis for means is expressed in the following z-score formula:

$$z = \frac{\bar{x}_1 - \bar{x}_2}{\sqrt{s_1^2/n_1 + s_2^2/n_2}}$$ (UNIT 18)

☐ The significance level, by specifying the minimum P value for the rejection of the hypothesis under test, thereby determines the probability of *false rejection* of that hypothesis. In assessing how good a test is, the risk of false rejection (*Type I error*) must be balanced against the risk of *false acceptance* (*Type II error*). To do this, we determine the *power* of a test: High power means a low chance of false acceptance. Therefore if two different tests have the same significance level (the same chance of false rejection), the one with the higher power is preferred since it will have a lower chance of false acceptance. The power of a proposed test of a hypothesis can and should be appraised before the collection of data bearing on that hypothesis is begun, and the data collection should be guided accordingly. Failure to do so risks the possibility that the hypothesis under test will be accepted no matter what the validity of the hypothesis may be. The z-score test is of broad applicability. As a one-sample test applied to proportions, the z score tests the hypothesis $P = P_H$ by using a sample proportion p in the formula

$$z = \frac{p - P_H}{\sqrt{P_H Q_H/n}}$$

Here, and with all discrete variables, the *continuity correction* should be used in determining the significance level (which, in turn, determines the *critical value* of a sample proportion). The continuity correction is necessary here because this z score is based on a discontinuous variable but is referred to the normal table, which assumes a continuous variable. As a two-sample test of the null hypothesis $P_1 = P_2$, the z score uses two sample proportions in the formula

$$z = \frac{p_1 - p_2}{\sqrt{pq(1/n_1 + 1/n_2)}}$$

where p and q are obtained by pooling both samples. For testing $\tilde{\mu} = \tilde{\mu}_H$ a sample median $\tilde{x}$ is used in the formula

$$z = \frac{\tilde{x} - \tilde{\mu}_H}{\mathrm{SE}_{\tilde{x}}}$$

The simple z-score test can be used whenever two conditions are met: (1) The estimate for the parameter has a (nearly) normal distribution centered at the population value and (2) a reliable SD or SE of the estimate can be obtained. (UNIT 19)

☐ When the two conditions for the use of a z-score test cannot be met, there are available several alternative tests. Where n is small, with the result that the SE is not very stable, we can instead use the *t test* (*Student's t*), where

$$t = \frac{\bar{x} - \mu_H}{s/\sqrt{n-1}}$$

provided that the population is normal. The quantity $n - 1$ is known as the *degrees of freedom (df) of the test*. The instability of s, characteristic of small samples, prohibits the use of the normal table to describe the distribution of t (and hence to determine the corresponding P values). Therefore separate distributions of t for each df must be calculated. These distributions are available in tabular or (in STATLAB) in graphic form, and they permit the determination of the P value that corresponds to any obtained t. Thus the t test differs from the simple z test (aside from the use of $n - 1$ rather than n) only in that it obtains its P values by reference to its own special distributions, whereas the z is referred to the normal distribution. Otherwise the P values, etc., have the same meanings. Although the t test is more powerful than any other for testing the location of a normal population, it fails when dealing with a nonnormal population. For the latter instance there are available a number of *nonparametric* tests, so called because they do not assume a population distributed in a form specified by its parameters μ and σ. Among these nonparametric tests the *sign test* for the hypothesis $\mu = \mu_H$ is the simplest and makes the least constraining *symmetric assumption* for the population distribution. This test determines (by the z-score test using the continuity correction as discussed in Unit 19) whether the proportion of the sample values that fall above μ_H is significantly different from $\frac{1}{2}$. The z-score test can be used here since we have converted all our sample observations into a simple proportion and it no longer matters whether the original observations came from a normal or nonnormal distribution or whether we can calculate a stable s for those original observations. The sign test by doing this does not make full use of the available evidence. The *Wilcoxon test,* however, uses both the sign and the size of the deviations from μ_H and thus provides us with a more powerful nonparametric test. With the Wilcoxon we rank the sample values in order of increasing magnitude of deviation from μ_H, attach the appropriate signs, and sum the ranks to obtain both V^+ and V^-. The exact form of the distribution of V can be worked out for any given n, and from these distributions the appropriate

P values can be obtained. However, a good *approximation for the P value* can be calculated by converting V into a z score (applying the continuity correction) and then referring the z to the normal table. This is possible because V is nearly normally distributed and formulas for its mean and its standard deviation have been worked out. The formula for converting V into a z score is

$$z = \frac{V - \frac{1}{4}n(n+1)}{\sqrt{\frac{1}{24}n(n+1)(2n+1)}}$$
(UNIT 20)

☐ The t and Wilcoxon tests can also be used to test the null hypothesis in the two-sample case. The *two-sample t test*, however, requires not only that the two populations both be normal but also that $\sigma_1 = \sigma_2$. These two assumptions (where valid) permit the pooling of the observations from the two samples in estimating a common σ. This gives us the following formula for t:

$$t = \frac{\bar{x}_1 - \bar{x}_2}{s\sqrt{1/n_1 + 1/n_2}}$$

where

$$s = \sqrt{\frac{n_1 s_1^2 + n_2 s_2^2}{n_1 + n_2 - 2}}$$

The value of the t thus determined may then be referred to t distributions (given in tabular or graphic form) for obtaining the appropriate P value, but now df $= n_1 + n_2 - 2$. Many variables are not normally distributed, however, and the t test cannot be recommended in such cases. Recourse can be had to the *Wilcoxon two-sample test*, which makes the nonparametric assumption that the two populations have *a common, but not necessarily normal, distributional shape* and that $\sigma_1 = \sigma_2$. To test the null hypothesis that $\mu_1 = \mu_2$, the Wilcoxon test ranks all the observations of both samples as if they were combined (permitted because of the two assumptions listed above) but sums the ranks of the observations in only one sample to obtain W. The exact distribution of W (and hence of the corresponding P values) can be worked out. However, since W is approximately normally distributed (no matter what the shape of the distribution of the original observations of the two samples), a good approximation of the P values can be obtained by using the familiar procedures of expressing W in a z score for reference to the normal table. Substituting the formula for the mean in the usual z-score formula, we obtain the following formula for converting W into a z score:

$$z = \frac{W - n_2 \dfrac{n_1 + n_2 + 1}{2}}{SD_W}$$

(UNIT 21)

☐ Statistical hypotheses involving qualitative or categorical data can be assessed by the *chi-square test*, applicable for both *simple hypotheses* and *composite hypotheses*. The simple hypothesis specifies the proportion of cases expected in each category, and χ^2 tests the agreement between these expected numbers and those observed in the sample. The statistic χ^2 (a modified sum of z^2's) is determined by the formula

$$\chi^2 = \frac{(np_1 - nP_1)^2}{nP_1} + \frac{(np_2 - nP_2)^2}{nP_2} + \cdots + \frac{(np_C - nP_C)^2}{nP_C}$$

For the composite hypotheses discussed in STATLAB, the population is classified in two different ways, thus permitting the construction of a *contingency table*. The sample is classified according to this table, yielding an observed number in each category. For a *test of independence,* once the sample has been drawn and categorized the expected number can be calculated. From then on the chi-square test proceeds as in the simple-hypothesis case. Since the P values of χ^2 depend only on the number of degrees of freedom, it is possible to present a table (or, in STATLAB, a set of curves) for P values corresponding to different χ^2's for different df's. The df's, however, are calculated differently for the simple and composite hypotheses. Since all deviations are relevant to χ^2, the test has low power in detecting a specific sort of deviation. (UNIT 22)

A final word. *Each of the nine tests of statistical hypotheses we have examined has its own set of assumptions, area of applicability, virtues, and limitations. These must be thoroughly mastered, since a main task of the statistician is to choose the test most appropriate for the problem and for the population. This choice should be made before the data are collected because (1) such a choice can be of crucial help in specifying how the data collection is to be carried out and (2) "fitting" a test to already available data can result in badly biased and erroneous analyses.*

BIVARIATE DISTRIBUTIONS AND CORRELATION

In our work so far we have been mainly studying *univariate* statistics: statistical methods useful in analyzing a single variable, conventionally denoted by x. We have seen how to organize a sample of x values into a graph or a frequency table and how to use the sample to make inferences (estimates, tests) about the x values in the population from which the sample came.

In this unit and the two that follow we shall take up *bivariate* methods: methods that are required when we want to consider two variables simultaneously, conventionally denoted by x and y. One often wants to know whether two variables are related to each other, and if so how and to what extent. Is smoking (x) related to blood pressure (y)? Are sunspots (x) related to world wheat yield (y)? Is there an association between money supply (x) and inflation in the following year (y)? Do high school grades (x) predict success in college (y)? It is obvious that you could not approach any such question by merely examining the variables x and y one at a time. They must be looked at *simultaneously* to see how they relate to each other, if indeed they do.

Our presentation of the bivariate methods required to handle two variables simultaneously will be more or less parallel to the treatment of univariate methods in earlier units. In this unit we shall show you the most useful graphic techniques for visualizing bivariate data; extend the idea of the simple frequency table to the two-variable case; and introduce a very widely used numerical measure of the degree of association of two variables. Unit 24 will be devoted to tests for deciding whether an apparent relation between two variables is real or just an accident of sampling. Finally, Unit 25 will show you how the relation between two variables can be exploited to make inferences from one variable to the other. In each unit, of course, you will be asked to use the techniques yourself in an investigation based on your own bivariate sample drawn from STATLAB.

THE SCATTER DIAGRAM

The slogan of Unit 5, that "one graph is worth a thousand numbers," is even more true when you want to grasp the significance of a sample of bivariate data. Let us illustrate this important fact. Suppose that we are interested in how the variables height x and weight y are related in the population of 648 STATLAB boys. To find out, we draw a random sample of $n = 20$ and for each boy look up x and y in the Census. The results are shown in Table 31. Spend a

TABLE 31 Height in inches (x) and weight in pounds (y) of a sample of 20 STATLAB boys

ID no.	x	y
43-42	49.0	60
52-33	51.8	87
52-66	50.4	60
54-51	61.1	65
54-36	56.3	98
64-54	58.1	96
61-25	52.8	60
53-53	51.9	62
52-42	52.6	65
53-22	54.7	68
41-35	50.2	58
62-34	54.9	82
61-62	53.1	62
54-22	48.9	52
41-21	50.8	59
63-42	61.4	94
51-35	55.8	76
66-12	52.9	93
42-36	58.4	86
53-42	50.6	62

These data are also depicted in Fig. 36 as a scatter diagram.

240

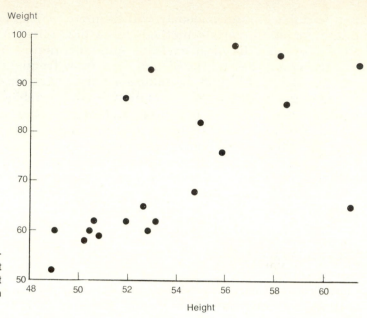

Weight

FIGURE 36 Scatter diagram of a sample of $n = 20$ STATLAB boys by height (inches) and weight (pounds). Each dot represents one of the boys listed in Table 31.

Height

little time examining these numbers. Is there a relation between x and y? If so, how much relation and of what nature?

Even a skilled statistician would have trouble extracting sense from this jumble of numbers in their present form. But turn to Fig. 36. This presents the same information as Table 31, but in graphic form. Each dot represents simultaneously *both* the x and the y values for one of the 20 boys. For example, the first boy in the table has $x = 49.0$ in and $y = 60$ lb. He is represented by a point to the far left of the diagram, where the vertical coordinate at $x = 49.0$ intersects the horizontal coordinate at $y = 60$. Such a depiction of bivariate data—the 20 points taken together—is known as a *scatter diagram* (sometimes referred to as *scatter plot*).

See how compellingly the scatter diagram tells you the main facts about this sample of bivariate data. We see at once which boy is tallest and which is shortest, which is heaviest and which is lightest. Clearly height and weight tend to be related, as you would expect, with the taller boys being on the average also heavier than the shorter boys, and the heavier boys on the average also taller than the lighter boys. The relation, however, is not a perfect one: Some boys are both taller and *lighter* than other boys, and some are heavier and *shorter* than others.

We can easily reach some quantitative conclusions as well—again merely by looking at the scatter diagram. As represented by this sample, one could venture the estimate that STATLAB boys who are 50 in tall will have a mean weight of about 58 lb; the 58-in boys, about 91 lb. Thus to an 8-in increase in height there corresponds a 33-lb increase in mean weight, or a rate of $\frac{33}{8} = 4.1$ added pounds, on the average, per added inch. Looked at the other way, the 60-lb boys have a mean height of about 51 in while the 95-pounders average about 57 in; mean height increases at the rate of $\frac{6}{35} = 0.17$ added inches, on the average, per added pound. These considerations will be pursued in Unit 25.

The merest glance at the scatter plot calls attention to one boy who lies away from the general pattern: boy 54-51 with $x = 61.1$ and $y = 65$. The mean weight of boys of his height is about 95 to 100 lb; he is at least 30 percent lighter than the average boy of his height. One of the main uses of the scatter diagram is to identify such *outliers* for closer examination. Not infrequently the examination reveals an error of plotting or transcription, so as always the graph helps to check the work. If the outlier is correctly plotted, the scatter plot thereby flags a case that is unusual or interesting in some regard.

THE BIVARIATE NORMAL DISTRIBUTION

Leaving aside the outlier, the remaining points of the scatter diagram form a (more or less) coherent pattern, shaped like an oval or ellipse. This elliptical pattern is characteristic of plots of bivariate data for which both x and y have a (more or less) normal distribution. The ellipses may be slanted up or down (from left to right), and they may be fat or thin. These two clearly visible characteristics of the bivariate distribution reveal both the *direction* and the *degree* of association between x and y. Figure 37 shows scatter diagrams for samples of $n = 20$ from five different bivariate normal distributions (the r values will be discussed later). In Fig. 37a and b the ellipses slant *up:* They depict samples in which (as in our height-weight example) large values of x tend to go with large values of y—in such cases we say that x and y are associated *positively*. Figure 37d and e have ellipses that slant *down*, and these illustrate *negative* association: Large x tends to go with small y. Figure 37c is the intermediate case: There is no slant; i.e., there is little or no association at all, and y is as likely to be big when x is big as when x is small. In Fig. 37a and e the strength of *association* is high—for these bivariate distributions, if we know the value of any single x we can almost make a reliable prediction about the value of its associated y and vice versa. In Fig. 37b and d the association between x and y is much weaker, and the value of a single x does not tell us with any great precision what the value of its associated y may be. Looking at Fig. 37a to e, the order of strength of association (disregarding the *direction* or sign of association) is, from lowest to highest, c, d, b, e, a.

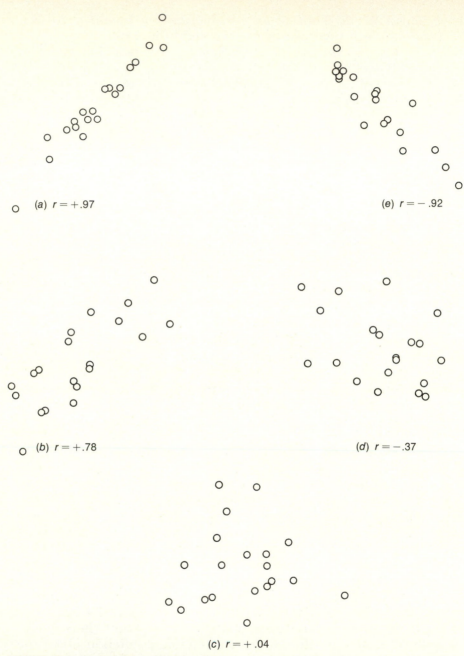

(a) *r* = +.97

(e) *r* = −.92

(b) *r* = +.78

(d) *r* = −.37

(c) *r* = +.04

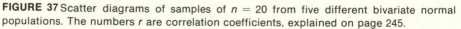

FIGURE 37 Scatter diagrams of samples of *n* = 20 from five different bivariate normal populations. The numbers *r* are correlation coefficients, explained on page 245.

A very great deal, indeed, can be learned about a bivariate collection of data from a careful inspection of its scatter diagram: We can judge whether the *x* and *y* variables have approximately *normal distributions;* we can *identify the extreme values* (highest and lowest x and y values); we can determine the *direction of association* which may exist between x and y; where there is an association we can spot the *individuals who deviate* markedly from the general trend; we can make a very good stab at estimating the *rate of increase, or decrease, of x relative to y;* and we can make comparisons of the *strength of associations* among different bivariate distributions (e.g., we can say, after visual inspection, that the data of Fig. 37*a* show a stronger degree of association than do those of 37*b*). And, finally, all this can be done more quickly and more surely as our experience with scatter diagrams increases. The making and contemplation of scatter diagrams is a most rewarding exercise for those who seek understanding of bivariate phenomena.

PEARSON'S COEFFICIENT OF CORRELATION

Although the scatter plot permits us to make quite good qualitative statements about degrees of association between x and y (and even enables the experienced statistician to make respectable *quantitative* estimates), it cannot do the quantitative job completely satisfactorily. For that, we need *numerical* measures of direction and strength of association. A variety of such measures is available, just as there are numerous different quantitative measures of location and dispersion. The most widely used measure of association is the *coefficient of product-moment correlation,* proposed in 1896 by Karl Pearson as a refinement of a somewhat similar measure suggested earlier by Francis Galton.

The development and logic behind the product-moment correlation coefficient can be very quickly and simply outlined. Recall from Unit 9 that the variance of x is the mean of the n values of $(x - \bar{x})(x - \bar{x})$; this quantity, of course, measures how widely x varies. Similarly, the variance of y is the mean of $(y - \bar{y})(y - \bar{y})$, and it measures how much y varies. If we want to see to what extent x and y vary *together,* it seems natural to take one factor from each of these products and use the mean of $(x - \bar{x})(y - \bar{y})$. Accordingly, we define the *covariance* of x and y to be

$$\text{Cov} = \frac{\Sigma(x - \bar{x})(y - \bar{y})}{n} \qquad \textbf{FORMULA 43}$$

Suppose that the scatter diagram of a sample looks like Fig. 37*a*. It is clear that for the points on the upper right, both $(x - \bar{x})$ and $(y - \bar{y})$ will be positive and large since these x and y values are much larger than the mean x and y values. Therefore the product $(x - \bar{x})(y - \bar{y})$ is a *large positive* number for these points. The points on the lower left have x and y values which are much smaller than the x and y means, and therefore multiplying the two large negative values will again give us *large positive* numbers, $(x - \bar{x})$ $(y - \bar{y})$. There are a few points for which $(x - \bar{x})$ and $(y - \bar{y})$ have opposite signs, and the products of these cases [multiplying a positive $(x - \bar{x})$ by a negative $(y - \bar{y})$ or vice versa] will be *negative* and also *small*. Therefore when we average the values of all the $(x - \bar{x})(y - \bar{y})$ products, we are averaging many large positive numbers with a few small negative numbers, so the result, the covariance, is *large and positive*. By similar reasoning, the data of the scatter diagram of Fig. 37e would have a *large negative* covariance [since the points on the upper left and lower right will have $(x - \bar{x})$ and $(y - \bar{y})$ values of *opposite* signs, but *large*]. Figure 37b will have a positive covariance, but smaller than that of Fig. 37a. Figure 37d will have a small negative covariance. Finally, Fig. 37c will have a covariance near zero, since the signs of $(x - \bar{x})(y - \bar{y})$ are as likely to be positive as negative and will tend to cancel each other when added up.

Here then, in the covariance, is a way to calculate a numerical value of the direction and strength of associations between x and y. It suffers, however, from a serious disadvantage: Its value depends on the units of measurement. If, for example, we were to have used centimeters instead of inches in measuring the heights of STATLAB boys, each of our x values would have been 2.54 times as large as it now is. As a result, each $(x - \bar{x})$ would be 2.54 times as large, as would each $(x - \bar{x})(y - \bar{y})$ and hence Cov itself. Similarly, if we had measured weight in kilograms, Cov would be reduced by the factor 2.2.

Now surely if someone asks "How closely associated are the heights and weights of STATLAB boys?" he does not want the answer to depend on whether we use metric or English units of measurement. He wants an answer that is "dimensionless."

Karl Pearson saw that a dimensionless measure of association would result if he divided Cov by the standard deviations of x and of y (as we shall demonstrate in a moment). He therefore defined his *correlation coefficient*, designated by the letter r, as

$$r = \frac{\text{Cov}}{s_x s_y} = \frac{\Sigma(x - \bar{x})(y - \bar{y})}{n s_x s_y}$$ FORMULA 44

If you change from inches to centimeters, Cov gets multiplied by 2.54—but so does s_x (see Formula 5, Unit 9). The net effect is that the factors cancel and leave r unchanged. Similarly, change from pounds to kilograms will divide both the numerator and denominator of r (Cov and s_y) by 2.2, and r again remains the same.

We can write r in a different form that helps to reveal its nature. Since Cov $= (1/n)\Sigma(x - \bar{x})(y - \bar{y})$, when we divide by $s_x s_y$ we can take this divisor inside the summation and write r as follows:

$$r = \frac{1}{n} \Sigma \left(\frac{x - \bar{x}}{s_x} \right) \left(\frac{y - \bar{y}}{s_y} \right)$$

But the factors in parentheses, as you can quickly recognize, are just x and y converted to the z scores by Formula 8 of Unit 12:

$$z_x = \frac{x - \bar{x}}{s_x} \qquad z_y = \frac{y - \bar{y}}{s_y}$$

Making the substitutions, we can now write the formula for Pearson's product-moment correlation coefficient in the form

$$r = \frac{\Sigma z_x z_y}{n}$$

FORMULA 45

That is, r is just the mean of the products of the standard scores of x and y for all the cases in the sample.

We have calculated the value of r for the data represented by each of the diagrams in Fig. 37. For Fig. 37a, the r has the very high value +0.97, reflecting the strong positive association of x and y in this case. The value $r = -0.92$ of Fig. 37e reflects the negative association of x and y in this diagram—very strong, but less so than the positive association of Fig. 37a. Figure 37c, where x and y seem nearly unrelated, gives the value of $r = +0.04$, near zero. It can be easily shown that r can never go above the value 1.00 nor below the value −1.00. These extreme values are attained by r only when x and y have a perfect straight-line relation (positive slope for +1.00 and negative for −1.00).

Before reading further, please stop to compare Fig. 36 with Fig. 37a to c. Try to judge which one of the five scatter plots of Fig. 37 shows a little stronger association between x and y than does Fig. 36 (positive or negative) and which one shows a little weaker association. Using the r values of Fig. 37 to guide you, guess the r value of Fig. 36. *Jot your guess down on the Fig. 36 scatter diagram.* (The wise statistician always guesses the value of a correlation by looking at the scatter diagram before he computes it. This is one of the easiest and most useful graphic checks of a computation in which it is easy to blunder.) Having made your guess, now let us calculate r. Turn to Table 32. The first two columns repeat the data of Table 31. In the last three columns we have used Formula 45 to calculate r for our sample of boys' heights and weights. By the familiar methods of Units 7 and 9, we first found the mean $\bar{x} = 53.78$ in and standard deviation $s_x = 3.64$ in of the 20 heights. Using these values we converted each boy's height into a standard score z_x. Thus for the first boy

TABLE 32 Calculation of the correlation of height and weight for the boys of Table 31 by the direct method

x	y	z_x	z_y	$z_x z_y$
49.0	60	−1.31	−0.83	1.09
51.8	87	−0.54	1.00	−0.54
50.4	60	−0.93	−0.83	0.77
61.1	65	2.01	−0.49	−0.98
56.3	98	0.71	1.75	1.24
58.1	96	1.19	1.62	1.93
52.8	60	−0.27	−0.83	0.22
51.9	62	−0.52	−0.70	0.36
52.6	65	−0.32	−0.49	0.16
54.7	68	0.25	−0.29	−0.07
50.2	58	−0.98	−0.97	0.95
54.9	82	0.31	0.66	0.20
53.1	62	−0.19	−0.70	0.13
48.9	52	−1.34	−1.38	1.85
50.8	59	−0.82	−0.90	0.74
61.4	94	2.09	1.48	3.09
55.8	76	0.55	0.26	0.14
52.9	93	−0.24	1.41	−0.34
58.4	86	1.27	0.94	1.19
50.6	62	−0.87	−0.70	0.61
				12.74

The x and y columns reproduce the data of Table 31. Then each x and y value is converted into a standard score, using these 20 boys as the reference group. Finally, the standard scores are multiplied in pairs and the products are added up.

$$z_x = \frac{49.0 - 53.78}{3.64} = -1.31$$

Similarly, we found $\bar{y} = 72.25$ lb, $s_y = 14.70$ lb, and calculated

$$z_y = \frac{60 - 72.25}{14.70} = -0.83$$

The final column gives the products of the standard scores: $-1.31 \times (-0.83) = 1.09$ and so on. When the sum of these scores is divided by $n = 20$, we get $r = 12.74/20 = +0.64$. How close did your guess come?

The seven steps in calculating the product-moment correlation coefficient r by this direct method may be summarized as follows (a shortcut method much more efficient for large samples will be presented later):

1 Construct a scatter diagram and, by visual inspection, guess the sign and value of r. Jot it down.
2 Compute (by the most convenient methods available): $\bar{x}, s_x, \bar{y}, s_y$.
3 Using Formula 8 for z scores, convert *each pair* of x and y into z_x and z_y pairs.

4   Multiply z_x by z_y of each pair to get n products $z_x z_y$.
5   Sum these $z_x z_y$ products to get $\Sigma z_x z_y$.
6   Divide $\Sigma z_x z_y$ by n. This will give you the calculated r.
7   Check your calculated r with your guessed r. If they are too widely off, look for an error—either in the construction of your scatter diagram or (more likely) in your calculations.

THE BIVARIATE FREQUENCY TABLE

One can also make a scatter diagram for an entire bivariate *population* if the x and y values are known, but the labor of plotting a great many points is great and often is laborious labor lost. When there are too many points, the scatter plot is hard to comprehend. A better sense of the data can then be given by means of a bivariate frequency table, which combines the advantages of tabular presentation with many of the virtues of graphic representation.

The construction of a bivariate (simple) frequency table is modeled on the univariate method of Unit 4. For each variable separately, the range is divided into a reasonable system of class intervals of equal width. The individuals in the population are then *cross-tabulated* according to their class interval for both variables, and the frequency f is recorded for each "cell" of the cross-tabulation.

Table 33 illustrates all this for the heights and weights of the 648 STATLAB boys. We chose a 2-in width for the height x and used the phasing 46.0–47.9 in, 48.0–49.9 in, and so forth, which is convenient for tabulation. This gives us eight class intervals for the entire population. For weight y we chose $w = 5$ lb, with phasing 45–49, 50–54, and so forth. This gave us 15 class intervals, not counting the outliers. As in constructing univariate frequency tables, it is important to give some thought to the choice of the class intervals in constructing bivariate frequency tables. Using too many intervals makes an unwieldy table; using too few imposes too crude a grid on the data. As a rule of thumb, there should be at least eight intervals for each variable, not counting outliers.

Once the intervals have been chosen, the number of cases in each cell must be tabulated. We tabulated the 648 boys' heights and weights, and the results are shown in Table 33. For example, the entry $f = 22$, in the cell corresponding to height 54.0–55.9 and weight 75–79, means that just 22 of the 648 boys had a height between 54.0 and 55.9 in (inclusive) and also a weight between 75 and 79 lb (inclusive). (One of these happened to be drawn in our sample: boy 51–35 with height 55.8 in. and weight 76 lb.) The frequency or f values should be summed by column (giving at the bottom margin of the table the *marginal* distribution of height) and by row (giving at the right margin of the table the marginal distribution of weight). The fact that the marginal distributions add up to

TABLE 33 Bivariate frequency table for the height and weight of 648 StatLab boys

Weight class (pounds)	Height class (inches)								Total
	46.0–47.9	48.0–49.9	50.0–51.9	52.0–53.9	54.0–55.9	56.0–57.9	58.0–59.9	60.0–61.9	
125–129						2			2
120–124					1				1
115–119							1	2	3
110–114						2	4		6
105–109						1		1	2
100–104					2	4	2		8
95–99					3	4	3	1	11
90–94			1	3	4	6	5	2	21
85–89			1	3	13	14	5	2	38
80–84				8	18	18	6		50
75–79			5	10	22	12	1		50
70–74		1	9	44	51	10			115
65–69		2	19	61	34	3		1	120
60–64		3	51	58	11	2			125
55–59		14	34	12	2				62
50–54		15	15						30
45–49	2	1		1					4
Total	2	36	135	200	161	78	27	9	648

Each of the 648 boys is assigned to the appropriate height class and weight class and then tabulated in the "cell" at which these two classes intersect. These same cell frequencies appear on Fig. 41.

648 assures us that we have not missed nor duplicated any cases in our tabulation.

A bivariate frequency table can be viewed, comprehended, and used in much the same way as a scatter diagram. A quick visual inspection of Table 33 identifies the outliers: three boys very heavy for their heights and one boy very tall for his weight—they are boys 63–34, 64–34, and 66–55, who are presumably very obese; and boy 54–51, who must be a real string bean. Aside from the outliers, the data show the typical elliptical pattern. A close look shows the skewness of weight as an elongated or heavy "skirt" on the upper side, compared to the short skirt below. The same sort of analysis we devoted to the sample shows that, on the average, weight increases about 3.8 lb per inch of increased height while height increases about 0.17 in per pound of increased weight. The correlation of a frequency table like Table 33 can be estimated by comparison with diagrams such as Fig. 37*a* to *e*, in the same way we asked you to do with Fig. 36. After you have gained considerable experience in making such estimates, you will find that you have "internalized" a set of idealized comparison figures like Fig. 37*a* to *e*. Then you will be able to look at any scatter diagram or bivariate frequency table and compare it with your private set of mental "comparison figures" in estimating the r.

The simple frequency table can be portrayed graphically (Fig. 38) by the bivariate analog of the histogram. As you see, the cases tend to pile up in the center and fade away at the margins. This is characteristic of a near-normal distribution.

SHORTCUTS IN CALCULATING CORRELATIONS

The method used to calculate r in Table 32 becomes very laborious when applied to a large sample or to a population. There are various machines—from mechanical desk calculators to electronic devices —that can do the work for you. These machines will calculate r even if the data have not been tabulated in an orderly form. They are programmed to take each pair of (x,y) values and operate on these raw values according to the following formula—known as the *machine formula*—in which there are no terms except the original x and y values and n:

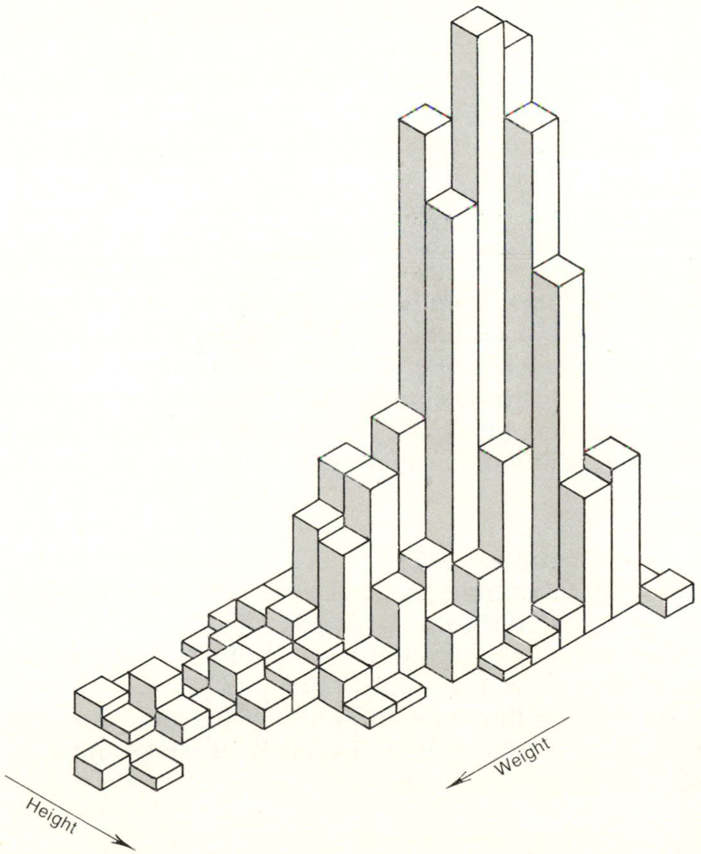

FIGURE 38 Graphic representation of the height-weight distribution of the 648 STATLAB boys. Compare this figure with Table 33, where the same data are given as a bivariate frequency table. Some of the columns are hidden behind others.

$$r = \frac{n(\Sigma xy) - (\Sigma x)(\Sigma y)}{\sqrt{(n\Sigma x^2) - (\Sigma x)^2}\,\sqrt{(n\Sigma y^2) - (\Sigma y)^2}}$$

FORMULA 46

However, being saved from laborious calculation by this *deus ex machina* is costly: You are deprived of all sense of the data, and even egregious mechanical errors may escape your detection.

Fortunately you can avoid both the laborious calculation and the heavy price by taking advantage of the tricks of Unit 9 which helped so much to reduce the labor of computing s. In fact, with the use of those shortcuts, the calculations of r—even where thousands of cases are involved—can be completed by any numerate person in very short order once the bivariate frequency table has been constructed. (And such a table should always be constructed whenever you deal with a large bivariate sample—even if you do not plan to calculate the correlations.)

When calculating r from such a table we must, to begin with, take the frequency f into account in the formula for covariance:

$$\text{Cov} = \frac{\Sigma f(x - \bar{x})(y - \bar{y})}{n}$$

Next the same bit of algebra used to simplify Formula 5 of Unit 9 applies here, to give

$$\text{Cov} = \frac{\Sigma fxy}{n} - \bar{x}\bar{y}$$

Finally, we may reduce the work greatly by using coding variables x' for x and y' for y. We do not even have to decode at the end if the same coding variables are used for s_x and s_y, because r is dimensionless and does not alter when the units are changed.

All this can be made clear by computing r for the 648 STATLAB boys' heights and weights. Table 34 reproduces the frequencies f of Table 33, but with convenient coding variables x' and y'. If we apply the methods of Unit 9 to the marginal distributions, we easily find that $\bar{x}' = -0.659$, $s_{x'} = 1.292$, $\bar{y}' = -1.191$, and $s_{y'} = 2.574$.

All that remains to compute is $\Sigma fx'y'$. This we do by first writing fy' after each f: For example, the cell at $x' = -2$ and $y' = -3$ has $f = 51$, so we write after it the entry $fy' = 51 \times (-3) = -153$. Each column of fy' values is totaled, giving the second line from the bottom, labeled $\Sigma fy'$. These entries are multiplied by x' to produce the values $x'\Sigma fy'$ in the bottom line, and when those are totaled on the right we have $\Sigma x'\Sigma fy' = \Sigma fx'y' = 2082$. Divide by $n = 648$ and subtract $\bar{x}'\bar{y}'$ to find

$$\text{Cov} = \frac{\Sigma fx'y'}{n} - \bar{x}'\bar{y}' = \frac{2082}{648} - (-0.659) \times (-1.191) = 2.428$$

TABLE 34 Calculation of the correlation of height and weight for the 648 StatLab boys by the short-cut method

y'	x' -4 f	x' -4 fy'	-3 f	-3 fy'	-2 f	-2 fy'	-1 f	-1 fy'	0 f	1 f	1 fy'	2 f	2 fy'	3 f	3 fy'
10										2	20				
9									1						
8												1	8	2	16
7										2	14	4	28		
6										1	6			1	6
5									2	4	20	2	10		
4									3	4	16	3	12	1	4
3					1	3	3	9	4	6	18	5	15	2	6
2					1	2	3	6	13	14	28	5	10	2	4
1							8	8	18	18	18	6	6		
0					5	0	10	0	22	12	0	1	0		
−1			1	−1	9	−9	44	−44	51	10	−10				
−2			2	−4	19	−38	61	−122	34	3	−6			1	−2
−3			3	−9	51	−153	58	−174	11	2	−6				
−4			14	−56	34	−136	12	−48	2						
−5			15	−75	15	−75									
−6	2	−12	1	−6			1	−6							
$\Sigma fy'$		−12		−151		−406		−371			118		89		34
$x'\Sigma fy'$		48		453		812		371	0		118		178		102

$$\Sigma x'\Sigma fy' = 2082$$

The bivariate frequencies f are the same as those shown in Table 33. The fy' values for column $x' = 0$ are omitted since they are not needed.

The final step is to divide by $s_{x'}$ and $s_{y'}$ to obtain

$$\frac{\text{Cov}}{s_{x'}s_{y'}} = \frac{2.428}{1.292 \times 2.574} = 0.73$$

This is the correlation of x' and y' and is also (because of the dimensionless nature of r) the correlation of the original variables x and y.

INTERPRETATION OF THE CORRELATION IN A SAMPLE

The coefficient 0.73 we have just computed is the correlation of x and y in the entire population of 648 boys. It is therefore a population parameter, often denoted by ρ (the Greek letter rho, which corresponds to the English letter r). The coefficient $r = 0.64$, computed in Table 32 for a sample of 20 boys, may be regarded as an estimate of the true population ρ. The law of large numbers assures us that if n is big enough, the r of a sample will probably be close to the ρ of the population from which the sample came. Because we have a sample of such modest size, it is not surprising that our estimate r

= 0.64 differs considerably from the true population value $\rho = 0.73$.

Having taken to heart our many exhortations about standard errors, you will now feel it necessary to attach a standard error to r as an aid in interpreting its accuracy. Unfortunately this is usually not possible, for the same reason that one cannot usually attach a standard error to the estimate s (see Unit 15). Like s, the r involves terms of the second degree: $(x - \bar{x})^2$ in the case of s, $(x - \bar{x})(y - \bar{y})$ in the case of r. These second-degree terms are very sensitive to the population values far out in the tails or skirts of the distribution. Since in practice you will seldom know the precise nature of the distributional shape in its periphery, it is seldom possible to know how variable r will be and hence not possible to attach a reliable standard error to r. This leads to another problem. Often we would like to test statistical hypotheses involving correlation coefficients. But since we cannot give a trustworthy value for SE_r, we cannot use the z score for testing a hypothesis that specifies a numerical value for ρ. (There is one important special case in which the z-score test *can* be used, however, as explained in the next unit.)

☐ You are now ready to try out these methods in a small investigation of bivariate data. We shall ask you to begin an inquiry into the question: What is the relation between the heights of the mother and father of the STATLAB families? Before going on, pause a minute to speculate about this question. On the basis of common sense and everyday observation, what do you think the answer might be? Do tall women tend to marry tall men? Or do opposites attract? And if so, is the tendency strong? Or are there so many exceptions as to yield only a negligible association? A scientist about to embark on such a study would always begin by wondering and speculating about its outcome.

WORK TO BE DONE

STEP 1   Before proceeding further, answer item 1 of the inquiry.

STEP 2   On page 1 of the worksheet make a scatter diagram of the heights of the mothers (x) and fathers (y) of the first 20 families of your sample of 40 STATLAB families, as recorded in File B-1 of the Databank. Use unrounded heights, recorded to tenth inch.

STEP 3   Before proceeding further, answer item 2 of the inquiry.

STEP 4   On page 3 of the worksheet, convert your scatter diagram into a bivariate frequency table. Use class interval width $w = 2$ for both variables, with phasing 60.0–61.9, 62.0–63.9, and so on.

STEP 5 Assign coding variables x' and y' to your table, and compute $\bar{x}'$, $s_{x'}$, $\bar{y}'$, $s_{y'}$, and r by the coding method, using the blanks on pages 2 and 3 of the worksheet.

STEP 6 Record in Databank, File J-1, the following values: C_x, $\bar{x}'$, $s_{x'}$, C_y, $\bar{y}'$, $s_{y'}$, and r (from step 5).

STEP 7 Complete the inquiry.

TESTS FOR INDEPENDENCE IN BIVARIATE DISTRIBUTIONS

In the preceding unit you became acquainted with Pearson's correlation coefficient, which is the most widely used measure of the degree to which two quantitative variables are associated. If you have in hand a sample of bivariate data, it may appear from their scatter diagram that they are indeed associated, and it is easy enough to compute r as the measure of the association. But after all, what you have is only a sample, and it is conceivable that the appearance of association is only the result of chance. Perhaps if you examined the entire population the two variables would be quite unrelated. We need a test of significance to assess the reality of an association that appears to hold in a sample.

If two variables have in reality no relation to each other, then their population correlation coefficient ρ will be equal to zero. This fact suggests that what we want is a test of the hypothesis that $\rho = 0$. Statisticians first approached the problem in this way, and only slowly realized that this formulation did not permit a satisfactory answer in most instances. The better approach, as we shall see, takes off instead from the concept of independence that we have already encountered in Unit 22 in the study of categorical data.

254

In this unit we shall explore the connections between the hypothesis that the population correlation between x and y is zero and the hypothesis that x and y are independent. Next we shall present two frequently used tests for the hypothesis of independence. One of these tests is based on the sample r; the other is based on a procedure which first converts the x and y values into ranks. Finally, as is usual when we have more than one test to do a job, we shall discuss the question of how to choose intelligently between the two available tests. In your work, you will have an opportunity to try out both tests on your own data. When the instructor reveals the true facts about the STATLAB population—facts which you will have been trying to infer from your bivariate sample—you will be able to see for yourself which test did the better job.

ZERO CORRELATION AND INDEPENDENCE

In Unit 22 we introduced the idea that two categorizations of a single population might be independent of each other, using the example of party affiliation and attitude toward capital punishment. If the distribution of attitude is the same within each of several parties, we say that attitude is independent of party. When this is so, finding out someone's party affiliation will tell you nothing about how likely he is to favor capital punishment; and, conversely, knowing his attitude will offer no clue as to his party.

The same notion of independence may also be applied to quantitative variables x and y. To see what independence means for a bivariate population, let us proceed as follows: We first classify the individuals of the population into classes according to their x values. The members of each x class will usually have somewhat different y values. In other words, within each x class there will be some distribution of y values. If these y distributions are the same in *all* the x classes (the y distributions having the same means, the same standard deviations, and the same shapes), then we may say that *y is independent of x.* Clearly, in that case, learning that an individual is in this or that x class (has this or that x value) adds *nothing* to your knowledge about what his y value is likely to be; and, conversely, the value of y tells nothing about what to expect for x. Let us take a simple example: The students in your college might be measured for resting pulse rate (x) and SAT verbal score (y). We would expect that the distribution of verbal scores among those students with a pulse rate of 60 per minute would turn out to be about the same as among those with a pulse rate of 65 or 70 per minute and so forth. Similarly, the distribution of pulse rates would be nearly the same among the high scorers as among the low scorers. If we are right about this, then we could conclude that SAT verbal scores (y) and pulse rate (x) are essentially independent of each other; they have little or nothing to do with each other.

If x and y are independent, then it turns out that the product-moment correlation between x and y has to be equal to zero. This is not too difficult to prove.

The argument goes like this: Consider the group of individuals having a particular value of x. Because y is independent of x, these individuals have the same y distribution as every other x group and hence the same y distribution as the whole population. Therefore, for the members of our x group the deviations $y - \bar{y}$ add up to zero. Since all these individuals have the same value of $x - \bar{x}$, this is just a constant factor in the product $(x - \bar{x})(y - \bar{y})$, so these products also add up to zero for our particular group. But the same is true of each other x group, so the grand total of $(x - \bar{x})(y - \bar{y})$ in the whole population is zero. That is, the covariance of x and y is zero and hence so is the correlation ρ—see Formulas 41 and 42.

But (and this may be a bit surprising) the *product-moment correlation coefficient between x and y can also turn out to be zero even if x and y are dependent.* Figure 39 illustrates two distributions in which y does depend on x, but in which the product-moment correlations are zero. In Fig. 39a the dependence of y on x is very strong: Knowing x almost tells you the precise value of y. Similarly, knowing y tells you that x has approximately one or the other of two values. In Fig. 39b the mean y value is the same in each x group, but the *dispersion* of y depends on the value of x. In this example, individuals with a small value of x have a much tighter y distribution than do the individuals with a large x value. [Can you see why in each of these distributions the *mean* value of $(x - \bar{x})$ $(y - \bar{y})$ is zero, so that the product-moment correlation is zero?]

FIGURE 39 Scatter plots of two distributions with zero correlation between x and y, although y depends strongly on x.

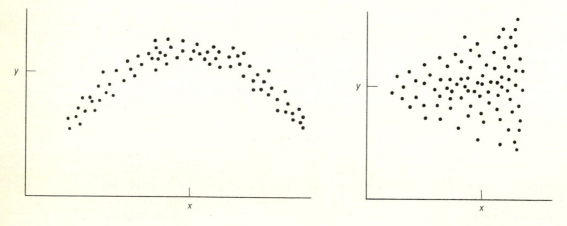

(a) Value of y almost determined by x

(b) Dispersion of y varies with x

The hypothesis of independence is *stronger* than the hypothesis of zero correlation, in the following sense: If x and y are independent, then the correlation between x and y *must* be zero; if the correlation between x and y is zero, then x and y *may or may not be* independent. To say that x and y are independent is to say that there is *no* sort of association or relation between x and y. To say that they have zero correlation is merely to say that on the *average* $(x - \bar{x})(y - \bar{y})$ is zero, but this does not rule out the possibility of various kinds of relations existing between x and y. This second statement is a much weaker statement since it permits various sorts of dependence relations between x and y, as illustrated in Fig. 39.

Statisticians have come to regard the stronger notion of independence as the proper way to make precise the intuitive idea that "x and y are unrelated." This preference is no doubt due in part to the fact that several good tests have been found for the stronger hypothesis.

THE z-SCORE TEST FOR INDEPENDENCE BASED ON r

At the end of the preceding unit we remarked that the correlation r of a sample is the natural estimate of the correlation ρ in the population from which the sample was drawn. Since r has the nature of an average (see Formulas 43 and 44), the central limit theorem leads us to expect that r will be normally distributed about ρ if n is not too small. Now we have just seen that when x and y are independent, then $\rho = 0$. This immediately suggests that we could test independence with the z-score test by seeing whether our sample r departs significantly from zero. This requires, however, that the deviation of r from zero be expressed in standard units. That is, we need the standard deviation of r.

As we warned you at the end of Unit 23, it is not generally possible to give SD_r, since its value depends on details of the bivariate distribution that are in practice not known. Fortunately, however, where $\rho = 0$ (under the strong hypothesis of independence) it has been found that SD_r is approximately equal to $1/\sqrt{n-1}$ regardless of the shape of the bivariate distribution.

Accordingly, if n is not too small the hypothesis of independence can be tested by the appropriate variant of our z-score test:

$$z = \frac{r - 0}{1/\sqrt{n-1}} = r\sqrt{n-1} \qquad \text{FORMULA 47}$$

The resulting z is then referred to the normal table. As a first illustration, let us recall the example on page 246 of Unit 23. We found the correlation $r = 0.64$ for a sample of $n = 20$ heights (x) and weights (y) of STATLAB boys. How significant is the apparent association? For testing the hypothesis that x and y are independent, we calculate

$$z = r\sqrt{n-1} = 0.64 \times \sqrt{19} = 2.79$$

and read the P value of 0.0026 from the normal table. The association is thus highly significant. (This is of course the one-sided P value; if you had decided in advance to use a two-sided test—we cannot imagine why—the P value would have been $2 \times 0.0026 = 0.0052$.)

Of course nobody would seriously entertain the notion that height and weight are independent in a population of boys. As a more realistic example, let us ask: Are the Peabody and Raven scores independent among the STATLAB girls? The Peabody test, like most tests of mental ability, emphasizes verbal facility. The Scottish psychologist J. C. Raven set out deliberately to measure a quite different sort of mental ability, related to geometrical intuition and spatial organization. The Raven test uses no words, only geometrical figures. The everyday notion that verbal and mathematical abilities are separate (e.g., the inarticulate engineer and the poet who cannot make change) has recently been reinforced by the discovery that different halves of the brain are apparently involved: The left half does the talking, while the right half does the spatial visualizing. Thus it is reasonable to ask whether Raven succeeded in making a test of aspects of mental functioning that are independent of the verbal aspects measured by the Peabody test.

As always, before looking at the data we should pause to think about the relevant alternatives so as to be able to choose between a one-tailed and two-tailed test. If these mental tests are not independent, would we expect them to be positively or negatively associated? Many psychologists think that, in addition to various special mental abilities, there is a *general* intelligence factor called g that is involved in all sorts of problems, including the verbal and spatial problems. According to this theory people have varying amounts of this general factor to which they add varying amounts of special abilities, such as verbal or mathematical talent. To the extent that this is so, we would expect a positive association between our two tests because the scores on both the verbal and the spatial tests would reflect the influence of this common factor g. If we are particularly interested in this alternative, we would decide in our test of independence to regard positive values of r as the relevant ones and to use a one-sided test.

Now we are ready to look at data. We draw a sample of $n = 10$ STATLAB girls and record in Table 35 their Peabody (x) and Raven (y) scores. When plotted (see Fig. 40), these data seem to suggest

TABLE 35 Peabody (x) and Raven (y) scores of a sample of $n = 10$ STAT-LAB girls

ID no.	x	y	$C_x = 80$	$C_y = 30$	$x'y'$
26-21	80	28	0	−2	0
14-54	77	23	−3	−7	21
16-42	64	22	−16	−8	128
21-35	74	29	−6	−1	6
31-12	86	20	6	−10	−60
21-26	84	39	4	9	36
23-32	92	37	12	7	84
33-44	89	38	9	8	72
16-35	81	26	1	−4	−4
36-61	69	21	−11	−9	99
			−4 =	−17 =	382 =
			$\Sigma x'$	$\Sigma y'$	$\Sigma x'y'$

$$\bar{x}' = -0.4 \qquad s_{x'} = \sqrt{\tfrac{700}{10} - (-0.4)^2} = 8.36$$

$$\bar{y}' = -1.7 \qquad s_{y'} = \sqrt{\tfrac{509}{10} - (-1.7)^2} = 6.93$$

$$r = [38.2 - (-0.4) \times (-1.7)]/(8.36) \times (6.93) = 0.65$$

The correlation of the two mental test scores for these 10 girls is calculated in the last three columns with the aid of coded variables and the shortcut method.

a rather weak positive association. (Jot down your guessed r for these data before reading on.) How significant is the apparent relation? Using coding variables $x' = x - 80$ and $y' = y - 30$, we compute the sample correlation $r = 0.65$ in the final columns of Table 35. (How close was your guess?)

The z score is

$$z = r\sqrt{n-1} = 0.65 \times \sqrt{9} = 1.95$$

to which corresponds the (one-sided) P value of 0.026. At the 5 percent level of significance, the data thus establish a positive association of Peabody and Raven scores. Let us repeat our reasoning: A P value of 0.026 says that it is not very probable (about 1 chance in 40) that, if the population ρ were really zero, our sample r would be as high as 0.65. Therefore we are justified in rejecting the hypothesis that $\rho = 0$. And since we know that if x and y are independent, ρ *must* equal zero, we can now properly drop the notion that x and y are independent.

Although the z-score test is easily carried out, we have doubts about its reliability. Is $n = 10$ big enough to justify treating r as normally distributed and to justify using the formula $SD_r = 1/\sqrt{n-1}$? Probably not. Fortunately, there is a test of independence that is valid even with very small samples.

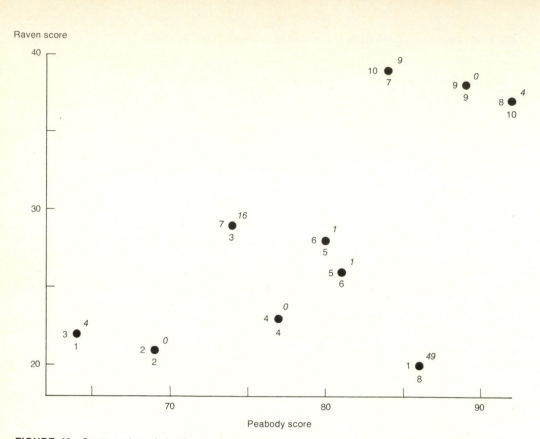

FIGURE 40 Scatter plot of the Peabody and Raven scores of the sample of $n = 10$ STATLAB girls listed in Table 35. The numbers under and to the left of the dots are ranks. The italic numbers are the squared rank differences required for the Spearman statistic.

THE TEST BASED ON SPEARMAN'S RANK CORRELATION

The idea that the relation between two variables could be assessed by using *ranks* was suggested by the British psychologist Charles Spearman in two remarkable papers published in 1904 and 1906. As a young man Spearman had gone into the army because he thought that the military profession offered more contemplative leisure than any other. After fighting in the Burmese and Boer Wars, he began to see the advantages of academic life and accepted a chair in psychology created for him at University College, London. He is responsible for the concept of a general factor of intelligence, mentioned above. His psychological researches led to an interest in measuring the association between variables.

Spearman strongly urged the advantages, theoretical and practical, of replacing the actual measurements in a sample by their ranks—a proposal now widely followed by statisticians, as the two Wilcoxon tests indicate. He pointed out that if x and y have a strong positive association (as for example in Fig. 37a), then an individual's rank on x and his rank on y would be close (e.g., the individual who was first in x would also be near the top in y; an individual with a middling x rank would tend to have a middling y rank; etc.). As a result, the difference d of the x rank and y rank of an individual would tend to be small and hence the sum $\Sigma|d|$ of the absolute values of these differences would tend to be small. Spearman was mainly interested in devising a new coefficient of association based on $\Sigma|d|$ (and in using this coefficient for a z-score test of independence), but he noted in passing that Pearson's coefficient r was much easier to compute if the actual measurements x and y were replaced by their ranks. When this is done, Formula 44 can be reduced by algebraic manipulation to the form

$$r' = 1 - \frac{6\Sigma d^2}{n(n^2 - 1)}$$ FORMULA 48

To find r' it is only necessary to compute Σd^2, which (like $\Sigma|d|$) would tend to be small if there is a strong positive association. To see how easy it is to compute Σd^2 and hence r' (which is usually referred to as Spearman's rank-order correlation coefficient), turn again to Fig. 40. We have written under each dot the x rank of its x coordinate; these ranks merely number the dots from left to right across the page. Similarly the y ranks, written to the left of the dots, number the dots from bottom to top. Finally, to the upper right of each dot is written d^2, the square of the difference of the x rank and the y rank. For the three girls whose x and y ranks are equal (ID numbers 14-54, 33-44, and 36-61), $d = 0$ so that $d^2 = 0$. Two girls (26-21 and 16-35) have $d = 1$ or $d = -1$, so $d^2 = 1$ for them. Only one girl (31-12) has a large value of d^2. Adding d^2 for all the dots gives

$$\Sigma d^2 = 4 + 0 + 16 + 0 + 1 + 1 + 9 + 49 + 0 + 4 = 84$$

$$r' = 1 - \frac{6 \times 84}{10 \times (100 - 1)} = 0.49$$

The whole calculation can be done with pencil and paper in a few minutes.

As so often happens with discoveries ahead of their time, Spearman's pioneering ideas were not followed up and the possibility of using the sum of the squares of the differences between ranks (Σd^2) to test the independence of x and y went unexplored for 30 years. In 1936 statisticians Harold Hotelling and Margaret Pabst returned to Spearman's work of 1906. They pointed out that his statistic Σd^2 provides a test of independence, with small values of Σd^2 relevant if you are interested in detecting positive association between x and y. And they noted that his test gives exact P

values, even for a very small n, without making any assumption about the form of the bivariate distribution.

The reasoning seems clear: If y is independent of x, then knowledge of x (or of its rank) tells you nothing about y (or its rank). It is as if the y ranks were randomly arranged with respect to the x ranks.

To see how this fact provides a P value for Σd^2, let us consider the simple case of a sample of $n = 4$. If the four points are arranged in order of increasing x rank, from 1 to 4, there are 24 possible *and equally likely* orders for the four y ranks. These 24 possibilities are shown in Table 36. At the top we list the y ranks in the order (1, 2, 3, 4) that indicates perfect positive association with the x ranks. Here each rank difference d is zero, so $\Sigma d^2 = 0$. At the bottom we list the y ranks in the order (4, 3, 2, 1) that indicates perfect negative association with the x ranks. Here the rank differences d are $(-3, -1, 1, 3)$, so that

$$\Sigma d^2 = (-3)^2 + (-1)^2 + (1)^2 + (3)^2 = 9 + 1 + 1 + 9 = 20$$

The other 22 possibilities are listed in between, in order of increasing Σd^2. Check a few values of Σd^2 in Table 36 to be sure you understand how it is computed.

Suppose that we are interested in detecting a positive association, so that the small values of Σd^2 are the relevant ones. We draw a sample of $n = 4$, arrange the points in order of increasing x rank (1, 2, 3, 4), and find that the y ranks then have the order (2, 1, 4, 3), represented by the fifth line of Table 36. In this case $\Sigma d^2 = 4$. There are just five rows in Table 36 with $\Sigma d^2 \leq 4$. Since the 24 rows are equally likely if x and y are *independent,* the chance of getting $\Sigma d^2 = 4$ in this case is just $\frac{5}{24} = 0.208$. This is the P value of these data.

This P value may also be read from Table 37, which shows the P values for sample sizes up to $n = 11$, worked out by considering all possible permutations of the y ranks as equally likely. In our illustration of $n = 10$ Peabody and Raven scores, we found $\Sigma d^2 = 84$. A glance at Table 37 shows that the P value is 0.0667 for $\Sigma d^2 = 80$ and 0.0956 for $\Sigma d^2 = 90$. Interpolation gives the P value of 0.078 at $\Sigma d^2 = 84$. In other words, the chance that we would get a Σd^2 of 84 or less if x and y are independent is about 1 in 13. The data appear considerably less significant by the rank test than they did by the z-score test based on r. However, the rank test's P value is entirely reliable whereas the P value of the z-score test is suspect with so small a sample.

Several comments should be made about the test of independence based on Spearman's statistic Σd^2. First, the P values of Table 37 are for the one-sided test of independence against the alternative that x and y have a *positive* association. By means of a simple trick,

TABLE 36 The 24 possible orders of the y ranks when the points are arranged by order of increasing x rank, with the value of Σd^2 for each of the 24 cases

x rank				
1	2	3	4	Σd^2
1	2	3	4	0
2	1	3	4	2
1	3	2	4	2
1	2	4	3	2
2	1	4	3	4
1	3	4	2	6
2	3	1	4	6
1	4	2	3	6
3	1	2	4	6
1	4	3	2	8
3	2	1	4	8
2	4	1	3	10
3	1	4	2	10
2	3	4	1	12
4	1	2	3	12
2	4	3	1	14
4	1	3	2	14
3	2	4	1	14
4	2	1	3	14
3	4	1	2	16
4	3	1	2	18
4	2	3	1	18
3	4	2	1	18
4	3	2	1	20

Each of the 24 rows shows one possible ordering of the four y ranks, as associated with the ordering (1, 2, 3, 4) of the x ranks listed at the head of the table. If x and y are independent, the 24 orderings are equally likely.

however, the same table may be used if you are interested in a one-sided test against the alternative that x and y are *negatively* associated: Just rank the y values from large to small instead of from small to large. Then a small value of Σd^2 is evidence of negative association, and Table 37 may be used as before.

Second, suppose you wish to test independence with a sample of $n = 9$ in a situation where *both* negative and positive association would be relevant. The scatter diagram suggests that the association is negative, so you assign the y ranks from top to bottom and find $\Sigma d^2 = 38$. As usual, the P value of a two-sided test is just double the corresponding one-sided value. When the entry 0.0252 in Table 37 is doubled, you have the two-sided P value of $2 \times 0.0252 = 0.0504$. The data would just fail to be significant at the 5 percent level.

TABLE 37 One-sided P values for Spearman's statistic Σd^2, used to test independence with a bivariate sample of size n. For values of Σd^2 larger than 22, linear interpolation may be used. P values above 0.25 have been omitted.

Σd^2	n 3	4	5	6	7	8	9	10	11
0	.1667	.0417	.0083	.0014	.0002	.0000			
2		.1667	.0417	.0083	.0014	.0002	.0000		
4		.2083	.0667	.0167	.0034	.0006	.0001		
6			.1167	.0292	.0062	.0011	.0002	.0000	
8			.1750	.0514	.0119	.0023	.0004	.0001	
10			.2250	.0681	.0171	.0036	.0007	.0001	
12				.0875	.0240	.0054	.0010	.0002	
14				.1208	.0331	.0077	.0015	.0003	.0000
16				.1486	.0440	.0109	.0023	.0004	.0001
18				.1778	.0548	.0140	.0030	.0006	.0001
20				.2097	.0694	.0184	.0041	.0008	.0001
22				.2486	.0833	.0229	.0054	.0011	.0002
26					.1179	.0347	.0086	.0019	.0003
30					.1512	.0481	.0127	.0029	.0006
34					.1978	.0661	.0184	.0044	.0009
38					.2488	.0854	.0252	.0063	.0014
42						.1081	.0333	.0087	.0020
46						.1337	.0429	.0117	.0027
50						.1634	.0540	.0153	.0037
60						.2504	.0888	.0272	.0072
70							.1348	.0441	.0126
80							.1927	.0667	.0201
90							.2603	.0956	.0304
100								.1316	.0438
110								.1744	.0609
120								.2241	.0817
130								.2801	.1070
140									.1365
150									.1708
160									.2091
170									.2517

The P value is the chance that Σd^2 will have a value as small as the tabular entry. In computing Σd^2, the x values are ranked from small to large. The y values are ranked from small to large if positive association is relevant and from large to small if negative association is relevant. We assume that the sample is drawn from a bivariate population in which the two variables are independent.

Third, for cases not covered by Table 37 there is available a z-score approximation:

$$z = \frac{6\Sigma d^2 - n(n^2 - 1)}{n(n + 1)\sqrt{n - 1}}$$

FORMULA 49

Here, as in all our other similar instances, the z is referred to the normal table for an approximate P value. Unfortunately, the approximation is not very accurate for small P values. For example, with $n = 11$ and $\Sigma d^2 = 60$, for which the correct P value is 0.0072 as given by Table 37, the normal approximation gives

$$z = \frac{(6 \times 61) - (11 \times 120)}{11 \times 12 \times \sqrt{10}} = -2.29$$

Here we have used the continuity correction. The values of Σd^2 are always even integers, so we are distinguishing $\Sigma d^2 \leq 60$ from $\Sigma d^2 \geq 62$. Splitting the difference gives $\Sigma d^2 = 61$ as the value to enter in Formula 49. For this value of z, the normal table shows the P value 0.0110; this is 53 percent larger than the correct P value. The normal approximation works better for data that are less highly significant, and it improves as n is increased. For $n \geq 20$ it is satisfactory.

Fourth, we have assumed that there are no ties in the ranks, and as always with rank tests you should try to avoid ties by making the original measurements of continuous variables with precision. In practice, however, ties often do occur, and as usual they can be handled by mean ranks. For example, if girl 36-61 in Table 35 had received Raven score 22 instead of 21, she would have been tied with girl 16-42 for the y ranks 2 and 3. Each would be given y rank 2.5, and the d^2 values for these two girls would have been $(1 - 2.5)^2 = 2.25$ and $(2 - 2.5)^2 = 0.25$. This would change Σd^2 from 84 to 82.5. Remember, however, that the values in Table 37 are for untied data and that the P value 0.074, obtained by interpolating in Table 37, would in this case be only approximate. Formula 49 can also be used to give a reasonable approximate P value when there are ties— at least if there are not too many of them and if the n is not too small.

CHOICE OF A TEST FOR INDEPENDENCE

In choosing between the z-score test based on r and the nonparametric test based on Σd^2, an important consideration is that the test based on r is the more powerful one when the population has a normal distribution. The sacrifice of power involved in using Σd^2 with normal data is not a large one, but it is substantial enough that we recommend the use of the r-based test if you think your data are nearly normal and if n is large enough to lend validity to the P value. As a rule of thumb, the test based on r needs n to be at least 20 and preferably at least 30.

For small values of n, say $n < 20$, the test based on Σd^2 begins to be attractive. It provides accurate P values with small samples, at least if there are not too many ties in the data. The test based on Σd^2 is also much easier to use than the r test when n is small. Once

BOX 8

ASSOCIATION, CAUSATION, AND ANDERSON'S LAW

A universal (and valid) belief among those who are concerned with the problem of inference and statistics can be expressed in the dictum: Association is not Causation. Yet the scientist is often interested in association only because he seeks causation. Whether a high degree of association exists between smoking and lung cancer is of interest, for example, because we want to know whether smoking causes lung cancer. Statisticians tend to take one of two ways in getting around this problem. One is that the pure statistician would do well to mind his p's and q's, leaving the problem of causation to others.

The second position is exemplified by the work of J. Yerushalmy, an American biostatistician who believed that the statistician and the researcher should join forces in seeking causal relations.

In 1957 the *American Journal of Obstetrics and Gynecology* reported that women who smoke cigarettes during pregnancy give birth to many more low-birth-weight infants than do women who do not smoke (a low-birthweight infant being defined as one weighing 2,500 grams or less). In the following fifteen years over thirty additional studies confirmed this association. This now very clear association had serious implications since it is well known that low-birthweight infants run more than 20 times as high a risk of dying in the first month of life than do heavier infants. It was therefore vitally important to know whether smoking caused an increase in low-birthweight incidence. This causal inference, furthermore, seemed reasonable on the hypothesis that cigarette smoking interfered with the intrauterine development of the fetus. Yerushalmy set about to check this out. The data came from his Child Health and Development Studies, the same source that provided our STATLAB Census.

A first analysis confirmed the previously reported association. "But," Yerushalmy writes, "we explored the problem further." In one study he examined 5466 pregnancies among 3422 women, all of whom had given birth at age 25 or less. Some of these women smoked, some did not. Instead of doing the usual analysis, he attempted a finer differentiation among the women. He divided the women into four groups and checked out the incidence of low birthweights among them: (1) the non-smokers, women who had never smoked; (2) the past and future smokers, women all of whose pregnancies occurred during periods of smoking and who were still smoking when interviewed for the study; (3) the future smokers, women who began to smoke only after they had given birth; and (4) the past smokers, women who quit smoking after they had given birth. The results are given in the following table.

	No. of births	Low-birthweight infants (%)
Nonsmokers	2529	5.3
Past and future smokers	2076	8.9
Future smokers	210	9.5
Past smokers	651	6.0

Comparing the nonsmokers with the

past and future smokers yields the expected association: Smoking mothers are associated with high incidence of low-birthweight infants. But comparing the future smokers with the nonsmokers immediately throws doubt on this association, since the mothers who did not smoke until *after* their babies were born also showed a high incidence of low-birthweight infants. (This difference is significant at the 0.01 level with $\chi^2 = 6.69$.) Finally the past smokers, during the period *before* they quit smoking, gave birth to significantly ($\chi^2 = 5.61$, $p > 0.02$) fewer low-birthweight infants than did the past and future smokers (the women who did not quit) and fewer even than the future smokers. In this last instance the usual association seems to be *reversed.*

Surveying these and other data, Yerushalmy concludes that "the evidence appears to support the hypothesis that the higher incidence of low-birthweight infants is due to the *smoker*, not the 'smoking.'" He is here suggesting that the "smokers" represent a group of women who, because of certain biological factors or because of a certain mode of living, would give birth to the same number of low-birthweight infants whether they smoked or not. Smoking does not *cause* low birthweights; both these variables are *effects* of a third set of factors, and *that* is why they seem to be related.

This hypothesis, in turn, implies all kinds of other associations, implicates other kinds of variables, and suggests that we look elsewhere for possible causal factors. And this is precisely what Yerushalmy did. To take an example: In a scouting study Yerushalmy sought to find some leads to differences that might exist between smokers and nonsmokers. He compared, on a number of characteristics, two groups of pregnant women: one group of over 7000 nonsmokers and another group of about 4000 smokers. He found that the smokers, as compared to the nonsmokers, were less likely to use contraceptive methods, less likely to plan their babies, more likely to drink whiskey, beer, and coffee (the non-smokers were more likely to drink tea, wine, and milk), and more likely to drink to excess. The smokers, it appears, tend to be more extreme in the pleasures they take and more carefree in their mode of life than the nonsmokers. And, to complicate the picture, he found that the women who were destined to become smokers in their adult life were, as girls, more precocious in the development of their reproductive system: More of them began to menstruate before age 12 than did those who were not to be smokers as adults.

The moral seems to be that a simple association, no matter how reasonable and strong, will not necessarily develop upon further examination into a sturdy causal connection; instead it could very well collapse into a puzzling array of intricately tangled associations and knotty problems. This should not dismay us, for it is precisely in this way that we may arrive at greater understanding. As Anderson's law reminds us: *There is no problem, no matter how complex, which, upon careful analysis, does not become more complex.*

The studies referred to are The Relationship of Parents' Cigarette Smoking to Outcome of Pregnancy: Implications as to the Problem of Inferring Causation from Observed Associations, by J. Yerushalmy, *American Journal of Epidemiology* **93:** 443–456 (1971); and Infants with Low Birth Weight Born before Their Mothers Started to Smoke Cigarettes, by J. Yerushalmy, *American Journal of Obstetrics and Gynecology* **112:** 227–284 (1972). The eponymous Anderson, whose epigram we think deserves the status of a law, is Poul Anderson, the science fiction writer.

you have made the scatter plot of a small sample (and this ought to be done in any case), it takes only a few minutes to calculate the P value of Σd^2 by the method illustrated in Fig. 40.

SUMMARY

Whenever we find, for a bivariate sample, that r differs from zero, it is incumbent upon us to test the significance of the implied association between x and y. The best available way to do this is to determine the probability that a value of r so far from zero would occur if x and y are in fact independent. If we can reject the independence hypothesis, then we can accept the implication that x and y are indeed associated; if we cannot reject the independence hypothesis, then it is prudent to assume that the obtained value of r is merely a chance deviation from a ρ of zero.

There are two tests we can use. In the first of these (z score based on r) we determine whether the sample r departs significantly from the hypothetical value zero of the population ρ. The logic of the test rests on the categorical imperative of the independence hypothesis: When x and y are independent, then ρ *must* equal zero. To obtain the normal table P values for these z scores requires the assumptions that r is normally distributed and that $\mathrm{SD}_r = 1/\sqrt{n-1}$. These assumptions are without merit if n is too small.

The second test, which makes no assumption about the form of the distribution and which is applicable for small n's, is based on Spearman's rank-difference methods. It can easily be shown that the size of the statistic Σd^2 reflects the degree of dependence of x and y such that with a small Σd^2 there is a positive association between the two variables. The Σd^2 distribution, in the case where x and y are independent, can be precisely determined for each n. Therefore the P value can be determined for any observed Σd^2. This provides us with the information needed to reject or accept the independence hypothesis.

In choosing between the two tests such considerations as the following must be borne in mind: the differential power of the tests, the normality of the population, the size n of the sample, and the ease of calculation.

Finally, we must issue a caveat before leaving the matter of association between variables. Scientists and others usually search for associations between variables in order to find *causal connections*. If smoking is associated with blood pressure, is it because smoking *causes* a rise in blood pressure? If there is an association between money supply and inflation in the following year, is it because an increase in money supply *causes* inflation the following year? And here is our caution: We must resist the temptation of arguing that because x and y are highly associated, one therefore *causes* the other. The association-causation relation is not that simple. Make a quick leap from association to cause and you may come a cropper. Much more information than a high correlation between x and y is needed before we can make a causal statement. And when we do dig out that information, we frequently find that our reasonable and logical causal relation between x and y has vanished—as Box 8 so nicely illustrates.

☐   You are now ready to test for independence in your own data. We shall ask you to inquire into the significance of the apparent association between mothers' and fathers' heights shown in the sample you drew and analyzed in the preceding unit. Is the evidence for an association you saw there real, or merely a chance result of the sampling? How strong is the evidence against the hypothesis of independence?

WORK TO BE DONE

STEP 1   Answer items 1 and 2 of the inquiry.

STEP 2   Retrieve from Databank, File J-1, the value of r you computed for the first 20 mothers' and fathers' heights in your sample of 40 STATLAB families and enter this r on page 1 of the worksheet. Use r to test the hypothesis that the heights are independent. Find the P value of the two-sided test, and also the one-sided test, when positive association is relevant.

STEP 3   Plot the *unrounded* height data of the first 20 mothers and fathers (to the nearest tenth inch) as recorded in Databank, File B-1, on the worksheet as a scatter diagram. Label each point with its x and y ranks (or mean ranks) as in Fig. 40 of the text. Calculate Σd^2 and find its associated P values by the normal approximation, both for the two-sided and for the one-sided tests.

STEP 4   Complete the inquiry.

REGRESSION

Once it has been established that two variables x and y are associated with each other, what use can be made of the relation between them? Often the very *fact* of association is what matters. Take, as an example, the relation between fluorine in water and cavities in teeth. Once a scatter plot of towns had been made, according to fluorine content of the water supply (x) and rate of dental caries in children (y), it became clear that there was a substantial negative relation: The towns with high fluorine had a much lower caries rate. This was an important discovery, and it led (after appropriate experimental checking) to the fluoridation of water supplies. Similarly, the Australian physician who noted the strong positive association, year by year, between incidence of rubella among pregnant women (x) and birth defects in infants born later in that year (y), had therewith made the essential finding leading toward preventive measures.

In other cases, the relation between x and y needs to be expressed quantitatively to enable one to use an observed value of *x* to *predict* what the value of *y* will turn out to be. For example, a quantitative expression of the relation between spring rainfall and autumn wheat yield will permit the prediction of food supplies months in advance. Similarly, SAT scores may be used by an admissions officer to predict college grades, or an insurance company may use physiological measurements to predict the life span of an applicant for a policy.

Quantitative analysis of a relation may also be used to establish *norms* with which individual cases can be compared. You have probably been told at some time the average weight for a person of your age, sex, and height as an indication of what your weight should be. The grammar school teacher is told how many words students of a given age should be able to read, on the average, as a standard of comparison for his students. Such norms are in part the result of studies along the lines we shall explore in this unit.

Sir Francis Galton, who was the first person to develop an extensive methodology for bivariate data, published in 1889 a study of the heights of fathers (*x*) and of their adult sons (*y*). He was mainly interested in finding scientific laws governing the inheritance of physical traits—an interest aroused by the evolutionary theories of his cousin Charles Darwin. His work also established norms that would help a physician to diagnose cases of abnormal growth and could be used by a pediatrician to predict the height a male child will achieve when fully grown.

Galton found, to the surprise of absolutely nobody, that tall fathers had tall sons. But within this obvious finding there was a surprise: The sons were less tall than might have been expected. The class of fathers who were themselves all 6 in above mean fathers' height had sons who were (on the average) only about 3 in taller than the mean height of all sons. Similarly, the fathers who were 4 in shorter than the mean had sons who averaged only about 2 in shorter, rather than 4. On the basis of this and similar findings, Galton announced a law of "regression toward mediocrity": *With extreme values of x are associated values of y that tend to deviate less extremely than the x values.* The *y* values seem to turn back (regress) toward their mean. The name stuck, and the curve which depicts the way the average *y* varies with *x* is still called *the regression of y on x.*

In this unit we shall explain just what a regression is, and then we shall show you how to estimate a regression from bivariate sample data. The results stored in your Databank from the work you have already done on your sample of mothers' and fathers' heights will permit you very easily to regress the height of each parent on the other. Then, by a not implausible use of the data, you will be able to predict the height of your spouse.

REGRESSION IN A POPULATION

Suppose that a STATLAB boy has been chosen at random and you are asked to predict what his weight will turn out to be. You know (Tables 33 and 34 of Unit 23) the weight distribution of these 648 boys, and (from the coded values given on page 250) you can compute that their mean weight is $\mu_y = 71.0$ lb. If it is necessary to guess the boy's weight, knowing only that he is randomly chosen from this population, 71.0 lb is perhaps the most reasonable guess. His actual weight y will of course deviate more or less from this predicted value; the magnitude of the deviation will reflect the standard deviation $\sigma_y = 12.9$ lb, since this quantity measures the extent to which y varies about its mean value.

Let us change the conditions of the problem. After choosing the boy at random, it is found that he is 49 in tall. Now what weight would you predict for him? It no longer makes sense to predict 71.0 lb; that is the mean weight of the entire population. But we now know that this boy is unusually short and hence probably of below-average weight (there being a positive association between height and weight). Turning again to Table 33 of Unit 23, we see displayed there the weight distribution of the 36 boys with heights in the interval 48.0–49.9 in. These data, reproduced here as Table 38, are much more pertinent to our new problem than is the entire y distribution. Our boy is one of the 36 boys whose weights are given in Table 38. The mean weight of these boys, which is 56.0 lb, is a reasonable guess for the weight of our boy. Of course, his actual weight y may deviate from this mean value, the magnitude of the deviation reflecting the standard deviation, 5.1 lb, of the distribution of weights given in Table 38.

TABLE 38 Weight distribution of the 36 STATLAB boys in the 48.0–49.9 in height class

Weight class (pounds)	f
70–74	1
65–69	2
60–64	3
55–59	14
50–54	15
45–49	1
	$36 = \Sigma f$

These frequencies constitute one column of the bivariate frequency distribution of Table 33.

TABLE 39 Frequency, mean, and standard deviation of the distribution of weight within each height class

Height class (inches)	f	Mean weight (pounds)	Standard deviation of weight (pounds)
60.0–61.9	9	95.9	15.1
58.0–59.9	27	93.5	11.2
56.0–57.9	78	84.6	12.5
54.0–55.9	161	74.7	9.4
52.0–53.9	200	67.7	7.0
50.0–51.9	135	62.0	6.9
48.0–49.9	36	56.0	5.1
46.0–47.9	2	47.0	0
	648		

Each row of this table corresponds to one column of Table 33. The distribution of Table 38 is represented here by the next-to-last row.

In a similar way, if a randomly chosen STATLAB boy is 59 in tall, the best guess we can make for his weight would be 93.5 lb, which is the mean weight of the 27 boys with heights in the 58.0–59.9 class interval. We show in Table 39 the mean weights for the boys of each height class interval. These same results are displayed graphically as the solid line of Fig. 41. This line represents the *regression of weight on height*. From it, you can see at a glance how much a STATLAB boy of any given height may be expected to weigh on the average. The regression rises as it goes from left to right, at a more or less steady rate in spite of local irregularities. The regression shows clearly that average weight increases at the rate of about 3.7 lb per added inch of height—a refinement of the rate of 3.8 lb per inch which we read off from Table 33 by inspection of the height-weight frequency table.

In an exactly analogous way one can compute the *regression of height on weight*. The data are given in Table 40 and are depicted as the dashed line of Fig. 41. This regression shows, for each weight, the mean height of boys having that weight. If a randomly chosen boy weighs 92 lb, for example, the best guess for his height is 56.6 in since this is the mean height of the 21 boys in the 90- to 94-lb weight class. The regression of height on weight also increases more or less steadily, at the rate of about 0.16 in of added height, on the average, per added pound of weight. (At the upper extreme of the distribution, the steady rise of the regression of height on weight breaks down under the influence of the few obese outliers.)

When the two regressions are considered together, certain paradoxical features appear. For example, boys who are 59 in tall have a mean weight of 93.5 lb; yet 93.5-lb boys do not have a mean

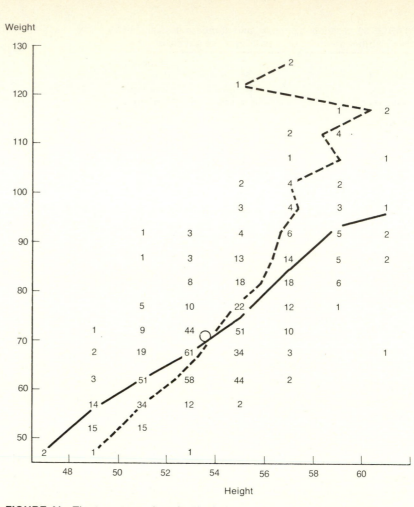

FIGURE 41 The two regressions for the heights and weights of the 648 STATLAB boys. The numbers, presented also in Table 33, give the bivariate distribution. The solid line is the regression of weight on height, giving the mean weight in each height class (see Table 39). The dashed line is the regression of height on weight, giving the mean height in each weight class (see Table 40). The circle marks the center of the bivariate distribution at $\mu_x = 53.6$ in and $\mu_y = 71.0$ lb.

height of 59 in, but only 56.8. These facts, of course, are reflections of Galton's discovery of "regression toward mediocrity." Extremely tall boys are heavy, but not extremely so; heavy boys tend to be only moderately tall. Note that the two regressions cross at a point near the center of the bivariate distribution. We show as a circle on Fig. 41 the central point, with x coordinate equal to the mean height $\mu_x = 53.6$ in and y coordinate equal to the mean weight $\mu_y = 71.0$ lb.

Figure 42 gives, as a second illustration, the two regressions associated with the bivariate distribution of Raven and Peabody

TABLE 40 Frequency, mean, and standard deviation of the distribution of height within each weight class

Weight class (pounds)	f	Mean height (inches)	Standard deviation of height (inches)
125–129	2	57.0	0
120–124	1	55.0	0
115–119	3	60.3	0.9
110–114	6	58.3	0.9
105–109	2	59.0	2.0
100–104	8	57.0	1.4
95–99	11	57.3	1.9
90–94	21	56.6	2.7
85–89	38	56.3	2.1
80–84	50	55.8	1.8
75–79	50	54.7	1.9
70–74	115	54.0	1.6
65–69	120	53.3	1.7
60–64	125	52.3	1.5
55–59	62	51.0	1.5
50–54	30	50.0	1.0
45–49	4	49.0	2.4

Each row of this table corresponds to one row of Table 33.

scores for the 648 STATLAB girls. As you see, in a general way this figure resembles Fig. 41. The two regressions look even more like straight lines in this case (except for the distortion due to the out-lier, girl 11-53, who has a Peabody score of 51 and a Raven score of 40), again crossing near the central point. A glance at these regres-sions shows that, for example, a girl who scores 42 points on Raven would, on the average, score 81.8 points on Peabody.

In both our examples, the distributions are more or less similar to the normal (weight less so than the other variables), and the fea-tures we have perceived in Figs. 41 and 42 are characteristic of nearly normal distributions. With perfectly normal distributions, the two regression curves become perfectly straight lines, and in this situation the regression is said to be *linear*. Furthermore, with normal distributions it turns out that the two regression lines cross exactly at the central point (μ_x, μ_y). We suggest that you turn to Fig. 37, page 242, and try to visualize what the two regression lines would look like for each of those five bivariate normal distributions.

USING A SAMPLE TO ESTIMATE REGRESSION

So far we have been dealing with an entire population. In practice, of course, you will usually know the x and y values for only a sam-

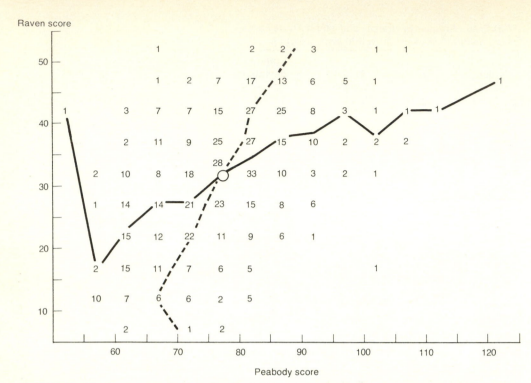

FIGURE 42 The two regressions for the Peabody and Raven scores of the 648 STATLAB girls. The numbers show the bivariate frequency distribution in score classes of width $w = 5$. The solid line is the regression of Raven score on Peabody score; the dashed line is the regression of Peabody on Raven. The circle locates the mean scores ($\mu_x = 77.2$, $\mu_y = 31.5$).

ple from the population. The problem then is to use these sample values to estimate the true regression in the population. Consider for example Fig. 36 in Unit 23, which shows the heights and weights of our sample of $n = 20$ boys. We now wish to use these 20 points to estimate the regression of weight on height for the 648 STATLAB boys.

This is a tall order. A sample of moderate size is likely to have very few points in any particular x class—perhaps none at all—and this is the case in our illustration. How then can we hope to estimate the y average in that class with reasonable precision? It is only possible to do this if we make some simplifying assumptions. The usual practice is to assume that the regression curve is a straight line and that this line goes through the point (μ_x, μ_y). These assumptions are exactly true for normal populations and are reasonable if the population from which the sample came is reasonably close to the normal shape.

The central point (μ_x, μ_y) of the population can be estimated, of course, by the point ($\bar{x}, \bar{y}$), so it makes sense to require our estimated regression line to go through that point. Once that decision has been made, all that remains is to determine the *slope* of the line, that is, the rate at which it will rise per unit change in x. If we denote the slope of the estimated regression line by b, and denote by y^* the value predicted for y for a given x, the equation of the regression line will have the familiar form of an equation for a straight line:

$$y^* = \bar{y} + b(x - \bar{x}) \qquad \text{FORMULA 50}$$

To see whether this equation does what we want, let us try two checks. Suppose x happens to be equal to the mean value $\bar{x}$; then the term $b(x - \bar{x})$ becomes zero and y^* then equals $\bar{y}$. That is, for $x = \bar{x}$, $y^* = \bar{y}$ and thus this line does go through the point $(\bar{x}, \bar{y})$ as we wanted it to do. To check the slope, note that for each unit increase in x, $(x - \bar{x})$ also will increase by one unit. This means that for each unit increase in x, $b(x - \bar{x})$ will increase by the amount b; therefore y^* will increase by b. Since the slope is defined as the rate at which the line will rise per unit change in x, the line does have slope b, as desired.

Now how should the value of b be chosen? Clearly, we want the line to fit as nearly as possible to the observed points (x,y) of the sample. In other words, the deviations between y and y^*, that is, $y - y^*$, should be as small as possible; or, as most statisticians prefer, the *squared* deviations $(y - y^*)^2$ should be as small as possible. When the slope b is made to meet this requirement, we say that the regression line has been "fitted by least squares." A bit of straightforward algebra will show that the least-squares fit can be achieved very simply; all we need do is to use the following formula:

$$b = \frac{rs_y}{s_x} \qquad \text{FORMULA 51}$$

where r is the correlation between x and y.

Here is the bit of algebra required to prove all this. The squared deviation for the point (x,y) is found from Formula 50 to be

$$(y - y^*)^2 = [(y - \bar{y}) - b(x - \bar{x})]^2$$
$$= (y - \bar{y})^2 - 2b(x - \bar{x})(y - \bar{y}) + b^2(x - \bar{x})^2$$

The mean of the squared deviations (MSD) for all n points is accordingly

$$\text{MSD} = \frac{1}{n} \Sigma (y - \bar{y})^2 - 2b \frac{1}{n} \Sigma (x - \bar{x})(y - \bar{y}) + b^2 \frac{1}{n} \Sigma (x - \bar{x})^2$$
$$= s_y{}^2 - 2b \, \text{Cov} + b^2 s_x{}^2$$

This quantity will not be changed in value if we simultaneously subtract and add $\text{Cov}^2/s_x{}^2$, and by doing this we can "complete the square":

$$\text{MSD} = \left(s_y{}^2 - \frac{\text{Cov}^2}{s_x{}^2}\right) + \left(\frac{\text{Cov}^2}{s_x{}^2} - 2b\,\text{Cov} + b^2 s_x{}^2\right)$$

$$= \left(s_y{}^2 - \frac{\text{Cov}^2}{s_x{}^2}\right) + \left(\frac{\text{Cov}}{s_x} - bs_x\right)^2$$

Remember that our least-squares problem is to choose b so as to make MSD as small as possible. The first term $[s_y{}^2 - (\text{Cov}^2/s_x{}^2)]$ of MSD does not depend on b. The second term $[(\text{Cov}/s_x) - bs_x]^2$ is a square and therefore cannot be less than zero; it achieves its minimum value of zero when $\text{Cov}/s_x = bs_x$ or when $b = \text{Cov}/s_x{}^2$. But recall from Formula 44 of Unit 23 that $\text{Cov} = rs_x s_y$. Therefore, the best choice for b is

$$b = \frac{\text{Cov}}{s_x{}^2} = \frac{rs_x s_y}{s_x{}^2} = \frac{rs_y}{s_x}$$

just as claimed in Formula 51. Further, with this choice MSD is reduced to its first term, so its least value is

$$s_y{}^2 - \frac{\text{Cov}^2}{s_x{}^2} = s_y{}^2 - \frac{r^2 s_x{}^2 s_y{}^2}{s_x{}^2} = s_y{}^2(1 - r^2)$$

It will be helpful at this point to summarize what we have found by combining Formulas 50 and 51 into the final equation for the regression line. If the true regression of y on x is linear, and if this true regression line goes through the center $(\mu_x.\mu_y)$ of the population (conditions which hold perfectly for normal populations), then a reasonable estimate for the regression line is

$$y^* = \bar{y} + \frac{rs_y}{s_x}(x - \bar{x}) \qquad\qquad \text{FORMULA 52}$$

Here the quantities $\bar{x}$, s_x, $\bar{y}$, s_y, and r are to be calculated from the values in a sample drawn randomly from the population.

As an illustration let us use our sample of 20 boys' heights and weights from Unit 23, the scatter plot of which is reproduced for this sample in Fig. 43. As we found for this sample, $\bar{x} = 53.8$, $s_x = 3.6$, $\bar{y} = 72.2$, $s_y = 14.7$, and $r = 0.64$. Therefore the estimated regression line is

$$y^* = \bar{y} + \frac{rs_y}{s_x}(x - \bar{x}) = 72.2 + 2.61(x - 53.8)$$

Let us see how to plot this line. To begin with, it must go

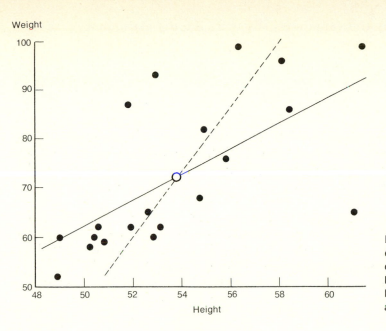

FIGURE 43 Sample regression lines of weight on height (solid) and height on weight (dashed) for the $n = 20$ STAT-LAB boys of Table 31, Unit 23. The circle locates the sample means, $\bar{x} = 53.8$ in and $\bar{y} = 72.2$ lb.

through the center, that is, through the point $x = \bar{x} = 53.8$ and $y = \bar{y} = 72.2$. We now need a second point, preferably not too close to $\bar{x}$. We choose $x = 60$ and compute

$$y^* = 72.2 + 2.61\,(60 - 53.8) = 88.4$$

The straight line through the two points determined by

$$x = 53.8 \qquad y = 72.2$$

and

$$x = 60 \qquad\qquad y = 88.4$$

is the estimated regression line for weight on height. It is shown as a solid line on Fig. 43.

To see how this regression line can be used for prediction, suppose that a boy is 50.6 in tall. From the solid line on Fig. 43, you can read his predicted weight (on the y axis) as being a bit less than 64 lb. As an alternative to this graphic method, we can make the prediction by substituting $x = 50.6$ in the formula for this regression line and compute

$$y^* = 72.2 + 2.61\,(50.6 - 53.8) = 63.8 \text{ lb}$$

Exactly the same methods will give the estimated regression line

$$x^* = \bar{x} + \frac{rs_x}{s_y}(y - \bar{y})$$

for the regression of x on y. Using this formula with the values of $\bar{x}$, s_x, $\bar{y}$, and s_y given above, we computed the estimated regression line of height on weight shown as the dashed line on Fig. 43. We urge you to check our plotting. You should in addition compute the regression lines for our Raven-Peabody example (where, as given in Table 35 of Unit 24, $\bar{x} = 79.6$, $s_x = 8.36$, $\bar{y} = 28.3$, $s_y = 6.93$, and $r = 0.65$). Sketch both lines on Fig. 40, page 260. The regression lines computed from sample data should always be plotted on the scatter diagram as a check to see that they do follow the trend of the dots. Do your lines agree reasonably with the dots of Fig. 40? Indeed, to make the check properly the regression lines should be sketched by eye *before* you compute the line from Formulas 50 and 51.

THE ERRORS OF PREDICTION

Once the sample estimates $\bar{x}$, s_x, $\bar{y}$, s_y, and r have been computed, it is very easy to write down the two regression lines as given by Formulas 52 and 53. And from these lines it is very easy to calculate the predicted y^* for a given x or the predicted x^* for a given y. But how much reliance may be placed on these predictions? There are three distinct sources of error, which must be considered separately. We shall discuss them for the regression of y on x; similar remarks apply to the regression of x on y.

First, the equations are based on the assumption that the true regression is linear. This is correct for the ideal case of normal populations, but in practice no true regression is exactly linear. And in some cases the true regression may be very nonlinear. Consider, for example, the regression of y on x in Fig. 39a of Unit 24. The assumption of linearity of regression can lead to very major errors in such a case. Figure 41 of this unit shows that the regression of height on weight becomes quite nonlinear for weights over 100 lb.

Second, even if the true regression is linear, the lines computed from the sample are only estimates of the true population regression line. Our line starts at $(\bar{x}, \bar{y})$ rather than at (μ_x, μ_y). This is not usually as serious a source of error as is the fact that the slope of the true regression has to be estimated. As you know, it is not usually possible to say how accurate are the estimates r, s_x, and s_y. Consequently it is not surprising that we cannot say how accurate is the estimated slope b, which depends on these quantities (see Formula 51). What we *can* say is that an error in the slope becomes more serious as we move away from the sample center.

Look for example at Figs. 41 and 42. As it happens, in this case our slope $b = 2.61$ estimated from the sample is too shallow. By the time we reach $x = 59$, the resulting prediction $y^* = 85.8$ is not very close to the true population regression, which is 93.5.

To be sure you understand the relation between a true population regression and the regression line estimated from a sample, you should plot both the true and the sample regression lines for the height-weight example on Fig. 41 and for the Peabody-Raven example on Fig. 42. Compare the sample regression lines with the true regression curves. Note the separate errors resulting from the nonlinearity of the true regression and from the misplacement of the line, especially because of errors in the slope.

Third, even if your estimated y^* is close to the true population regression for that x, your prediction is subject to errors because of the variability of y about the true mean in its x class. There is available a formula which may be taken as a sort of standard deviation of prediction:

$$s_y \sqrt{1 - r^2} \hspace{4cm} \text{FORMULA 54}$$

It is, in fact, the square root of the minimized MSD (see the last formula given in the small-print section on page 278). It must be emphasized that this is an *average* value, taken over all points in the sample. In the height-weight illustration, for example, $s_y = 14.7$ and $r = 0.64$; therefore the standard deviation of predicting weight from height is $14.7 \sqrt{1 - (0.64)^2} = 11.3$. However, an inspection of Table 39 shows that the variability of y about the true regression differs substantially from one x class to another. Here is a good illustration of why Formula 54 should be viewed as giving only a rough average guidance as to the likely deviations of y from its predicted values. (It is only for precisely normal populations, for which the variability of y is the same in each x class, that Formula 54 applies to each x class individually.)

☐ You are now ready to try out various regression calculations on your own sample of mothers' and fathers' heights and to make a number of predictions—about STATLAB people and about your own family. You have already, in Unit 23, done the basic calculations for your sample, so the computational work of this unit will be easy.

WORK TO BE DONE

STEP 1   Plot on the worksheet the scatter diagram of the mothers' and fathers' heights (unrounded) of the first $n = 20$ families, as stored in Databank, File B-1.

STEP 2 Draw on this figure, as light dashed lines, the two regressions, fitted by eye without the aid of calculations.

STEP 3 Retrieve from Databank, File J-1, the mean and standard deviation of mother's height and father's height, and their correlation, for your sample of $n = 20$.

STEP 4 Calculate the two regressions, and draw them on the scatter plot as heavy solid lines. Label all regression lines.

STEP 5 Record in Databank, File J-1, your two regressions.

STEP 6 Complete the inquiry.

With this self-test review you take your last formal glimpse of STATLAB. *Perhaps as you work through the following items you will be impressed with how far you have come since you first threw* STATLAB's *red and green huckle-bones in accordance with Galileo's reckoning. Or perhaps you will merely agree with one of our favorite glosses of Shakespeare: "All's well that ends."*

☐   As with other statistical problems, graphics can play an extraordinarily important role in dealing with *bivariate* data. This becomes clear when we consider the uses of the *scatter diagram* (or *scatter plot*) and the *bivariate frequency table*. These tell us much about the *direction* and comparative strengths of *association between x and y* in a *bivariate distribution* through indicating the direction and degree of slope created by the simultaneous portrayals of both variables. We also need, in addition to these graphic portrayals, *numerical measures of direction and strength of association*. Covariance of a *bivariate sample* as determined by

$$\text{Cov} = \frac{\Sigma(x - \bar{x})(y - \bar{y})}{n}$$

is one such measure, but it suffers from the fact that its value varies with changes in units of measurement. Pearson's *product-moment correlation coefficient* remedies this by dividing Cov by s_x and s_y, thus producing a *dimensionless measure of association:*

$$r = \frac{\Sigma(x - \bar{x})(y - \bar{y})}{ns_x s_y}$$

This formula can also be expressed in standard scores, where r becomes the *mean of the products of the z scores of x and y for all the cases in the sample.* With the use of a bivariate frequency table and various shortcuts (including coding), the calculation of r becomes relatively quick and simple—even for large n's. There are also available machine formulas that permit the calculation of r directly from the raw data. It is usually not possible to determine a reliable SE_r to aid in estimating the population correlation ρ, because r involves *terms of the second degree* which are very sensitive to the population values far out in the skirts of the distribution. This also means that (except for one special case) we cannot use the z-score test for a hypothesis that specifies a numerical value for ρ. (UNIT 23)

☐ To assess the reality of an association between two variables that appears to hold in a sample, we may test either the *hypothesis of zero correlation* or the *hypothesis of independence*. The latter is the *stronger* of the two since *if x and y are independent, then the correlation between x and y must be zero;* whereas *if the correlation between x and y is zero, then x and y may or may not be independent.* Under the strong hypothesis of independence, SD_r is approximately equal to $1/\sqrt{n-1}$. This permits the testing of the hypothesis of independence by referring

$$z = \frac{r-0}{1/\sqrt{n-1}} = r\sqrt{n-1}$$

to the normal table: However, this z-score test is of doubtful reliability except where n is large enough. There is available a *nonparametric test of independence* based on Spearman's *rank-order correlation coefficient;* exact P values of this test are given for $n \le 11$. For larger n, one may use a variant of the z-score test:

$$z = \frac{6\Sigma d^2 - n(n^2-1)}{n(n+1)\sqrt{n-1}}$$

In choosing between these two tests of independence (z-score test based on r versus Spearman's test), such considerations as the differential power of the tests, the normality of the population, the size of the sample, and the ease of calculation must be borne in mind. If we can reject the independence hypothesis, then we can accept the implication of the sample finding that x and y are indeed associated—but we cannot determine, with precision, the value of ρ. Even if we reject the independence hypothesis, however, we must not assume that association between x and y necessarily implies a *causal* connection between the two variables. A final addendum: Spearman's original formula

$$r' = 1 - \frac{6\Sigma d^2}{n(n^2-1)}$$

yields coefficients which, like Pearson's correlation, *vary from +1.00 to −1.00* and measure the association between x and y in a bivariate sample in terms of the rank order of x and y within each univariate distribution. (UNIT 24)

☐ Once it has been established that x and y are associated with each other, there remains the problem of using an observed value of x to predict what the value of y will turn out to be. This can be done if one knows the *regression of y on x.* If the bivariate population is nearly normal, this regression may be estimated by the *regression line*, given by

$$y^* = \bar{y} + b(x - \bar{x})$$

where the *slope b* is fitted by *least squares* through the formula

$$b = \frac{rs_y}{s_x}$$

(It is, of course, also possible to discuss *the regression of x on y* in an analogous way.) There are, however, three sources of error for such predictions:

1 The equations are based on the assumption of *linearity*, and where this assumption is invalid, major errors of prediction can result.
2 The computed regression lines are estimates of the true *population regression lines*, and since it is not usually possible to say how accurate r, s_x, and s_y are, we cannot determine the accuracy of the estimated slope b.
3 The prediction is also subject to error because of the variability of y about the true mean in its x class. Although the quantity

$$s_y\sqrt{1 - r^2}$$

is usually given as the *standard error of prediction*, this holds precisely only for normal populations. In all other instances, this value is to be understood as only an *average value taken over all points in the sample*—it gives us a rough average guidance as to the likely deviation of y from its predicted value. (UNIT 25)

☐ You are now ready.

DATABANK

UNIT 1

Inquiry Item 10: I would guess that the average STATLAB mother's height is about _____ in.

Inquiry Item 11: I would guess that the average length at birth of the STATLAB girls was about _____ in, and that of boys was about _____ in.

Inquiry Item 12: It would appear that of the two sexes, blood type O is more common among the STATLAB (boys)(girls).

UNIT 6

Inquiry Item 5: I would guess that the STATLAB population median for mothers' heights is _____ in.

UNIT 7

Inquiry Item 6: On the basis of my sample mean I would guess that the mean height of STATLAB mothers is about _____ in.

ID no.	UNIT 2 Work Step 4: Mother's height		UNIT 3 Work Step 5: Father's height		ID no.	UNIT 2 Work Step 4: Mother's height		UNIT 3 Work Step 5: Father's height	
	to tenth inch	to inch	to tenth inch	to inch		to tenth inch	to inch	to tenth inch	to inch
—					—				
—					—				
—					—				
—					—				
—					—				
—					—				
—					—				
—					—				
—					—				
—					—				
—					—				
—					—				
—					—				
—					—				
—					—				
—					—				
—					—				
—					—				
—					—				

FILE B 1

FILE B-1

UNIT 2

Inquiry Item 1: I drew _____ duplicates; I would guess that _____ percent of the class drew one or more duplicates.

UNIT 3

Inquiry Item 2: Of my mothers' heights, _____ were rounded up and

_____ were rounded down.

Inquiry Item 5: Of my rounded mothers' heights, _____ are even.

Inquiry Item 6: Of my mothers' heights, _____ ended in .0; of my fathers'

heights, _____ ended in .0.

UNIT 4

Work Step 7:

HEIGHT OF MOTHER (rounded to nearest inch)

$w = 1$			$w = 2$			$w = 4$				
x	f	F	Class	f	F	Class	f	F	$f\%$	$F\%$
——	——	——	– ——	——	——	– ——	——	——	——	——
——	——	——	– ——	——	——	– ——	——	——	——	——
——	——	——	– ——	——	——	– ——	——	——	——	——
——	——	——	– ——	——	——	– ——	——	——	——	——
——	——	——	– ——	——	——	– ——	——	——	——	——
——	——	——	– ——	——	——					
——	——	——	– ——	——	——					
——	——	——	– ——	——	——					
——	——	——								
——	——	——								
——	——	——								
——	——	——								
——	——	——								
——	——	——								
——	——	——								

Inquiry Item 1: The minimum rounded height is _____; the maximum

is _____.

Inquiry Item 2: The range is _____; I guess that the largest range is

_____.

Inquiry Item 3: The (mode is)(modes are) _____.

Inquiry Item 4: The median is near $x =$ _____.

UNIT 6

Work Step 5: Mothers' calculated median height (to tenth inch) for

$w = 1$ _____.

UNIT 7

Work Step 6: Mothers' mean heights, $n = 10$: _____.

For $n = 40$: $w = 1$ _____, $w = 2$ _____,

$w = 4$ _____.

UNIT 13

Inquiry Item 11: The 68 percent confidence interval for μ (mothers'

heights) = _____ $\pm$ _____.

UNIT 5

Work Step 5:
Mothers' heights—sample, $n = 40$

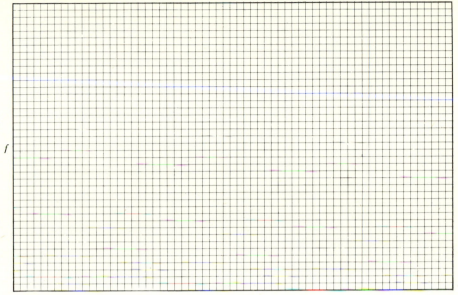

Simple frequency polygon

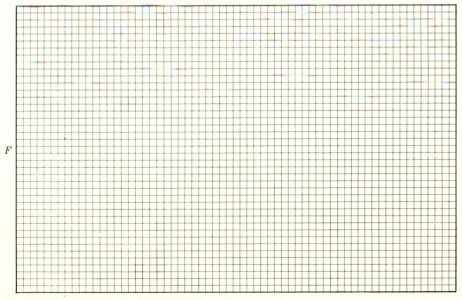

Cumulative frequency polygon

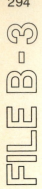

FILE B-3

Boys' heights—population

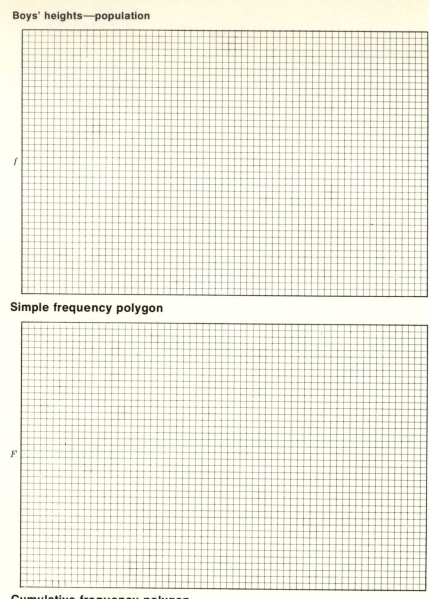

f

Simple frequency polygon

F

Cumulative frequency polygon

Inquiry Item 5: From my cumulative frequency polygon for mothers' heights, the median is _____ in.

Inquiry Item 6: From my population frequency curve for boys' heights, the median is _____ in.

UNIT 7

Note, page 53: Class distributions provided by instructor for the means $\bar{x}$ and medians $\tilde{x}$ of samples of 40 mothers' heights.

Average height	For $\bar{x}$			For $\tilde{x}$		
	f	F	$F\%$	f	F	$F\%$
63.1–63.2	——	——	——	——	——	——
63.3–63.4	——	——	——	——	——	——
63.5–63.6	——	——	——	——	——	——
63.7–63.8	——	——	——	——	——	——
63.9–64.0	——	——	——	——	——	——
64.1–64.2	——	——	——	——	——	——
64.3–64.4	——	——	——	——	——	——
64.5–64.6	——	——	——	——	——	——
64.7–64.8	——	——	——	——	——	——
64.9–65.0	——	——	——	——	——	——
65.1–65.2	——	——	——	——	——	——
65.3–65.4	——	——	——	——	——	——
65.5–65.6	——	——	——	——	——	——
65.7–65.8	——	——	——	——	——	——

FILE C-1

UNIT 8

Work Step 6:

ID no.	Income at test (to nearest $1000)	ID no.	Income at test (to nearest $1000)	Class	f	F
–		–		0–1		
–		–		2–3		
–		–		4–5		
–		–		6–7		
–		–		8–9		
–		–		10–11		
–		–		12–13		
–		–		14–15		
–		–		16–17		
–		–		18–19		
–		–		20–21		
–		–		22–23		
–		–		24–25		
–		–		26–27		
–		–		28–29		
–		–		30–31		
–		–				
–		–				
–		–				
–		–				

The frequency table header: **Frequency table family income, $w = 2$**

$\bar{x}$ (to nearest $500) _____

$\tilde{x}$ (to nearest $500) _____

UNIT 9

Inquiry Item 3: The range of income for my sample is _____.

UNIT 13

Inquiry Item 12: The 68 percent confidence interval for μ (income) is

_____ $\pm$ _____ .

UNIT 15

Inquiry Item 11: The standard error of my estimate $\tilde{x}$ is $_____.
The 68 percent confidence interval for $\tilde{\mu}$ (income), based

on my sample, runs from $_____ to

$_____ .

FILE C-2

UNIT 8

Note, page 69: Class distributions provided by instructor for means and medians of samples of 40 families' incomes (to nearest $500).

Average income	For $\bar{x}$			For $\tilde{x}$		
	f	F	$F\%$	f	F	$F\%$
12,000	——	——	——	——	——	——
12,500	——	——	——	——	——	——
13,000	——	——	——	——	——	——
13,500	——	——	——	——	——	——
14,000	——	——	——	——	——	——
14,500	——	——	——	——	——	——
15,000	——	——	——	——	——	——
15,500	——	——	——	——	——	——
16,000	——	——	——	——	——	——
16,500	——	——	——	——	——	——
17,000	——	——	——	——	——	——
17,500	——	——	——	——	——	——
18,000	——	——	——	——	——	——
18,500	——	——	——	——	——	——
19,000	——	——	——	——	——	——

Inquiry Item 9: I would say that (the two statistics, $\bar{x}$ and $\tilde{x}$, are about equally precise)(the _____ is more precise).

UNIT 9

Note, page 81: Class distributions provided by instructor for the s of samples of 40, for

Mother's height (to nearest tenth inch)				Family income (to nearest $500)			
s	f	F	$F\%$	s	f	F	$F\%$
1.7	___	___	___	4,500	___	___	___
1.8	___	___	___	5,000	___	___	___
1.9	___	___	___	5,500	___	___	___
2.0	___	___	___	6,000	___	___	___
2.1	___	___	___	6,500	___	___	___
2.2	___	___	___	7,000	___	___	___
2.3	___	___	___	7,500	___	___	___
2.4	___	___	___	8,000	___	___	___
2.5	___	___	___	8,500	___	___	___
2.6	___	___	___	9,000	___	___	___
2.7	___	___	___	9,500	___	___	___
2.8	___	___	___	10,000	___	___	___
2.9	___	___	___	10,500	___	___	___
3.0	___	___	___	11,000	___	___	___
3.1	___	___	___	11,500	___	___	___
3.2	___	___	___	12,000	___	___	___
3.3	___	___	___	12,500	___	___	___
3.4	___	___	___	13,000	___	___	___

FILED-1

Work Step 4: The s for my sample of mothers' heights, $w = 1$, is _____,

and for $w = 2$ is _____.

The s for my sample of family income, $w = 2$, is _____.

UNIT 15

Inquiry Item 1: The standard error of my sample s of mothers' heights is

_____ in. The 68 percent confidence interval for σ

runs from _____ to _____.

Inquiry Item 2: Using Formula 15 on my sample, family income, the

68 percent confidence interval for σ runs from $_____

to $_____.

UNIT 10

Work Step 6:

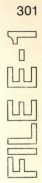

ID no.	Peabody score	Raven score	ID no.	Peabody score	Raven score
—			—		
—			—		
—			—		
—			—		
—			—		
—			—		
—			—		
—			—		
—			—		
—			—		
—			—		
—			—		
—			—		
—			—		
—			—		
—			—		
—			—		
—			—		
—			—		

UNIT 10

Work Step 6:

GROUPED FREQUENCY TABLES, $w = 5$

Peabody scores				Raven scores			
Class	f	F	$F\%$	Class	f	F	$F\%$
–				–			
–				–			
–				–			
–				–			
–				–			
–				–			
–				–			
–				–			
–				–			
–				–			
–				–			
–				–			

$\bar{x}$, Peabody scores: _____ s, Peabody scores: _____

$\bar{x}$, Raven scores: _____ s, Raven scores: _____

Inquiry Item 4: Of my ungrouped Peabody scores there are _____ (or

_____ percent) in the interval $\bar{x} \pm s$; and _____

(or _____ percent) in the interval $\bar{x} \pm 2s$.

Inquiry Item 5: Of my ungrouped Raven scores there are _____ (or

_____ percent) in the interval $\bar{x} \pm s$; and _____

(or _____ percent) in the interval $\bar{x} \pm 2s$.

UNIT 13

Inquiry Item 9: My 68 percent confidence interval for μ, Peabody scores,

runs from _____ to _____; my 95 percent confi-

dence interval runs from _____ to _____.

UNIT 14

Inquiry Item 2: _____ percent of my sample earning Peabody scores

over 80 is _____ $\pm$ _____.

Inquiry Item 3: The 95 percent confidence interval for this percentage

runs from _____ to _____.

FILE E-3

Work Step 6:
Cumulative Frequency Polygon, Peabody

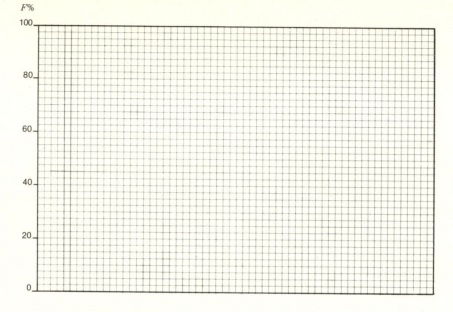

Cumulative Frequency Polygon, Raven

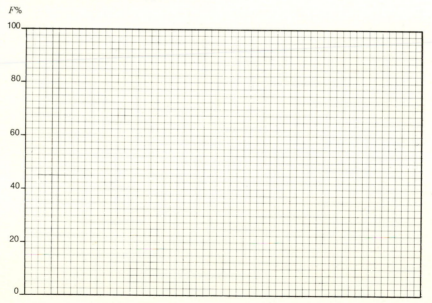

UNIT 14

Work Steps 2 and 3:

ID no.	ID no.	ID no.	ID no.
___ - ___	___ - ___	___ - ___	___ - ___
___ - ___	___ - ___	___ - ___	___ - ___
___ - ___	___ - ___	___ - ___	___ - ___
___ - ___	___ - ___	___ - ___	___ - ___
___ - ___	___ - ___	___ - ___	___ - ___
___ - ___	___ - ___	___ - ___	___ - ___
___ - ___	___ - ___	___ - ___	___ - ___
___ - ___	___ - ___	___ - ___	___ - ___
___ - ___	___ - ___	___ - ___	___ - ___
___ - ___	___ - ___	___ - ___	___ - ___
___ - ___	___ - ___	___ - ___	___ - ___
___ - ___	___ - ___	___ - ___	___ - ___
___ - ___	___ - ___	___ - ___	___ - ___
___ - ___	___ - ___	___ - ___	___ - ___

My proportion p of fathers with $x = 1$ (codes 0, 1, or 2) is _____; SE_p

is _____.

Inquiry Item 1:   My 68 percent confidence interval for P runs from _____

to _____.

FILE G-1

UNIT 16, WORK STEP 1 AND
UNIT 18, WORK STEP 3

	BOYS				GIRLS	
Unit 16 Work step 1 ID no.	Unit 18 Work step 3 Raven scores	Unit 16 Work step 1 ID no.	Unit 18 Work step 3 Raven scores	Unit 16 Work step 1 ID no.	Unit 18 Work step 3 Raven scores
—	_____	—	_____	—	_____
—	_____	—	_____	—	_____
—	_____	—	_____	—	_____
—	_____	—	_____	—	_____
—	_____	—	_____	—	_____
—	_____	—	_____	—	_____
—	_____	—	_____	—	_____
—	_____	—	_____	—	_____
—	_____	—	_____	—	_____
—	_____	—	_____	—	_____

BOYS' RAVEN SCORES			GIRLS' RAVEN SCORES		
Unit 18 Work step 4 Class ($w = 5$)	f	F	Unit 18 Work step 4 Class ($w = 5$)	f	F
_	___	___	_	___	___
_	___	___	_	___	___
_	___	___	_	___	___
_	___	___	_	___	___
_	___	___	_	___	___
_	___	___	_	___	___
_	___	___	_	___	___
_	___	___	_	___	___
_	___	___	_	___	___
_	___	___	_	___	___

UNIT 16

Inquiry Item 1: The difference in mean birthweights between boy and

girl infants in my sample is _____ ± _____ lb. I estimate a 95 percent chance that the true difference between

STATLAB male and female infants lies between _____

lb in favor of the (boy)(girl) infants and _____ lb in favor of the (boy)(girl) infants.

FILE G-1

Inquiry Item 3: _____ percent of the boy infants and _____ percent of the girl infants had blood type O. The difference in percentage of boy and girl infants with blood type O is

_____ ± _____ .

Inquiry Item 4: I estimate a 99 percent chance that the true difference lies

between _____ and _____ percent.

UNIT 18

Work Steps 2; 3 (Raven statistics)

Boys: $\bar{x}_1$ _____ , s_1 _____

Girls: $\bar{x}_2$ _____ , s_2 _____

Inquiry Item 4: Using the historical value for σ_1, I found a z value of

_____ .

Inquiry Item 5: This z value yielded a P value of _____ .

Inquiry Item 7: With s_1, I found a z value of _____ and a P value of

_____ .

Inquiry Items 12 and 13: For the second hypothesis, I found $z =$ _____

and P value $=$ _____ .

Inquiry Item 20: My boys' sample (does)(does not) seem reasonably normal; my girls' sample (does)(does not) seem reasonably normal.

UNIT 20

Work Step 2:

Computed s, boys' Raven scores _____ .

Work Step 6:

The number of sets of ties in ranks in my sample is _____ .

Inquiry Item 6: My P values were _____ for the t test, _____

for the sign test, and _____ for the Wilcoxon test.

Inquiry Item 8: I guess that, of the three tests, the _____ test
has worked best.

UNIT 21

Inquiry Item 2: My P value for the t test is _____ , and my P value

for the Wilcoxon test is _____ .

FILE H-1

UNIT 17

Work Step 5:

ID no.	ID no.	Weekend birth	Father Q at test
____-____	____-____	$p_B =$ _____	$p_B =$ _____
____-____	____-____	$SE_B =$ _____	$SE_B =$ _____
____-____	____-____	$n_B =$ _____	$n_B =$ _____
____-____	____-____	$p =$ _____	$p =$ _____
____-____	____-____	$SE_p =$ _____	$SE_p =$ _____
____-____	____-____	$n_C =$ _____	$n_C =$ _____
____-____	____-____	$p_C =$ _____	$p_C =$ _____
____-____	____-____	$SE_C =$ _____	$SE_C =$ _____
____-____	____-____		
____-____	____-____		
____-____	____-____		
____-____	____-____		
____-____	____-____		
____-____	____-____		
____-____	____-____		
____-____	____-____		
____-____	____-____		
____-____			

Inquiry Item 1: I found it (easy)(hard) to arrive at my p_B; (easy)(hard) at my SE_B.

Inquiry Item 3: I guess that _____ percent of the class will cover true P.

Inquiry Item 4: I think that $(p_B)(p)(p_C)$ will be closest.

Inquiry Item 5: I found it (easy)(hard) to arrive at my p_B; (easy)(hard) at my SE_B.

Inquiry Item 7: I guess that _____ percent of the class will cover true P.

Inquiry Item 8: I think that $(p_B)(p)(p_C)$ will be closest.

FILE I-1

UNIT 19

Work Step 3:

ID numbers of sample, $n = 45$				
ID no.	ID no.	ID no.	ID no.	ID no.
__ - __	__ - __	__ - __	__ - __	__ - __
__ - __	__ - __	__ - __	__ - __	__ - __
__ - __	__ - __	__ - __	__ - __	__ - __
__ - __	__ - __	__ - __	__ - __	__ - __
__ - __	__ - __	__ - __	__ - __	__ - __
__ - __	__ - __	__ - __	__ - __	__ - __
__ - __	__ - __	__ - __	__ - __	__ - __
__ - __	__ - __	__ - __	__ - __	__ - __
__ - __	__ - __	__ - __	__ - __	__ - __

Worksheet page 3:

Number of younger mothers, n_1 _____; number of older mothers,

n_2 _____.

Percentage of churchgoing younger mothers, p_1 _____.

Percentage of churchgoing older mothers, p_2 _____.

Percentage of churchgoing mothers (total), p _____.

Inquiry Item 3: I (do)(do not) reject the hypothesis at the 10 percent level, and so I (accept)(reject) the speculation that 50 percent of STATLAB mothers attend church regularly.

Inquiry Item 4: My standard deviation of $p_2 - p_1$ is _____. I (accept) (reject) the hypothesis at the 15 percent level.

Inquiry Item 5: I find for the hypothesis $\tilde{\mu} = \$16,000$ the two-sided P value

of _____.

UNIT 22

Inquiry Item 2: I found $\chi^2 = $ _____, with a P value of _____.

Inquiry Item 4: My one-sided P value is _____.

Inquiry Item 6: I find $z = $ _____, with one-sided P value =

_____.

Inquiry Item 8: My $\chi^2 = $ _____; my P value is _____.

UNIT 23

Work Step 6:

For my sample $(n = 20)$ of mothers' (x) and fathers' (y) heights: $C_x = $

_____, $s_{x'} = $ _____, $C_y = $ _____, $s_{y'} = $ _____, $r = $ _____.

Inquiry Item 1: I speculate that the relation in the STATLAB population will be (negligible)(positive)(negative), with ρ about

_____.

Inquiry Item 2: I now guess r will be about _____.

Inquiry Item 6: On the average, husbands' heights (rise)(fall) about

_____ in per 1-in increase in mothers' heights.

Inquiry Item 7: On the average, wives' heights (rise)(fall) about _____ in per 1-in increase in fathers' heights.

Inquiry Item 8: I now speculate that the relation in the STATLAB population will be (negligible)(positive)(negative), with ρ about

_____.

UNIT 24

Inquiry Item 3: The two-sided P values by test based on r is _____,

and by test based on Σd^2 is _____.

Inquiry Item 8: My value of $r' = $ _____.

314

FILE X

CENSUS

The STATLAB Census covers 1296 member families of the Kaiser Foundation Health Plan (a prepaid medical care program) living in the San Francisco Bay Area during the years 1961–1972. These families were participating members of the Child Health and Development Study conceived and directed by Jacob Yerushalmy, for many years Professor of Biostatistics in the School of Public Health, University of California, Berkeley.

On her first visit to the Oakland hospital of the Health Plan after pregnancy was diagnosed, each woman was interviewed intensively on a wide range of medical and socioeconomic matters relating both to herself and to her husband. In addition, various physical and physiological measures were made. When her child was born, further data about her and her newborn baby were recorded. Approximately 10 years later the child and mother were called in for follow-up testing, interviewing, and measurement. In some instances, the husband was also interviewed and measured.

The 1296 families of the STATLAB Census are divided into two equal subpopulations: 648 families consisting of a mother, father, and female child; and 648 families of a mother, father, and male child. The children were all born in the Kaiser Foundation Hospital, Oakland, California, between 1 April 1961 and 15 April 1963. The Census does not cover any other children who may also have existed in these families.

Of the multitude of available data, 32 variables have been selected for the Census. The following 36 Census pages list each of these 32 variables for each of the 1296 families. The first 18 pages cover the families with girls; the second 18 pages cover the families with boys. Within each of these two sets of pages the families are listed in order of mother's age, with the youngest mothers first and the oldest last.

The Census pages consist of printouts numbered in consecutive dice numbers (i.e., the Census pages are numbered 11, 12, 13, 14, 15, 16, 21, 22, . . . , 65, 66). Similarly the 36 families listed on each page are designated in consecutive dice numbers from 11 to 66. The identification number (ID no.) for any given family consists of two pairs of dice numbers, the first pair indicating page and the second pair indicating family on the page. To select a family purely at random from the population of 1296, it is necessary to throw the pair of dice twice. If, for example, the first throw gives a red 2 and a green 6, this selects page 26. If the second throw gives a red 5 and a green 4, this selects family 54 on that page. Thus the ID number for this family is 26–54.

KEY TO THE VARIABLES

The 32 variables pertaining to each STATLAB family are grouped by *child, mother, father,* and *family.* Certain of the data were collected at the time of birth (1961–1963) and certain other data at the time of test (1971–1972), thus resulting in seven clusters of data. The description and codes (where relevant) for each of the variables for each of the seven clusters, starting at the left of the printout, are as follows.

Child—Birth		Child—Test	
Variable	**Code/Definition**	**Variable**	**Code/Definition**
B	BLOOD TYPE 1 O — Rh-negative 2 A — Rh-negative 3 B — Rh-negative 4 AB — Rh-negative 5 O — Rh-positive 6 A — Rh-positive 7 B — Rh-positive 8 AB — Rh-positive 9 Unknown	HGHT	Height of child to tenth inch
		WGT	Weight of child to nearest pound
LGTH	Length of baby to tenth inch	L	Laterality: Combination of left or right handedness (H), with left or right eyedness (E), the latter being measured on two occasions (E_1 and E_2)
WGT	Weight of baby to tenth pound		
MO-D-HR	Month, day, and hour of baby's birth		H E_1 E_2 1 R R R 2 R R L 3 R L R 4 R L L 5 L R R 6 L R L 7 L L R 8 L L L
	MONTH 1 January 2 February ⋮ 12 December		
			For example, code 2 indicates that a *right-handed* child was *right-eyed* dominant on the first observation and *left-eyed* dominant on the second.
	DAY 1 Sunday 2 Monday ⋮ 7 Saturday	PEA	Score on the Peabody Picture Vocabulary Test
	HOUR 1 1 A.M. 2 2 A.M. ⋮ 12 12 noon 13 1 P.M. ⋮ 24 12 midnight	RA	Score on the Raven Progressive Matrices Test

	Mother—Birth		Mother—Test
Variable	**Code/Definition**	**Variable**	**Code/Definition**
B	BLOOD TYPE 1 O — Rh-negative 2 A — Rh-negative 3 B — Rh-negative 4 AB — Rh-negative 5 O — Rh-positive 6 A — Rh-positive 7 B — Rh-positive 8 AB — Rh-positive 9 Unknown	HGHT	Height of mother to tenth inch
		WGT	Weight of mother to nearest pound
		E	EDUCATION OF MOTHER 0 less than 8th grade 1 8th to 12th grade 2 High school graduate 3 Some college 4 College graduate
AG	Age of mother at last birthday before baby's birth	O	OCCUPATION OF MOTHER 0 Housewife 1 Office/Clerical 2 Sales 3 Teacher/Counselor 4 Professional/Managerial 5 Services 7 Factory worker 8 All other
WGT	Weight of mother (to nearest pound) at time pregnancy was first diagnosed		
O	OCCUPATION OF MOTHER 0 Housewife 1 Office/Clerical 2 Sales 3 Teacher/Counselor 4 Professional/Managerial 5 Services 7 Factory worker 8 All other	SM	SMOKING HISTORY OF MOTHER N Never smoked cigarettes Q Has now quit smoking, but did smoke at one time 01–99 Number of cigarettes currently being smoked per day
SM	SMOKING HISTORY OF MOTHER N Never smoked cigarettes Q Has now quit smoking, but did smoke at one time 01–99 Number of cigarettes currently being smoked per day		

Father—Birth

Variable	Code/Definition
AG	Age of father at last birthday before baby's birth
O	OCCUPATION OF FATHER 0 Professional 1 Teacher/Counselor 2 Manager/Official 3 Self-employed 4 Sales 5 Clerical 6 Craftsman/Operator 7 Laborer 8 Service worker
SM	SMOKING HISTORY OF FATHER N Never smoked cigarettes Q Has now quit smoking, but did smoke at one time 01–99 Number of cigarettes currently being smoked per day

Father—Test

Variable	Code/Definition
HGHT	Height of father to tenth inch
WGT	Weight of father to nearest pound
E	EDUCATION OF FATHER 0 less than 8th grade 1 8th to 12th grade 2 High school graduate 3 Some college 4 College graduate
O	OCCUPATION OF FATHER 0 Professional 1 Teacher/Counselor 2 Manager/Official 3 Self-employed 4 Sales 5 Clerical 6 Craftsman/Operator 7 Laborer 8 Service worker
SM	SMOKING HISTORY OF FATHER N Never smoked cigarettes Q Has now quit smoking, but did smoke at one time 01–99 Number of cigarettes currently being smoked per day

Family

Variable	Code/Definition
I–B	Income of family at time of birth in hundreds of dollars
I–T	Income of family at time of test in hundreds of dollars
C	CHURCH ATTENDANCE 1 Entire family attends fairly regularly 2 Mother and child attend fairly regularly 3 Child only attends fairly regularly 4 Sporadic attendance—anyone in family 5 Attend on Holy Days only—anyone in family 6 No one in family ever attends

StatLab CENSUS: Girls

	CHILD-BIRTH				CHILD-TEST					MOTHER-BIRTH					MOTHER-TEST					FATH-BRTH			FATHER-TEST					FAMILY		
ID	B	LGTH	WGT	MO-D-HR	HGHT	WGT	L	PEA	RA	B	AG	WGT	O	SM	HGHT	WGT	E	O	SM	AG	O	SM	HGHT	WGT	E	O	SM	I-B	I-T	C
11	2	20.0	6.6	3-4-4	55.7	85	5	85	34	2	17	119	0	Q	66.0	130	1	1	20	19	8	10	70.1	171	3	8	10	33	150	6
12	2	20.0	6.4	5-7-13	48.9	59	3	74	34	6	17	130	0	20	62.8	159	3	1	10	23	6	11	65.0	130	1	6	20	40	175	6
13	7	19.8	6.1	6-2-4	54.9	70	2	64	25	7	18	134	0	10	66.1	138	2	0	N	21	0	20	70.0	175	2	6	N	44	116	1
14	6	19.5	7.0	10-2-19	53.6	88	3	87	43	6	18	135	1	06	61.8	123	2	0	N	26	5	20	71.8	196	3	2	N	42	112	4
15	5	19.5	7.9	8-7-2	53.4	68	1	87	40	5	18	130	0	N	62.8	146	2	8	N	21	6	N	63.0	163	2	6	N	50	129	4
16	5	22.0	9.5	11-6-12	59.9	93	4	83	37	5	18	104	0	N	63.4	116	2	1	N	17	6	20	74.0	186	3	0	22	0	214	6
21	7	21.0	7.1	12-7-8	53.1	72	2	81	33	7	18	145	0	N	65.4	220	2	0	N	23	6	20	68.1	173	2	6	20	55	142	4
22	6	20.5	6.4	4-2-18	52.2	84	1	74	37	6	18	102	0	04	62.3	120	2	0	N	20	0	20	72.0	150	3	8	N	38	120	4
23	5	21.5	8.2	12-1-10	56.8	68	4	72	21	5	13	128	0	06	65.4	141	3	0	N	23	7	10	71.0	150	2	7	10	34	104	2
24	6	20.5	7.6	7-4-8	53.8	76	1	64	31	7	18	145	0	N	66.4	184	2	2	N	28	8	20	75.0	235	3	4	20	48	194	4
25	2	19.0	5.4	12-2-20	49.0	51	1	72	29	1	18	99	0	30	64.8	107	1	0	Q	22	6	20	70.0	145	3	5	Q	37	140	1
26	5	21.0	8.1	1-1-2	51.9	59	2	87	38	5	18	100	0	04	56.9	114	2	0	10	23	0	20	72.8	197	3	7	20	36	170	1
31	5	19.0	6.4	5-3-4	55.7	78	1	72	19	1	19	145	0	20	65.6	151	1	0	30	24	2	30	74.0	204	2	0	Q	46	125	6
32	5	20.0	6.4	5-1-11	53.4	73	3	78	27	5	19	112	0	N	63.6	123	2	5	N	22	6	30	71.0	220	3	6	50	50	170	4
33	5	20.0	6.9	12-5-24	50.3	60	1	77	35	5	19	120	2	06	65.4	143	2	1	N	23	4	09	65.0	150	2	4	22	67	176	6
34	1	21.8	7.9	9-1-2	55.8	70	1	75	29	5	19	115	1	0	63.7	150	2	0	Q	20	8	N	69.0	180	3	3	N	92	96	6
35	5	20.0	6.9	3-4-19	55.2	78	3	71	25	5	19	117	0	01	63.7	149	2	0	Q	25	8	N	69.0	170	3	7	N	39	84	1
36	5	20.0	7.4	10-6-24	58.8	106	1	60	20	5	19	125	5	02	64.3	205	2	8	02	27	6	20	72.0	235	2	0	20	68	124	6
41	6	20.0	8.1	7-6-4	53.6	76	1	79	36	6	19	100	1	0	58.3	121	3	0	20	20	0	25	75.0	190	3	4	20	27	150	4
42	5	20.0	6.2	8-4-15	53.3	64	4	63	32	5	19	122	0	N	63.3	147	2	1	N	30	6	N	70.0	190	2	0	N	36	28	1
43	5	20.5	7.4	7-6-18	49.9	50	4	80	42	1	20	128	1	N	65.3	134	2	1	N	25	0	03	68.5	160	3	0	N	48	216	6
44	8	19.5	6.1	9-5-18	53.8	56	1	66	28	7	19	106	0	20	66.4	117	2	1	N	33	6	20	65.5	130	1	7	15	68	108	1
45	6	20.5	7.1	10-3-10	53.9	68	1	65	38	6	19	135	0	N	65.6	151	3	1	N	22	6	N	69.0	160	3	6	N	51	89	2
46	5	19.5	6.6	11-6-8	53.1	68	1	69	15	5	19	129	5	02	63.8	150	3	4	10	23	5	05	73.5	154	2	6	Q	32	179	1
51	5	21.5	7.1	4-2-18	56.1	64	1	63	24	5	19	105	0	Q	62.0	110	2	1	02	21	8	N	70.8	183	3	4	N	35	180	1
52	5	19.0	5.3	3-1-9	51.7	56	1	70	14	5	20	97	0	N	59.6	138	2	1	N	21	6	05	64.9	169	1	2	05	50	100	6
53	1	18.0	5.4	4-7-23	48.0	64	4	51	40	1	20	113	1	N	65.1	128	2	4	N	21	0	N	67.0	146	4	4	Q	54	207	6
54	3	20.3	7.3	5-4-14	51.3	58	1	80	45	2	20	121	1	1	62.0	132	2	2	N	21	4	24	71.0	165	2	4	N	54	114	6
55	9	18.5	7.5	11-3-13	50.5	60	4	75	29	6	20	140	1	20	65.4	144	2	1	40	20	7	20	71.0	190	3	5	06	86	180	6
56	8	21.0	7.9	8-7-10	57.3	79	1	96	38	7	20	138	5	18	68.1	143	3	4	N	21	6	20	73.5	185	3	2	20	99	220	6
61	7	21.0	6.8	9-1-20	52.8	64	4	94	25	5	20	140	1	N	66.3	147	2	0	N	28	8	N	71.0	180	3	8	N	66	120	1
62	5	20.0	7.0	10-3-23	51.3	56	4	82	34	5	20	104	1	N	61.0	113	2	0	N	24	6	Q	71.0	149	2	3	N	88	120	2
63	5	18.5	4.6	11-2-6	50.1	56	1	82	39	3	20	97	1	Q	64.6	116	2	0	Q	21	8	20	66.0	145	3	8	20	50	140	4
64	1	19.5	6.8	12-2-8	48.7	50	1	74	35	6	20	122	0	Q	62.8	132	2	0	Q	22	6	20	67.8	166	3	5	40	46	120	1
65	5	20.0	7.8	12-5-24	57.7	89	4	90	35	5	20	123	0	Q	66.8	134	3	8	20	23	6	17	70.5	200	2	7	20	58	150	4
66	5	21.0	7.8	2-1-11	54.8	65	4	64	32	5	20	110	0	10	67.0	121	2	0	20	25	6	20	73.0	175	2	6	20	50	0	1

ID	CHILD-BIRTH				CHILD-TEST					MOTHER-BIRTH					MOTHER-TEST					FATH-BRTH			FATHER-TEST					FAMILY		
	B	LGTH	WGT	MO-D-HR	HGHT	WGT	L	PEA	RA	B	AG	WGT	O	SM	HGHT	WGT	E	O	SM	AG	O	SM	HGHT	WGT	E	O	SM	I-B	I-T	C
11	7	19.5	6.1	11-1-4	51.8	62	1	67	27	5	20	100	0	02	62.1	108	2	0	02	21	7	10	69.0	165	2	6	Q	45	83	3
12	5	21.0	7.1	4-7-24	52.1	61	1	66	27	5	20	110	1	10	64.1	113	2	8	10	21	6	20	69.0	160	2	2	20	88	133	1
13	5	21.0	8.1	5-4-9	56.9	65	1	88	31	7	20	125	1	N	67.1	130	2	0	10	23	6	20	72.5	163	2	7	20	76	156	6
14	9	20.0	7.2	4-5-19	49.4	58	1	76	25	6	20	95	1	10	62.7	109	2	5	20	23	8	10	71.0	180	4	2	20	43	114	3
15	5	19.5	6.3	4-4-2	53.0	63	1	65	28	5	20	130	5	N	65.4	154	7	7	N	27	5	N	68.0	170	2	5	N	35	122	1
16	5	20.0	8.2	5-5-21	51.1	57	4	80	35	5	20	109	1	10	60.8	111	2	0	10	20	5	20	68.0	154	2	0	20	51	168	4
21	5	22.0	9.7	6-6-1	55.3	83	1	84	30	5	20	145	0	N	69.0	200	3	0	N	24	0	N	72.3	194	4	4	N	54	120	3
22	5	20.0	6.3	5-3-14	52.4	66	1	86	45	6	20	102	1	Q	64.0	120	3	0	Q	21	0	20	70.5	150	3	3	30	89	150	6
23	5	20.0	6.4	7-4-16	54.0	74	1	75	41	6	20	102	1	20	63.8	121	2	1	Q	22	6	Q	66.0	151	2	6	N	96	126	2
24	6	20.0	7.1	4-1-14	48.5	51	2	60	18	5	20	130	0	10	65.2	146	2	0	20	23	2	20	72.0	180	4	0	20	37	144	3
25	5	20.5	7.3	10-2-19	49.0	56	4	72	21	5	20	110	0	20	60.9	126	3	2	20	22	6	10	70.0	175	2	3	30	90	180	4
26	6	20.0	8.3	9-5-5	55.3	80	5	64	23	5	20	167	5	N	66.0	140	3	4	20	22	6	18	73.0	190	2	7	10	50	172	3
31	5	19.0	7.0	8-7-22	52.0	63	4	78	37	1	20	134	0	09	64.1	154	1	0	01	23	5	N	71.0	140	2	6	02	75	72	6
32	9	19.0	6.4	1-3-23	50.3	61	1	74	24	5	20	110	1	20	63.6	126	2	2	04	30	6	Q	69.0	155	2	7	Q	65	108	4
33	5	20.0	7.3	9-7-6	49.9	62	1	66	35	6	20	112	0	20	63.1	121	2	1	10	20	2	15	68.0	140	2	2	20	44	144	1
34	5	19.8	6.5	10-7-7	50.9	59	1	62	28	6	20	118	0	20	63.4	125	2	5	20	21	8	09	71.0	201	1	6	N	70	130	1
35	2	20.0	7.8	5-5-11	49.9	51	3	70	31	6	20	120	0	N	61.2	131	2	0	20	23	4	N	73.0	185	2	4	N	66	140	1
36	3	20.0	7.7	9-7-23	56.3	86	4	75	17	5	21	120	1	10	66.5	129	3	0	10	29	3	15	67.0	178	3	3	Q	45	150	4
41	2	19.8	6.8	7-6-22	55.2	72	4	66	26	6	21	138	0	06	67.0	135	2	2	15	23	0	10	69.6	159	3	6	25	42	200	4
42	5	21.5	7.2	4-7-11	51.4	59	5	76	32	5	21	135	0	N	59.1	172	2	0	N	23	7	20	72.0	168	3	8	Q	40	75	6
43	2	19.5	7.4	6-3-9	57.0	102	1	94	40	3	21	130	5	N	61.9	135	3	0	N	23	0	N	67.0	180	4	2	N	72	120	6
44	5	20.5	7.8	5-5-17	53.3	64	1	70	17	1	21	125	0	N	67.0	137	3	2	N	23	7	N	69.9	208	2	6	20	50	190	4
45	5	18.8	6.1	12-7-2	53.4	78	4	63	8	6	21	135	0	06	59.6	146	3	2	07	25	5	11	69.0	150	2	2	12	64	180	1
46	8	20.5	6.6	6-3-19	53.8	74	1	76	18	6	21	135	0	06	63.5	145	2	5	06	26	6	05	70.5	177	3	3	15	80	97	4
51	1	21.0	8.9	5-4-16	57.9	69	1	74	33	2	21	201	0	20	66.4	205	2	2	20	32	7	20	72.0	184	2	6	Q	46	45	1
52	3	21.0	7.3	7-1-14	53.9	68	1	83	25	3	21	135	0	09	66.3	142	2	1	02	24	7	30	70.5	160	2	6	Q	41	152	6
53	6	19.0	5.6	6-4-20	56.8	80	4	84	42	1	21	124	0	N	63.8	139	4	0	N	22	6	N	72.0	170	3	6	N	37	100	6
54	5	20.5	8.6	9-4-20	57.6	78	4	74	20	7	21	134	0	10	68.1	143	3	4	Q	22	2	20	73.5	185	3	2	20	90	220	6
55	1	21.0	9.0	1-2-18	50.7	67	1	63	27	5	21	140	0	10	63.1	161	3	1	10	22	6	10	67.0	160	3	6	Q	50	213	2
56	6	21.0	6.8	2-1-10	52.2	65	6	93	26	6	21	120	0	N	65.4	128	3	0	N	23	6	03	70.0	168	4	2	10	61	148	6
61	5	22.0	7.8	12-4-11	54.8	70	1	84	39	5	21	155	8	20	67.7	156	3	4	30	24	0	N	68.5	165	4	0	N	54	180	6
62	6	19.0	7.1	8-2-17	55.6	82	4	77	27	6	21	135	0	20	67.0	158	3	0	20	24	0	20	69.0	170	4	4	Q	0	100	1
63	6	19.0	6.2	8-1-6	49.8	58	3	84	26	6	21	100	1	01	62.0	104	3	4	Q	39	8	20	67.0	175	3	3	25	102	177	4
64	6	18.5	5.8	4-4-22	52.7	52	1	83	33	6	21	127	1	N	65.4	132	2	0	N	28	8	Q	71.8	195	3	8	Q	125	140	4
65	6	18.5	6.8	5-5-21	47.9	50	1	60	16	5	21	95	8	03	56.8	110	2	0	01	20	6	20	65.5	148	2	7	40	74	91	4
66	5	20.0	6.1	9-4-13	53.5	55	8	75	25	5	21	118	0	20	62.4	114	2	0	40	22	6	20	72.0	185	1	2	40	80	150	1

	CHILD-BIRTH				CHILD-TEST					MOTHER-BIRTH					MOTHER-TEST					FATH-BRTH			FATHER-TEST					FAMILY		
ID	B	LGTH	WGT	MO-D-HR	HGHT	WGT	L	PEA	RA	B	AG	WGT	O	SM	HGHT	WGT	E	O	SM	AG	O	SM	HGHT	WGT	E	O	SM	I-B	I-T	C
11	9	20.5	7.1	10-1-15	52.6	63	1	78	40	5	21	135	1		66.0	145	2	0	06	22	0		71.0	197	4	0	02	45	162	1
12	5	20.0	6.4	9-7-7	50.5	63	3	71	21	1	21	98	0	18	62.9	100	2	0	12	22	6	20	72.0	160	2	6	20	80	104	4
13	6	17.0	4.2	7-5-5	51.1	51	1	73	35	6	21	96	0	10	61.8	103	3	0	15	24	6	09	70.0	160	2	6	15	48	74	6
14	2	20.5	6.8	1-7-24	52.1	69	4	67	30	7	21	115	4	20	62.4	134	2	0	15	22	6	10	72.0	180	2	6	Q	91	125	4
15	5	21.0	7.2	3-2-13	53.0	88	2	64	28	5	21	116	0	N	64.3	126	2	1	15	25	5	N	67.0	145	3	5	N	63	156	1
16	5	20.0	7.6	2-5-5	51.3	58	3	59	15	5	21	135	1	10	66.0	125	3	0	30	24	7	N	67.5	169	3	7	10	40	169	3
21	7	20.0	6.4	12-7-9	51.1	59	1	66	28	5	22	137	1	20	66.8	152	2	0	20	25	6	N	68.0	155	1	7	N	80	68	6
22	5	20.0	7.1	10-7-2	54.1	73	4	76	34	5	22	155	0	N	64.0	168	1	0	N	24	7	20	70.0	180	1	6	40	58	109	4
23	5	20.5	7.3	6-2-2	55.1	73	1	73	29	5	22	108	0	10	62.1	120	2	1	08	25	6	10	71.5	177	2	6	Q	68	133	6
24	7	19.0	5.8	4-2-24	51.7	75	2	69	37	4	22	125	1	20	65.5	135	2	1	20	27	8	20	72.0	165	2	4	20	70	172	6
25	5	20.0	6.4	7-7-9	53.4	70	2	84	32	1	22	127	0	07	65.8	137	3	0	10	23	6	20	65.0	137	3	0	Q	52	130	1
26	7	19.0	7.0	8-2-24	51.7	70	4	79	42	7	22	106	0	15	63.5	112	3	4	12	27	6	N	64.6	142	3	5	N	91	137	1
31	2	19.8	6.2	10-6-22	55.7	73	1	80	34	2	22	136	0	34	67.8	136	1	0	30	24	6		72.0	220	3	0	N	60	190	1
32	6	20.0	7.4	4-6-16	53.1	74	1	72	23	6	22	105	0	Q	61.6	111	2	0	Q	33	8	N	71.0	170	3	8	Q	76	120	1
33	6	19.0	6.1	1-3-6	52.3	59	1	71	17	6	22	121	1	N	63.6	138	2	8	N	22	6	N	71.0	190	2	8	N	100	121	6
34	7	19.5	6.6	11-5-22	51.8	66	1	64	20	6	22	122	0	10	66.0	124	2	7	10	24	4	11	67.0	195	3	6	22	49	144	6
35	8	19.5	6.9	4-3-1	51.4	60	4	64	14	8	22	118	0	N	63.1	159	2	0	N	23	6	30	68.0	185	3	6	60	50	121	6
36	9	19.5	8.1	6-1-19	53.5	82	4	86	40	7	22	135	0	Q	61.3	133	3	0	20	30	1	N	69.0	150	4	1	N	42	120	6
41	5	19.5	7.9	6-1-3	49.4	60	4	58	13	6	22	150	0	Q	64.5	205	3	5	N	26	8	10	68.0	150	2	7	10	30	150	1
42	2	20.5	8.1	6-4-21	54.0	61	7	83	37	2	22	126	5	Q	65.0	134	3	0	Q	25	0	05	74.0	180	4	0	Q	75	165	4
43	5	20.0	6.8	5-7-4	52.8	62	6	79	27	1	22	111	0	N	63.4	117	3	1	Q	28	8	N	68.0	125	4	2	N	40	120	6
44	5	20.5	6.6	8-3-9	51.4	58	4	86	46	1	22	117	1	03	63.6	121	3	1	06	30	5	01	68.0	150	3	2	N	103	192	4
45	7	18.0	6.1	6-1-3	52.1	73	1	76	27	5	22	130	0	Q	63.6	156	3	5	N	24	8	15	64.0	140	2	6	10	33	135	2
46	6	21.0	7.9	12-4-9	55.2	106	4	69	15	1	22	130	1	15	66.1	138	2	1	15	28	6	24	69.0	185	2	6	80	99	145	1
51	6	21.0	9.8	10-1-20	52.8	64	1	77	32	2	22	145	0	N	66.0	140	1	0	N	23	6	20	68.0	170	2	6	Q	48	103	3
52	5	20.0	7.1	11-1-4	52.1	60	4	72	11	5	22	120	0	Q	67.5	135	2	4	Q	27	5	N	69.6	170	3	2	N	40	180	6
53	5	20.5	7.6	11-4-5	52.4	68	4	61	23	6	22	140	1	N	63.9	197	2	1	N	36	7	07	66.0	155	0	7	Q	70	147	6
54	7	20.5	7.1	5-5-3	53.4	58	4	67	41	7	22	135	4	Q	68.2	143	3	5	N	26	6	N	73.5	220	3	6	N	62	175	4
55	6	19.5	6.0	5-5-23	52.8	76	4	71	23	5	22	105	1	20	62.8	112	2	0	20	24	8	10	71.0	215	4	1	40	65	165	6
56	1	20.0	7.1	4-4-9	52.4	81	1	67	18	1	22	160	0	15	65.9	174	2	0	20	26	6	20	71.0	220	2	2	20	40	193	2
61	5	19.5	6.0	9-4-19	50.8	52	1	81	23	5	22	101	1	20	60.9	111	2	0	30	23	6	N	69.0	150	2	2	N	96	120	1
62	6	21.0	7.3	7-1-17	55.2	71	5	86	34	6	22	120	0	10	66.8	117	3	0	N	29	1	20	73.0	190	4	1	40	34	160	1
63	5	20.0	6.6	6-6-18	52.1	70	4	75	23	5	22	126	1	N	66.0	150	3	0	Q	26	0	Q	71.0	165	3	7	Q	82	96	1
64	5	19.0	6.4	5-6-6	49.6	56	4	64	24	5	22	97	0	15	61.7	117	1	5	N	26	6	20	74.0	230	3	7	30	60	116	3
65	7	21.0	8.9	6-6-8	56.3	68	1	67	11	7	22	166	0	N	63.4	171	2	5	N	23	8	20	67.0	170	2	2	20	20	129	2
66	9	22.0	6.8	9-4-7	54.0	70	4	78	24	6	22	125	7	20	62.2	129	3	3	Q	28	6	N	75.0	185	2	3	N	60	256	1

ID	B	LGTH	WGT	MO-D-HR	HGHT	WGT	L	PEA	RA	B	AG	WGT	O	SM	HGHT	WGT	E	O	SM	AG	O	SM	HGHT	WGT	E	O	SM	I-B	I-T	C
		CHILD-BIRTH				CHILD-TEST					MOTHER-BIRTH					MOTHER-TEST					FATH-BRTH			FATHER-TEST					FAMILY		
11	7	20.0	7.6	8-1-2	53.3	70	4	62	13	5	22	116	0	N	63.1	130	2	7	N	30	7	20	67.0	147	3	7	10	150	132	1
12	5	21.0	7.8	1-2-18	49.8	56	1	82	44	6	22	177	4	N	62.5	175	3	0	N	25	0	40	72.0	137	4	1	30	42	120	1
13	5	19.0	6.0	1-7-	49.6	67	4	77	10	5	22	110	2	N	64.0	118	2	0	N	23	6	N	69.0	170	4	1	N	60	132	4
14	3	20.0	7.8	2-7-21	54.9	71	1	71	28	5	22	110	0	N	66.4	144	2	0	N	23	6	Q	66.6	165	2	6	N	46	144	1
15	7	20.0	7.2	1-5-3	54.9	74	1	70	33	5	22	145	1	N	65.6	146	3	0	N	26	5	N	74.0	205	3	6	20	59	190	2
16	5	19.5	6.1	2-2-3	51.9	74	4	63	16	5	22	140	1	10	65.8	140	3	4	20	25	6	20	73.0	165	3	0	20	73	260	3
21	5	18.0	4.9	3-4-6	55.3	67	4	67	24	5	22	132	8	Q	70.0	134	4	0	N	25	0	N	72.0	180	4	8	N	60	156	6
22	5	19.5	6.1	8-4-6	56.4	74	1	79	39	5	23	112	0	15	64.0	119	2	0	15	27	6	Q	69.0	155	3	7	20	60	96	1
23	6	21.5	8.3	12-5-11	53.6	64	1	80	39	6	23	125	0	N	68.4	120	3	1	N	32	4	N	70.5	170	4	1	N	97	167	6
24	6	21.0	7.6	6-4-14	55.1	74	1	80	27	5	23	139	1	05	67.5	162	3	0	02	27	2	20	73.0	180	3	8	22	149	148	6
25	5	21.0	7.7	12-6-14	52.2	58	4	82	38	5	23	119	0	05	65.3	126	3	0	05	31	0	20	67.0	139	3	0	18	75	118	6
26	5	20.5	7.3	7-2-11	53.7	64	1	65	14	6	23	145	1	10	68.0	158	2	5	03	24	6	N	76.0	204	2	6	N	58	112	4
31	6	19.5	6.9	5-6-19	51.7	74	3	62	12	6	23	185	0	10	65.1	184	3	0	18	31	6	Q	71.0	190	1	6	Q	70	200	6
32	5	20.5	6.4	6-6-21	52.1	63	1	64	14	5	23	120	0	20	67.2	124	2	0	Q	31	5	30	68.0	150	1	6	30	58	17	3
33	5	20.0	7.1	8-1-19	53.3	92	1	63	12	5	23	148	0	N	64.4	162	2	7	N	26	8	24	71.0	125	2	8	20	40	154	4
34	6	21.5	8.2	8-3-9	51.1	87	1	66	21	5	23	183	5	N	64.0	214	3	0	N	35	7	Q	68.0	175	1	6	N	84	100	1
35	6	21.0	9.6	4-6-17	51.1	60	1	70	15	6	23	114	0	09	64.9	114	4	0	10	28	5	05	68.0	145	4	0	N	52	120	1
36	9	19.0	5.9	6-6-17	50.8	60	8	91	43	7	23	123	0	10	62.9	130	2	0	Q	38	0	N	68.0	153	4	0	N	100	120	6
41	6	20.5	8.2	6-3-12	52.2	64	8	89	26	7	23	147	1	N	61.8	172	3	0	N	24	5	10	72.0	179	3	0	20	55	135	4
42	7	20.5	7.4	4-7-13	55.1	87	1	81	35	5	23	112	1	Q	60.7	154	2	0	N	23	2	N	73.0	205	4	3	N	66	125	1
43	6	20.3	9.1	5-1-19	57.0	80	3	78	43	6	23	129	1	Q	65.5	133	2	2	Q	25	8	25	71.0	169	2	0	Q	45	120	1
44	5	19.0	9.6	5-7-18	59.3	114	4	78	38	6	23	110	0	10	62.1	141	2	7	N	29	8	N	70.8	221	3	8	N	52	59	1
45	7	20.5	7.8	6-5-5	53.1	66	2	88	39	1	23	118	1	N	61.0	112	3	1	N	25	4	N	74.3	258	4	1	15	100	192	1
46	5	20.5	7.6	8-7-7	51.9	70	4	83	34	2	23	120	0	20	67.0	155	4	3	Q	27	5	20	70.8	138	4	1	15	36	216	6
51	6	20.5	7.4	11-7-16	50.0	54	8	84	33	5	23	114	0	07	63.4	121	3	1	10	31	5	Q	69.5	145	3	2	Q	78	221	4
52	7	19.5	6.6	10-1-23	55.6	80	1	77	44	2	23	119	3	Q	62.0	150	4	3	15	24	4	15	77.5	240	4	2	N	76	305	6
53	7	21.0	7.5	8-6-8	61.7	112	1	79	38	1	23	141	1	15	70.0	150	3	4	20	28	6	20	73.1	186	2	1	20	110	221	2
54	5	18.5	6.2	9-2-18	52.3	66	4	77	23	5	23	106	7	10	62.4	130	2	0	30	25	6	20	70.5	170	1	7	20	50	140	6
55	8	21.0	7.1	10-6-7	55.6	90	1	92	26	6	23	147	0	N	67.5	140	2	5	N	29	4	N	70.3	177	2	4	N	60	159	1
56	3	20.0	7.1	10-1-15	53.4	54	1	72	19	1	23	126	0	N	65.6	144	2	7	N	30	6	N	70.0	185	2	7	Q	50	100	1
61	4	20.0	7.6	9-5-16	52.9	75	2	91	41	7	23	185	1	15	68.6	247	3	0	20	25	0	20	68.0	161	3	0	20	53	96	6
62	5	17.0	3.1	11-5-2	52.9	67	4	82	41	1	23	110	0	N	64.3	115	3	0	N	24	0	Q	71.0	175	4	2	N	65	240	1
63	2	20.0	7.9	12-7-8	54.0	80	1	85	42	1	23	138	1	40	67.3	152	2	1	40	29	5	35	71.0	158	3	2	Q	70	190	3
64	5	19.0	5.9	4-1-8	50.6	69	4	90	27	6	23	115	1	20	60.0	115	3	4	12	25	4	30	70.1	168	3	0	20	89	224	5
65	5	20.0	7.6	3-3-24	54.4	82	4	71	33	5	23	110	7	N	61.6	140	3	7	N	23	7	24	72.0	156	3	2	20	57	201	4
66	7	21.0	7.8	6-6-5	52.4	54	1	85	21	5	23	132	0	N	65.9	166	2	0	Q	25	6	17	76.0	160	1	6	10	45	88	3

ID	B	LGTH	WGT	MO-D-HR	HGHT	WGT	L	PEA	RA	B	AG	WGT	O	SM	HGHT	WGT	E	O	SM	AG	O	SM	HGHT	WGT	E	O	SM	I-B	I-T	C
		CHILD-BIRTH				CHILD-TEST					MOTHER-BIRTH					MOTHER-TEST				FATH-BRTH				FATHER-TEST				FAMILY		
11	3	20.5	7.6	6-1-23	55.4	80	5	92	36	7	23	134	2	N	65.4	143	3	2	N	33	6	N	72.0	175	3	2	N	113	231	4
12	6	19.0	5.6	6-7-23	52.2	67	4	75	19	6	23	112	5	03	61.9	115	2	2	05	23	6	20	69.5	145	2	6	20	84	143	6
13	7	20.0	6.8	4-6-7	53.3	72	4	66	11	5	23	105	1	N	61.3	112	2	2	N	22	6	N	68.0	170	2	6	N	70	110	2
14	5	20.0	6.2	7-1-24	50.4	62	4	69	18	5	23	112	0	10	65.7	117	3	0	15	22	6	10	73.0	200	3	4	05	46	78	2
15	5	22.5	10.5	12-2-10	52.1	80	2	69	17	6	23	130	0	N	66.4	163	2	0	N	26	6	20	67.3	162	2	7	Q	50	88	4
16	5	21.0	7.7	7-5-14	50.5	52	1	83	45	5	23	143	0	N	63.9	128	3	2	N	27	0	N	65.5	130	4	0	Q	26	103	6
21	6	19.0	6.7	10-4-21	52.4	66	1	73	42	2	23	127	1	20	63.9	110	2	0	20	27	8	20	72.0	160	2	6	Q	59	140	1
22	9	21.5	9.7	1-4-10	53.4	66	4	85	44	6	23	137	1	Q	67.9	179	2	0	Q	28	6	N	71.5	160	3	0	N	110	140	6
23	6	19.0	6.8	9-2-2	50.3	64	4	64	17	5	23	124	0	N	65.0	192	2	0	N	38	7	20	69.0	162	3	6	Q	55	96	1
24	5	17.0	3.9	9-5-19	53.1	60	1	85	28	5	23	118	1	15	65.5	121	2	0	23	23	4	10	70.8	150	2	3	30	90	150	2
25	6	21.5	9.4	2-1-11	51.4	64	1	63	28	5	23	104	0	20	61.3	117	2	0	20	28	6	10	66.0	170	1	6	20	80	130	6
26	5	19.0	7.4	3-2-23	49.6	51	1	74	45	5	23	110	0	Q	63.8	124	2	5	N	26	6	Q	70.0	150	2	6	Q	70	192	3
31	6	17.5	4.1	1-7-14	50.7	69	1	91	30	8	23	104	0	20	64.3	108	3	0	20	29	8	30	71.9	175	3	8	20	68	132	1
32	7	19.8	7.1	12-1-18	52.8	66	4	64	28	7	23	127	1	06	65.0	180	1	0	Q	23	2	20	70.5	160	2	2	25	86	100	2
33	6	20.0	7.1	2-5-16	51.1	54	1	59	16	7	23	145	0	20	65.6	182	2	0	50	25	0	20	72.8	175	3	0	20	50	120	6
34	6	18.8	6.1	10-1-4	52.5	84	1	86	40	5	23	130	5	N	67.4	155	3	0	N	26	8	N	65.0	145	4	1	17	50	96	1
35	5	19.0	7.0	8-1-19	52.4	68	1	75	26	5	23	135	0	20	62.3	139	3	0	20	29	0	20	73.0	195	4	0	Q	80	260	6
36	1	21.0	7.6	11-2-2	56.3	73	1	68	50	1	24	125	0	03	66.1	133	2	0	01	33	3	20	67.0	168	3	6	40	70	102	1
41	1	19.8	6.1	11-7-4	52.7	56	1	72	23	1	24	120	1	N	63.0	144	3	0	N	30	6	N	69.0	190	2	3	N	90	300	1
42	9	21.5	8.3	10-1-13	49.8	54	1	84	36	5	24	115	0	Q	63.6	132	3	0	N	26	4	01	68.0	135	4	2	22	73	230	6
43	9	20.5	8.3	6-2-10	50.2	61	1	69	38	5	24	139	0	N	62.8	151	2	0	N	24	4	N	72.0	180	3	2	N	60	168	3
44	7	20.5	7.7	9-5-3	54.4	63	1	59	12	5	24	158	1	N	67.9	147	2	1	N	25	5	N	74.0	200	2	8	N	68	194	2
45	5	20.5	7.3	5-6-5	52.6	77	3	66	41	6	24	120	0	N	60.5	141	2	5	N	36	2	20	68.5	180	1	2	10	116	101	1
46	5	19.0	7.6	4-1-16	51.3	56	1	93	38	5	24	122	0	08	62.3	127	4	0	N	27	5	10	69.0	185	4	0	Q	92	170	1
51	6	20.5	6.9	3-7-8	55.9	91	4	85	34	6	24	123	0	N	64.9	131	2	0	N	26	5	N	72.0	175	3	0	N	72	150	1
52	9	19.5	7.2	4-5-19	59.6	96	1	100	45	5	24	118	1	10	67.9	125	2	2	15	24	6	20	77.0	230	2	3	30	78	185	4
53	1	21.0	8.6	5-6-4	56.9	75	2	97	46	6	24	126	1	N	65.5	130	2	1	N	26	0	N	69.0	150	4	3	N	76	200	4
54	7	20.8	9.6	5-5-1	55.1	82	1	74	17	4	24	130	1	N	64.0	132	2	1	N	25	6	N	69.9	177	2	6	Q	66	220	2
55	5	19.3	6.7	6-1-18	54.8	69	1	86	42	5	24	97	1	N	60.1	101	3	0	08	25	0	20	67.0	150	4	3	Q	36	280	1
56	6	22.0	9.8	7-1-11	57.1	88	8	96	39	6	24	145	0	N	67.0	157	4	4	N	30	5	14	74.0	240	4	0	Q	59	333	6
61	6	19.0	6.3	6-1-23	54.9	80	1	78	28	6	24	143	1	01	62.0	179	4	4	Q	22	8	12	65.0	135	2	8	Q	50	192	1
62	6	19.8	6.4	7-7-3	54.9	76	1	73	27	6	24	120	1	N	67.2	126	2	0	N	27	6	N	69.0	175	2	6	N	73	150	6
63	5	19.5	5.3	7-4-22	54.4	78	1	88	49	5	24	144	3	Q	67.6	147	4	0	N	26	0	10	70.5	165	4	0	N	92	140	1
64	5	20.0	6.9	8-5-17	57.4	94	8	88	48	6	24	151	0	N	61.7	170	2	0	N	31	5	06	68.7	168	2	7	08	35	96	1
65	6	21.5	9.3	5-7-8	54.1	75	4	81	43	1	24	125	0	N	64.0	146	2	3	N	29	0	24	69.0	190	4	1	Q	42	220	4
66	5	21.5	7.4	9-3-9	57.3	80	1	88	36	6	24	115	5	05	66.5	122	3	0	20	26	6	10	70.6	202	2	6		41	120	1

StatLab CENSUS: Girls

ID	CHILD-BIRTH B	LGTH	WGT	MO-D-HR	CHILD-TEST HGHT	WGT	L	PEA	RA	MOTHER-BIRTH B	AG	WGT	O	SM	MOTHER-TEST HGHT	WGT	E	O	SM	FATH-BRTH AG	O	SM	FATHER-TEST HGHT	WGT	E	O	SM	FAMILY I-B	I-T	C
11	6	21.8	9.0	2-1-18	50.0	54	4	76	34	5	24	121	0	N	62.8	118	3	0	N	31	6	N	69.0	235	2	6	N	55	110	1
12	5	18.5	5.8	5-3-12	55.0	60	1	74	23	5	24	105	5	N	63.1	122	2	5	Q	42	6	20	68.0	165	1	6	40	14	94	1
13	2	18.5	5.6	10-5-16	51.8	63	1	93	49	5	24	115	1	Q	64.5	118	4	0	15	24	2	N	70.0	165	4	2	N	100	150	1
14	5	20.0	7.3	11-3-3	55.6	94	6	70	33	5	24	115	4	N	62.6	142	4	4	N	26	2	N	73.0	240	3	2	N	107	230	4
15	5	21.0	8.3	12-1-16	53.5	82	1	79	33	5	24	102	2	Q	60.7	111	2	2	N	28	4	N	68.0	170	4	4	N	100	123	4
16	5	20.0	7.2	12-2-16	51.9	62	1	80	35	6	24	106	2	15	63.3	123	2	0	14	26	5	20	71.0	208	2	0	Q	70	140	1
21	9	19.5	6.5	10-3-2	56.8	96	1	97	40	6	24	110	4	10	64.2	110	3	4	10	35	8	15	66.5	155	2	8	Q	84	198	1
22	6	21.0	7.9	12-6-6	51.1	63	4	74	32	6	24	118	1	01	64.0	125	1	1	20	27	5	17	65.5	125	2	2	20	97	215	4
23	6	21.0	7.3	10-5-13	52.7	64	4	84	46	6	24	123	1	N	66.3	140	4	0	N	27	0	N	71.8	160	4	0	N	35	130	6
24	2	19.5	6.1	5-6-11	49.6	58	2	73	25	5	24	111	0	15	64.6	113	3	0	16	27	8	20	71.5	162	3	5	25	60	101	4
25	5	20.5	7.3	6-7-14	51.1	61	3	72	22	5	24	110	0	N	58.7	112	2	0	20	25	0	N	68.0	172	3	0	N	60	110	1
26	5	18.0	5.4	6-3-6	53.2	70	1	82	33	3	24	96	0	12	62.8	98	2	0	20	25	6	05	73.1	180	3	4	3	50	81	3
31	7	19.0	5.4	6-6-2	52.6	72	1	65	29	7	24	117	2	10	63.7	119	3	0	Q	24	0	20	69.3	151	3	0	40	54	120	4
32	5	19.0	6.3	6-7-6	49.0	54	4	76	30	5	24	98	1	10	62.0	106	2	2	20	25	6	05	69.0	174	2	6	10	87	140	1
33	5	20.5	6.4	8-2-7	52.5	63	8	64	24	5	24	122	4	20	64.3	136	2	0	20	24	0	20	72.0	200	2	8	15	40	120	1
34	4	19.0	6.7	7-1-6	52.8	60	4	94	47	6	24	130	1	N	66.6	146	4	0	N	29	0	N	70.0	180	4	1	Q	96	180	4
35	9	19.5	6.6	8-1-9	55.4	81	1	81	26	4	24	105	3	Q	60.0	115	4	3	N	25	1	10	72.4	195	4	0	Q	110	143	4
36	5	21.5	7.9	10-7-22	54.3	58	1	73	45	5	24	120	0	12	67.1	133	3	0	N	28	6	Q	71.5	156	2	0	Q	49	138	1
41	5	19.0	5.9	11-4-8	51.9	72	4	85	40	5	24	120	1	20	65.2	139	3	0	Q	25	0	Q	74.0	175	4	3	Q	54	300	4
42	5	20.5	7.3	1-4-18	53.3	59	1	64	22	5	24	128	0	10	67.0	122	2	7	15	24	8	24	68.1	178	3	6	07	17	135	4
43	2	19.0	6.8	12-4-18	52.0	71	1	72	31	5	24	142	0	20	65.3	159	2	0	20	24	6	30	71.8	187	1	7	22	38	96	1
44	9	20.0	6.8	1-7-23	55.9	88	2	90	45	3	24	127	1	06	68.4	125	3	0	10	27	0	20	73.0	188	2	3	10	67	120	6
45	2	21.0	7.4	2-5-11	50.1	49	2	90	37	2	24	140	1	Q	64.6	180	4	0	N	24	0	N	67.0	165	3	0	N	50	170	1
46	5	19.0	5.3	2-2-4	50.6	52	1	83	43	5	24	148	8	10	54.8	139	4	4	3	29	5	15	72.0	160	4	2	6	76	195	6
51	5	20.0	6.9	3-1-20	52.6	67	8	82	18	5	24	107	1	N	61.9	104	2	0	N	28	6	N	71.0	155	3	0	N	70	120	6
52	2	14.8	2.3	11-11-16	50.4	54	7	69	16	6	24	155	7	20	67.9	152	2	0	15	28	5	20	71.8	176	3	0	Q	80	102	4
53	1	21.0	7.5	6-6-4	57.3	85	2	82	26	5	25	145	0	Q	64.6	143	1	1	Q	31	6	15	74.5	175	3	3	N	54	220	2
54	5	20.5	7.8	8-1-2	49.9	55	2	80	35	5	25	176	0	N	64.3	170	2	5	N	27	4	20	64.0	200	2	4	N	85	61	1
55	6	19.5	5.9	7-1-15	57.1	69	1	81	47	6	25	131	0	20	68.8	123	3	0	20	33	0	N	73.5	190	4	6	N	46	130	6
56	5	20.0	6.3	3-4-14	53.5	74	4	76	20	6	25	109	1	20	63.8	110	2	0	30	32	2	N	72.0	190	3	1	N	80	61	3
61	6	19.5	5.6	1-6-13	51.3	61	4	59	32	7	25	110	1	N	64.4	125	2	0	N	24	6	20	72.3	147	2	6	20	79	114	6
62	6	20.5	7.3	8-5-3	54.7	80	2	83	43	6	25	145	8	02	67.0	200	1	0	08	26	0	06	69.0	175	4	0	Q	64	220	1
63	5	21.5	8.3	6-6-7	47.9	55	1	79	38	5	25	130	0	10	67.2	144	2	1	Q	33	0	Q	65.0	149	4	3	Q	100	140	4
64	6	20.0	8.3	5-5-18	54.3	72	4	91	39	6	25	125	0	05	65.3	127	2	1	05	31	2	22	69.0	165	3	2	Q	84	204	6
65	6	20.0	8.7	11-3-3	56.1	98	1	64	26	6	25	154	0	N	69.3	154	3	0	N	25	8	02	75.0	165	4	1	30	23	120	4
66	5	20.0	6.8	3-4-5	49.6	58	4	77	33	6	25	140	0	N	64.1	156	3	2	N	26	8	20	69.5	190	3	7	N	48	136	1

StatLab CENSUS: Girls

ID	CHILD-BIRTH B	LGTH	WGT	MO-D-HR	CHILD-TEST HGHT	WGT	L	PEA	RA	MOTHER-BIRTH B	AG	WGT	O	SM	MOTHER-TEST HGHT	WGT	E	O	SM	FATH-BRTH AG	O	SM	FATHER-TEST HGHT	WGT	E	O	SM	FAMILY I-B	I-T	C
11	6	20.5	6.9	4-5-18	52.4	68	1	69	15	6	25	135	0	10	63.8	148	2	0	Q	27	7	08	68.0	165	2	6	15	30	60	1
12	7	20.5	9.4	5-5-4	53.5	74	2	77	37	7	25	138	0	23	63.4	132	3	0	N	26	6	12	75.0	215	3	2	N	63	130	6
13	8	20.0	8.0	5-6-2	53.7	72	4	80	39	6	25	123	0	23	68.7	139	4	3	20	27	6	12	71.0	170	2	6	10	42	160	1
14	6	22.0	8.4	4-7-7	53.5	65	1	60	40	6	25	127	0	Q	65.5	133	4	0	20	27	0	25	71.0	169	2	2	N	66	125	1
15	5	18.5	5.8	6-6-22	54.6	65	4	63	33	6	25	110	0	N	66.0	129	3	0	N	26	6	N	65.9	145	2	6	N	68	78	3
16	6	20.5	7.5	7-6-23	55.5	128	1	60	20	6	25	180	0	5	66.0	165	3	3	N	28	8	20	66.0	190	2	8	20	41	108	2
21	6	20.0	6.9	4-1-12	55.5	69	1	81	12	5	25	135	0	12	67.1	138	1	0	10	40	0	16	70.5	175	1	6	Q	56	99	6
22	7	20.0	6.2	7-1-7	52.3	59	7	82	45	5	25	106	0	0	62.3	117	3	0	N	32	0	0	71.0	135	4	0	N	6	240	6
23	6	20.0	6.8	5-1-6	52.3	58	1	69	31	5	25	172	0	20	62.9	149	1	7	25	29	7	10	71.5	175	2	7	Q	64	173	4
24	6	21.0	8.2	8-2-8	51.5	68	1	83	48	5	25	120	0	10	66.5	122	3	0	20	27	6	20	70.6	202	2	6	Q	57	120	1
25	5	20.5	7.3	8-4-19	56.5	70	4	73	8	5	25	127	0	20	65.5	141	2	1	20	30	5	N	68.5	180	3	8	N	74	152	4
26	8	19.0	6.8	5-7-12	56.3	68	3	84	39	7	25	112	0	30	66.8	131	2	0	Q	27	0	30	72.0	168	3	0	Q	84	45	1
31	6	19.5	6.6	1-3-6	51.9	65	4	83	36	2	25	118	0	Q	66.3	153	2	1	Q	26	6	03	72.0	205	2	6	15	50	219	6
32	1	19.5	6.3	10-3-16	56.3	68	1	70	22	7	25	98	1	N	65.6	98	2	0	N	35	6	20	70.1	166	1	6	15	84	104	3
33	6	21.5	9.4	11-6-17	57.6	85	4	74	44	5	25	142	0	N	65.9	139	3	0	N	26	6	N	68.0	156	4	0	N	60	78	1
34	6	21.0	7.3	12-1-7	54.4	86	4	59	33	5	25	124	4	N	60.6	180	3	3	N	27	6	20	68.0	192	2	7	30	65	155	1
35	3	20.0	7.5	11-7-9	49.8	58	1	74	29	5	25	110	0	08	64.2	110	3	4	10	36	8	10	66.5	155	2	8	Q	50	198	1
36	5	22.0	10.5	3-7-12	56.0	86	1	86	38	5	25	178	0	N	66.6	173	2	0	N	26	4	25	70.0	170	2	4	05	77	150	1
41	7	21.0	8.1	5-3-6	53.5	64	1	68	26	8	25	129	0	N	60.8	133	1	0	N	30	6	24	70.5	165	3	8	10	77	120	4
42	5	20.0	6.6	6-1-21	46.2	52	1	78	47	7	25	145	0	20	62.7	130	3	4	20	30	0	30	67.0	140	4	0	50	100	275	6
43	5	20.0	7.3	5-7-3	49.6	59	4	87	50	5	25	115	3	04	62.2	124	4	0	01	25	0	20	68.0	160	4	3	N	96	300	4
44	6	20.5	6.8	7-2-10	57.3	80	1	76	19	5	25	114	1	N	66.4	142	4	3	N	31	6	N	71.3	177	2	6	N	113	180	1
45	2	22.0	9.4	12-6-13	52.9	60	1	86	41	5	25	127	0	Q	64.3	145	3	4	N	31	1	24	72.0	175	4	1	N	58	150	1
46	7	19.0	5.1	6-5-7	55.3	77	1	62	33	7	25	121	0	N	63.6	124	2	0	N	26	4	25	73.0	175	2	6	N	70	104	1
51	6	20.5	7.4	7-4-19	52.8	70	1	77	39	6	25	115	4	N	63.8	128	4	0	N	26	0	02	69.0	160	4	0	N	65	200	6
52	5	19.5	6.8	9-1-4	49.0	50	1	65	17	5	25	100	5	10	64.3	123	2	0	Q	27	2	20	69.3	143	0	6	30	29	68	1
53	5	18.5	5.9	7-5-18	50.3	49	4	62	9	5	25	127	5	N	65.3	156	3	0	N	28	0	N	68.0	150	2	6	N	74	90	1
54	6	20.5	7.5	7-1-12	56.4	71	4	87	41	6	25	130	1	N	67.8	142	4	0	N	27	0	N	73.0	170	4	0	N	135	200	6
55	5	19.0	6.0	8-5-13	50.4	62	1	93	30	5	25	119	1	10	61.8	123	2	1	18	29	2	20	71.0	191	3	6	30	101	180	6
56	5	19.5	6.4	8-3-14	49.9	57	2	86	42	6	25	103	5	N	60.8	108	4	0	N	30	2	20	71.0	157	3	2	30	58	132	6
61	5	20.0	7.6	5-2-23	50.2	63	1	63	34	5	25	128	0	N	63.0	123	3	0	N	31	8	20	75.5	195	4	0	N	68	168	1
62	5	18.5	5.6	8-6-19	53.6	80	1	58	12	5	25	107	1	20	65.3	107	3	1	20	29	5	20	76.0	200	3	2	30	105	142	3
63	5	21.0	7.4	10-4-5	52.9	66	4	81	24	2	25	135	4	15	65.4	129	4	4	20	26	7	20	69.0	180	3	7	30	60	138	1
64	6	20.0	7.8	6-6-3	52.9	64	1	74	31	5	25	130	0	20	64.2	153	2	0	Q	26	0	10	72.5	193	3	2	Q	35	130	1
65	5	18.8	6.2	11-4-4	52.8	58	4	68	24	5	25	142	1	08	67.0	145	3	0	Q	24	8	20	68.5	190	3	5	Q	76	84	1
66	5	19.5	6.4	1-1-6	52.4	59	1	68	27	5	25	115	8	N	62.2	119	2	1	N	27	7	N	66.5	145	2	2	N	72	206	3

StatLab CENSUS: Girls

	CHILD-BIRTH				CHILD-TEST					MOTHER-BIRTH					MOTHER-TEST					FATH-BIRTH			FATHER-TEST					FAMILY		
ID	B	LGTH	WGT	MO-D-HR	HGHT	WGT	L	PEA	RA	B	AG	WGT	Q	SM	HGHT	WGT	E	Q	SM	AG	Q	SM	HGHT	WGT	E	Q	SM	I-B	I-T	C
11	5	21.5	7.3	4-2-10	50.9	57	1	84	39	5	25	106	0	Q	62.0	105	3	0	0	34	1	Q	69.0	155	4	1	Q	60	85	6
12	9	20.5	6.9	1-3-1	52.9	53	1	82	31	6	25	110	0	N	63.7	118	3	0	N	27	1	N	74.4	284	4	1	N	72	180	1
13	7	20.0	6.2	4-2-16	52.2	70	1	83	21	7	25	120	0	N	64.9	127	3	0	N	25	8	N	69.0	170	3	6	N	39	106	1
14	6	19.5	6.7	3-3-2	51.9	74	4	71	25	5	25	113	0	20	58.4	111	1	0	0	33	6	04	67.0	135	3	0	N	50	120	4
15	7	20.0	7.0	9-6-14	49.6	50	4	56	12	7	26	148	0	09	66.1	169	3	0	Q	27	6	13	75.0	235	2	4	Q	52	120	1
16	7	19.0	5.7	9-2-16	52.1	58	1	66	41	5	26	110	0	05	61.8	106	3	0	10	32	0	10	70.0	135	4	0	20	83	275	6
21	6	19.5	7.0	9-5-1	53.5	68	4	66	44	2	26	140	0	Q	66.0	165	2	1	N	32	6	13	70.8	171	2	6	23	91	170	1
22	1	19.5	7.4	6-5-24	54.3	79	4	76	37	5	26	138	1	N	67.8	151	2	0	N	28	6	20	73.0	198	2	6	N	108	110	4
23	5	21.0	8.9	10-7-10	55.7	91	4	64	31	6	26	142	0	Q	66.3	160	4	3	N	26	5	20	75.0	220	3	2	N	49	200	1
24	7	21.0	7.4	11-2-8	48.8	56	1	71	26	5	26	127	0	18	63.7	150	2	0	N	27	6	N	67.0	155	3	6	Q	70	144	6
25	7	20.0	7.2	11-2-14	52.1	62	4	77	36	7	26	130	0	18	65.9	143	3	4	N	32	0	10	72.0	190	4	0	N	62	216	4
26	9	20.5	7.1	7-7-19	52.4	68	1	98	47	5	26	131	5	01	67.8	137	3	0	Q	29	6	Q	70.0	175	3	6	Q	84	150	1
31	7	21.5	8.1	11-6-12	49.4	60	4	64	33	5	26	140	8	N	63.4	141	2	0	N	38	6	20	66.0	145	0	6	Q	126	140	1
32	6	20.5	6.7	9-3-8	47.6	58	1	78	35	6	26	112	0	N	57.3	116	4	0	N	36	1	20	64.0	155	4	1	Q	72	180	1
33	5	21.8	7.8	9-2-13	54.0	72	4	81	41	6	26	150	0	Q	67.9	140	4	3	N	32	0	N	71.0	175	4	0	Q	80	200	1
34	5	19.0	5.6	4-5-13	58.1	110	4	80	30	8	26	114	0	N	63.2	115	2	1	N	31	6	N	67.5	180	2	0	N	42	142	1
35	1	18.5	5.4	5-6-10	51.4	68	4	79	35	1	26	101	4	12	60.3	100	4	0	10	28	0	10	73.5	200	4	1	Q	56	165	6
36	5	21.5	8.3	6-2-9	55.1	74	1	98	41	6	26	121	0	10	64.9	130	4	0	Q	26	1	N	73.0	185	4	1	N	31	100	6
41	7	20.0	6.7	4-2-5	53.2	108	1	73	23	7	26	137	0	Q	63.6	219	2	0	N	30	7	30	70.0	236	1	7	Q	65	96	1
42	5	20.5	7.3	6-3-19	57.0	82	1	93	50	5	26	172	3	N	68.7	159	4	3	N	39	0	20	74.0	180	3	2	Q	170	310	3
43	6	22.0	8.4	6-7-7	55.1	75	2	89	29	6	26	145	3	Q	65.2	145	4	0	N	26	0	N	69.0	185	4	0	N	105	300	1
44	5	21.5	10.4	4-7-16	57.4	114	1	80	32	5	26	216	0	12	66.6	239	2	0	Q	31	4	N	74.5	250	2	2	N	52	125	5
45	1	20.0	6.9	7-1-13	55.7	82	1	121	48	6	26	138	0	0	64.3	152	3	0	01	30	0	N	72.0	168	4	0	N	48	200	6
46	5	20.0	8.3	7-4-16	57.8	132	3	99	34	6	26	135	0	20	62.8	140	1	2	18	38	5	N	69.4	191	2	3	N	80	107	1
51	5	21.0	6.9	7-2-14	54.4	78	7	82	44	6	26	110	0	N	62.8	125	4	0	N	31	1	N	68.0	170	4	1	N	80	170	1
52	5	21.5	8.9	8-5-14	53.4	74	1	83	30	5	26	112	0	0	67.4	121	2	0	N	30	6	N	69.0	172	2	6	Q	64	127	1
53	5	18.0	4.6	8-4-1	51.4	65	1	71	14	6	26	131	0	20	65.4	145	2	0	20	29	1	N	67.0	150	4	1	N	90	160	6
54	6	20.5	6.6	9-4-7	52.6	72	6	82	35	5	26	113	1	20	65.0	122	3	0	Q	29	4	Q	70.0	165	3	4	Q	53	87	1
55	7	21.0	8.8	10-5-16	58.3	81	1	77	29	5	26	124	0	N	61.9	136	2	1	N	26	5	N	72.4	160	2	5	Q	30	136	4
56	6	21.0	6.6	12-1-15	52.9	62	1	86	44	6	26	134	0	Q	69.2	160	4	3	N	29	2	20	70.0	180	4	2	30	55	242	1
61	5	20.5	7.3	12-5-12	51.7	82	4	63	34	5	26	130	5	N	60.6	180	3	3	N	28	6	20	68.0	192	2	7	30	105	155	1
62	2	20.0	7.3	3-7-22	55.3	94	4	87	40	2	26	127	1	Q	65.3	165	4	1	N	26	2	N	73.5	240	3	2	N	98	224	6
63	6	20.0	7.1	11-3-3	54.8	73	1	69	32	5	26	126	0	10	66.4	148	2	0	20	29	6	20	74.0	245	2	2	10	94	146	1
64	5	18.0	4.8	12-6-8	54.3	62	4	89	41	5	26	103	0	20	64.7	170	4	2	N	29	0	20	72.0	170	4	2	20	94	350	1
65	7	18.5	5.1	12-4-10	54.1	78	1	73	21	5	26	107	2	10	63.3	115	3	2	N	30	7	20	73.0	200	2	7	22	56	130	2
66	7	20.0	8.7	4-7-6	50.2	51	1	75	31	5	26	125	0	06	62.9	137	2	5	10	32	5	20	72.0	175	2	5	20	54	167	2

	CHILD-BIRTH				CHILD-TEST					MOTHER-BIRTH					MOTHER-TEST					FATH-BRTH			FATHER-TEST					FAMILY		
ID	B	LGTH	WGT	MO-D-HR	HGHT	WGT	L	PEA	RA	B	AG	WGT	O	SM	HGHT	WGT	E	O	SM	AG	O	SM	HGHT	WGT	E	O	SM	I-B	I-T	C
11	9	20.5	7.6	4-1-21	50.9	80	1	76	25	5	26	134	0	15	65.2	149	3	0	Q	29	0	Q	70.0	200	4	0	Q	74	185	6
12	5	19.5	6.9	5-1-3	49.6	47	4	69	27	1	26	118	5	07	71.0	126	1	0	20	26	5	20	70.1	123	1	8	10	48	90	6
13	5	21.0	7.4	5-2-3	48.7	53	1	73	28	1	26	112	0	Q	63.9	123	1	0	Q	27	0	20	72.0	227	2	1	Q	72	150	4
14	5	20.0	7.6	4-4-20	52.4	67	1	83	32	5	26	122	7	15	63.9	131	2	0	20	34	7	10	70.0	160	3	6	Q	90	240	6
15	6	21.5	6.5	7-3-19	54.2	68	6	81	39	8	26	109	0	0	69.3	123	4	0	N	30	4	N	73.0	173	4	2	N	60	120	1
16	8	19.0	6.7	6-7-17	51.1	57	1	86	39	7	26	121	3	03	63.9	124	4	3	Q	28	1	Q	68.0	157	4	1	N	68	156	6
21	5	21.0	8.6	6-6-9	53.2	90	1	67	11	5	26	110	0	N	57.4	119	2	0	N	28	6	30	69.0	197	3	3	20	80	83	4
22	7	20.5	6.8	8-4-13	49.3	52	1	74	35	3	26	145	1	N	63.0	135	3	1	N	28	6	20	67.8	176	3	5	20	95	195	4
23	9	19.0	5.4	8-5-23	55.2	78	2	73	32	6	26	138	1	12	68.1	147	3	1	Q	27	6	20	71.0	205	2	6	20	102	164	1
24	6	20.0	6.6	9-7-14	51.9	70	4	79	43	6	26	140	0	07	68.0	159	3	0	08	31	1	N	72.0	215	4	1	N	80	145	1
25	9	17.5	5.6	11-3-12	49.8	57	2	81	33	9	26	120	4	12	66.9	147	3	0	Q	26	6	15	64.0	130	2	6	10	96	100	1
26	5	21.0	7.8	1-5-24	47.4	48	2	66	30	5	26	123	0	N	61.0	151	3	0	N	26	0	N	70.0	145	2	2	20	36	125	1
31	9	20.5	7.6	1-1-23	54.1	60	4	65	36	1	26	108	1	N	62.8	113	2	0	N	26	1	N	71.0	173	4	1	N	102	182	1
32	5	22.0	8.1	3-7-8	51.1	60	8	92	37	5	26	130	3	N	65.5	142	4	0	N	30	0	01	73.0	190	4	1	Q	114	120	4
33	5	20.5	6.9	1-3-7	51.9	56	4	60	21	5	26	142	0	N	62.0	175	2	0	N	36	6	20	67.0	165	2	7	20	61	84	1
34	5	19.5	6.4	4-5-22	53.2	84	1	68	38	5	26	121	1	10	65.0	120	3	0	20	33	6	10	70.3	188	3	6	Q	91	120	6
35	8	21.5	8.4	2-5-18	50.0	63	1	63	28	8	26	125	0	20	62.8	154	3	0	N	32	5	20	74.0	190	2	5	Q	50	100	3
36	5	20.0	7.1	12-6-18	53.6	68	1	77	42	5	27	140	0	20	66.5	137	4	0	20	33	0	20	70.0	148	4	0	20	75	200	2
41	1	20.0	8.1	4-2-7	56.5	76	1	68	24	5	27	110	0	04	65.3	113	2	5	05	30	7	10	74.0	195	1	7	10	41	44	2
42	5	20.5	7.4	12-1-23	58.5	93	8	73	27	5	27	165	0	35	66.3	184	3	8	40	36	4	40	73.0	200	3	0	N	58	103	3
43	8	20.0	7.1	2-2-3	50.4	56	1	76	43	8	27	110	0	0	65.1	114	4	0	Q	35	5	N	69.5	165	2	5	20	46	115	1
44	3	20.5	6.6	9-1-6	54.1	64	5	62	26	6	27	132	4	N	68.0	140	3	2	N	29	5	20	68.8	155	4	6	N	83	140	1
45	6	20.8	8.0	10-7-14	50.2	54	1	68	24	6	27	160	0	N	65.0	195	2	0	10	34	6	Q	70.0	165	1	6	Q	120	150	4
46	1	21.0	7.1	8-4-21	53.9	64	1	80	39	2	27	115	0	20	66.8	124	4	0	10	27	0	N	74.0	220	4	6	N	56	200	1
51	6	21.0	10.0	4-7-8	53.8	80	8	66	26	6	27	130	1	N	64.6	143	2	0	N	34	6	35	71.0	180	2	6	Q	48	140	4
52	5	20.5	7.6	5-4-14	56.0	84	4	75	36	5	27	137	0	10	62.6	146	3	8	20	35	0	10	68.0	160	4	0	15	54	137	2
53	6	21.0	7.6	4-3-13	55.3	78	4	73	37	5	27	122	2	13	63.0	122	1	0	20	30	0	13	71.0	200	4	0	N	82	160	2
54	9	20.5	7.3	6-2-21	51.9	66	1	79	30	6	27	111	0	01	62.9	128	3	0	20	31	8	25	69.8	165	2	3	70	56	141	1
55	6	20.0	5.8	5-5-7	51.7	54	1	75	37	5	27	119	0	N	62.6	157	4	0	N	27	0	N	69.0	150	3	0	02	19	84	4
56	6	20.5	7.7	7-6-16	49.1	46	2	85	46	2	27	109	4	Q	60.2	120	1	0	10	25	5	10	68.0	130	2	3	N	75	224	1
61	5	20.0	6.6	4-6-6	53.1	74	1	60	28	7	27	156	0	N	66.3	188	2	5	N	35	6	20	67.5	152	0	6	30	34	75	4
62	6	22.5	7.8	5-5-12	56.7	68	4	83	41	6	27	150	0	N	68.6	157	0	0	N	31	6	Q	77.0	230	2	6	N	51	74	6
63	9	20.0	7.2	7-4-4	55.4	94	1	81	41	5	27	145	0	03	65.9	160	4	0	N	27	0	01	72.0	205	4	0	N	57	200	4
64	5	22.0	9.0	8-5-20	57.6	98	1	64	28	6	27	126	0	Q	63.6	166	1	8	0	30	6	N	67.0	185	2	6	N	70	98	2
65	9	21.0	7.4	8-2-6	58.8	114	1	82	36	5	27	122	0	N	64.3	121	4	0	0	28	1	N	68.5	150	4	1	N	46	180	1
66	5	21.0	7.1	8-3-13	51.1	64	4	83	31	7	27	154	1	15	66.8	181	3	0	Q	37	0	20	70.0	175	4	2		91	120	1

ID	CHILD-BIRTH B	LGTH	WGT	MO-D-HR	CHILD-TEST HGHT	WGT	L	PEA	RA	MOTHER-BIRTH B	AG	WGT	O	SM	MOTHER-TEST HGHT	WGT	E	O	SM	FATH-BRTH AG	O	SM	FATHER-TEST HGHT	WGT	E	O	SM	FAMILY I-B	I-T	C
11	7	21.0	7.8	9-6-24	54.9	72	5	85	32	6	27	132	1	08	70.0	140	2	0	10	27	2	40	72.0	225	3	4	25	120	127	4
12	6	21.0	8.1	9-7-12	56.0	82	1	89	45	6	27	130	1	N	68.6	129	4	0	N	26	0	Q	69.0	155	4	1	Q	45	150	6
13	5	19.5	6.8	11-4-22	54.3	70	1	93	49	6	27	127	1	N	62.1	158	4	0	N	26	5	06	65.5	150	4	2	Q	50	185	1
14	6	18.8	5.9	11-1-4	51.6	52	1	81	43	5	27	116	1	Q	64.6	117	3	0	Q	34	2	30	71.0	185	3	3	40	130	240	1
15	1	18.5	5.7	3-5-6	53.6	72	4	85	47	2	27	125	0	N	62.6	145	4	0	N	27	0	02	71.0	165	4	3	Q	36	70	6
16	9	20.0	7.2	3-2-4	56.6	96	4	114	43	6	27	107	0	20	65.2	113	4	4	10	28	0	N	70.0	170	4	1	N	40	240	6
21	5	22.0	10.0	6-6-13	56.1	97	4	78	33	5	27	162	0	15	65.7	177	2	0	Q	31	6	N	70.0	180	2	0	N	95	150	2
22	6	22.0	9.8	1-1-11	53.1	57	1	84	52	6	27	125	0	N	64.3	145	4	0	N	28	1	Q	72.0	175	4	1	N	62	150	1
23	2	21.0	8.9	7-2-12	54.3	62	1	63	37	8	27	125	0	N	66.3	141	3	1	N	28	0	20	72.0	180	3	3	20	65	250	1
24	7	18.0	5.9	7-3-20	50.5	56	3	81	28	5	27	120	0	N	63.8	125	3	1	Q	33	1	N	69.0	185	4	1	N	50	238	1
25	2	20.0	7.5	8-3-1	53.4	78	1	92	42	6	27	185	0	Q	66.7	163	4	4	N	37	0	N	73.0	184	4	0	N	34	70	6
26	6	20.5	7.2	9-7-8	50.3	62	8	75	32	6	27	138	0	20	64.9	153	3	0	10	28	0	N	71.5	185	4	1	N	78	220	4
31	5	21.8	7.3	10-6-10	50.8	52	1	71	39	5	27	96	3	N	60.0	100	4	0	N	30	0	Q	67.1	144	4	0	Q	115	150	6
32	2	21.0	6.8	7-3-23	57.9	114	1	81	39	2	27	134	1	20	64.6	154	4	3	10	34	5	N	71.0	189	2	8	N	79	224	1
33	2	20.0	7.2	12-3-1	52.4	68	2	67	41	7	27	110	1	03	61.8	123	2	1	03	32	6	03	68.0	190	3	6	06	100	151	4
34	6	19.0	6.3	6-7-8	53.8	68	4	76	45	6	27	131	1	24	67.1	137	4	4	20	27	2	20	73.0	170	4	2	15	92	234	4
35	5	21.0	7.5	1-2-11	50.5	63	4	72	25	7	27	102	1	12	52.9	102	3	4	N	31	0	N	66.0	150	3	0	Q	54	128	6
36	6	21.0	7.8	2-3-15	52.3	66	1	85	38	6	27	111	3	Q	64.1	115	4	3	N	28	1	30	68.0	160	4	1	Q	103	290	1
41	5	19.5	5.8	2-3-21	56.1	93	2	64	35	1	27	130	1	Q	66.2	155	2	4	20	32	6	15	73.0	165	3	0	Q	103	208	3
42	7	21.0	7.1	12-5-11	53.1	77	1	85	35	5	28	125	0	20	64.0	151	3	1	20	29	3	10	72.0	175	3	0	15	80	210	1
43	8	19.5	6.9	12-2-8	53.6	68	5	76	29	2	28	105	1	N	60.8	100	1	2	N	31	6	03	66.4	153	3	3	Q	70	172	3
44	6	20.0	7.0	4-2-22	47.6	49	5	80	46	3	28	115	3	N	59.9	117	0	7	N	37	5	N	67.0	160	3	3	Q	114	125	1
45	6	22.5	9.3	11-5-9	52.5	54	4	81	22	6	28	125	0	N	63.6	144	2	2	N	29	6	10	72.0	185	3	6	Q	68	143	1
46	7	22.0	9.1	5-3-8	56.3	75	1	76	44	8	28	189	3	30	71.1	189	3	0	40	37	6	N	73.8	201	2	3	N	60	200	1
51	6	21.5	7.5	9-7-13	60.0	100	8	79	42	5	28	130	1	01	65.5	137	8	0	02	29	6	Q	72.6	193	2	0	Q	100	97	1
52	1	20.8	7.4	8-2-15	52.4	55	4	58	25	5	28	115	0	N	62.6	124	0	0	N	28	6	Q	67.0	160	0	7	Q	57	62	1
53	8	20.0	7.6	12-4-11	52.3	59	1	82	14	7	28	128	0	N	61.9	158	2	0	N	27	6	20	71.0	180	1	6	30	75	110	3
54	6	19.0	6.9	5-6-14	52.1	60	1	76	29	6	28	147	4	Q	65.3	183	2	0	Q	30	6	N	66.5	185	2	6	N	41	96	1
55	5	20.5	7.6	6-7-3	54.3	65	1	85	47	1	28	127	4	N	64.3	148	3	4	N	33	6	N	67.5	175	3	6	N	66	241	3
56	5	21.5	7.9	7-3-7	56.8	72	1	65	33	5	28	128	1	N	61.0	135	3	0	Q	29	0	N	68.7	207	3	0	N	50	280	6
61	5	19.0	5.6	8-5-21	55.3	68	4	79	38	5	28	129	3	15	63.4	139	4	0	Q	28	0	05	68.0	150	4	3	04	16	300	3
62	6	20.0	8.3	10-2-13	56.9	98	4	63	21	8	28	189	0	N	71.0	200	3	0	N	38	6	13	74.0	198	3	6	Q	65	84	1
63	5	21.0	9.7	10-7-3	57.0	78	4	68	41	5	28	140	0	Q	67.5	175	1	0	07	24	5	20	71.0	180	2	0	N	31	72	6
64	1	20.0	7.4	3-6-3	52.9	57	1	82	45	6	28	202	0	N	63.5	175	3	2	N	39	1	Q	72.0	155	3	0	Q	65	174	1
65	5	16.5	3.1	11-4-8	55.6	76	1	76	21	5	28	115	0	20	63.5	139	3	0	N	34	1	10	70.0	175	4	1	N	84	144	4
66	6	21.0	7.9	3-3-6	54.2	77	1	81	45	5	28	126	0	Q	66.4	139	3	0	Q	28	0	20	70.5	180	4	0	N	70	144	1

	CHILD-BIRTH				CHILD-TEST					MOTHER-BIRTH					MOTHER-TEST					FATH-BRTH			FATHER-TEST					FAMILY		
ID	B	LGTH	WGT	MO-D-HR	HGHT	WGT	L	PEA	RA	B	AG	WGT	O	SM	HGHT	WGT	E	O	SM	AG	O	SM	HGHT	WGT	E	O	SM	I-B	I-T	C
11	2	21.0	8.3	5-1-21	54.6	82	1	78	35	8	28	195	7	N	65.4	237	2	0	N	43	8	Q	70.0	210	2	8	Q	33	100	1
12	7	20.5	7.6	5-3-24	57.4	69	7	66	37	7	28	135	1	N	66.8	150	2	0	N	31	8	N	76.5	220	3	8	N	91	120	1
13	2	20.0	7.3	6-4-5	49.1	66	1	66	13	2	28	122	0	Q	63.9	145	1	1	Q	32	7	04	70.0	208	1	6	20	26	131	6
14	5	21.5	9.4	7-1-22	56.1	127	4	64	22	5	28	143	7	N	64.0	150	2	1	N	34	6	24	68.6	180	2	5	10	54	145	1
15	6	20.5	6.4	8-5-13	53.7	56	1	74	28	6	28	125	4	20	65.3	138	3	0	20	30	3	Q	72.0	190	2	3	Q	135	270	1
16	5	21.0	7.8	11-5-19	53.4	60	1	83	34	5	28	117	0	20	65.9	141	4	3	N	30	1	N	68.0	205	4	1	N	59	175	2
21	1	20.3	6.9	9-3-8	52.6	59	5	82	48	2	28	115	0	N	64.8	120	4	0	N	28	0	N	72.0	160	4	1	N	75	150	4
22	5	20.5	6.4	1-5-8	51.1	59	1	86	36	5	28	99	0	N	62.9	127	2	0	N	35	2	Q	63.5	134	3	0	N	50	90	4
23	5	22.0	8.6	3-1-14	52.3	93	1	77	33	6	28	115	0	N	62.1	141	3	2	N	28	7	N	66.5	210	2	6	N	80	112	4
24	7	21.0	8.3	12-7-5	54.1	77	8	83	40	7	28	122	0	N	61.3	152	2	0	N	38	0	20	70.5	210	4	0	N	61	120	2
25	7	18.5	6.4	2-4-10	52.8	86	1	84	40	7	28	107	4	20	62.2	119	3	4	20	33	6	15	66.0	145	2	2	10	125	208	4
26	5	20.0	6.6	3-5-5	51.7	57	1	77	40	5	28	110	0	15	61.8	88	3	1	15	28	6	20	69.0	160	2	2	20	55	151	3
31	5	18.0	5.1	4-7-9	50.3	60	3	61	43	5	28	125	0	Q	65.1	138	3	0	Q	30	6	20	70.0	183	3	6	40	75	190	1
32	7	21.3	7.3	4-4-16	49.5	45	1	86	47	5	29	105	0	N	63.0	110	4	4	N	33	0	N	65.1	115	4	0	N	40	315	6
33	5	20.0	7.1	8-1-21	56.1	78	1	88	44	5	29	147	0	07	68.7	152	4	4	06	32	0	Q	75.0	185	4	0	05	70	154	6
34	6	20.5	8.0	5-5-19	56.4	88	1	86	36	6	29	135	0	N	66.7	133	4	0	N	30	1	02	71.0	175	4	1	N	70	150	6
35	5	21.5	9.6	4-5-19	48.7	57	1	84	42	5	29	128	0	N	63.1	148	3	0	N	31	0	N	69.0	162	4	0	N	90	270	1
36	6	19.5	5.9	11-7-6	54.6	66	4	80	24	5	29	157	0	36	67.1	182	3	0	30	41	5	20	75.0	200	2	0	10	72	173	1
41	1	19.0	5.4	11-2-13	52.6	65	5	83	45	7	29	105	4	20	65.6	118	3	4	20	30	6	20	72.5	195	2	6	20	120	144	1
42	5	19.0	5.1	4-7-7	51.3	51	8	76	38	5	29	115	0	N	66.1	126	2	0	N	37	6	20	69.5	168	2	6	40	40	96	1
43	6	19.3	5.0	5-2-3	51.4	72	1	62	15	5	29	140	0	10	63.0	157	2	0	10	32	8	20	71.5	160	3	8	30	50	123	1
44	5	20.0	7.3	7-7-5	55.4	62	4	76	45	7	29	145	0	N	65.1	147	4	4	N	40	0	N	67.0	135	4	0	N	60	168	1
45	6	19.0	6.0	4-6-3	50.9	50	1	83	35	2	29	117	0	N	66.7	113	3	0	N	34	0	N	69.0	150	4	3	Q	52	200	6
46	3	19.0	6.6	6-5-9	55.6	72	2	76	31	3	29	108	0	05	60.3	106	3	0	30	31	4	N	71.0	152	2	3	N	80	160	3
51	4	21.0	8.3	7-2-15	54.4	82	5	86	36	6	29	153	0	Q	67.8	165	4	5	04	38	1	20	68.0	165	4	1	25	64	216	6
52	5	21.0	7.9	7-4-9	57.8	76	4	64	16	5	29	137	0	20	64.4	179	2	0	15	38	7	40	72.0	164	1	6	25	36	42	1
53	8	20.0	6.8	8-5-10	51.4	75	4	78	31	7	29	108	1	N	56.6	147	2	0	N	37	0	20	67.9	155	2	0	Q	110	38	6
54	6	21.0	7.8	9-1-10	58.3	73	1	92	32	7	29	135	3	N	67.0	139	4	0	N	38	0	25	73.0	260	4	0	Q	100	130	6
55	5	19.0	7.9	9-5-9	49.9	57	8	70	20	5	29	135	0	N	62.7	155	2	0	N	35	5	Q	62.0	130	3	2	N	40	120	1
56	5	19.3	7.1	10-6-1	51.4	64	4	88	51	5	29	135	4	N	64.3	128	4	4	N	27	0	Q	71.0	176	4	0	N	45	133	4
61	5	20.0	6.6	12-6-20	58.8	143	4	81	28	6	29	126	1	12	62.5	125	2	0	05	34	2	20	71.0	190	2	3	70	85	78	3
62	9	21.0	9.1	11-4-24	52.1	85	3	85	21	6	29	125	0	07	60.7	160	4	3	Q	35	1	15	68.5	153	4	1	13	78	218	1
63	9	18.0	6.8	4-1-13	50.6	54	3	82	25	6	29	109	0	15	66.3	154	4	0	10	32	0	15	67.0	145	4	0	Q	69	170	1
64	6	19.5	6.4	4-5-7	51.1	59	1	71	33	5	29	94	0	20	62.8	104	2	0	30	31	8	20	72.0	195	2	8	Q	70	110	1
65	6	19.5	6.6	3-3-20	53.5	63	1	69	25	5	29	128	0	12	64.6	141	3	2	20	30	8	20	69.5	168	3	8	Q	70	225	1
66	5	19.8	6.9	6-1-3	53.2	73	1	84	43	6	29	135	0	N	67.2	144	2	0	N	34	1	20	74.0	190	4	0	Q	86	185	1

StatLab CENSUS: Girls

ID	CHILD-BIRTH B	LGTH	WGT	MO-D-HR	CHILD-TEST HGHT	WGT	L	PEA	RA	MOTHER-BIRTH B	AG	WGT	O	SM	MOTHER-TEST HGHT	WGT	E	O	SM	FATH-BRTH AG	O	SM	FATHER-TEST HGHT	WGT	E	O	SM	FAMILY I-B	I-T	C
11	5	19.0	6.3	7-5-20	49.4	61	7	79	45	5	29	132	4	2J	65.2	142	3	0	20	27	1	30	64.5	170	4	1	Q	135	200	2
12	6	19.5	6.4	4-3-1	49.3	48	4	101	33	6	29	124	1	N	63.4	117	2	0	N	35	6	N	65.0	160	2	6	N	80	92	1
13	6	20.0	7.1	1-7-14	50.1	46	1	84	32	6	29	116	1	N	63.4	130	4	0	N	28	4	Q	70.0	196	3	0	N	61	186	1
14	9	19.0	6.3	10-3-10	57.8	100	1	77	28	9	29	120	3	20	65.4	117	4	3	30	33	1	30	76.0	230	0	4	Q	110	172	4
15	2	20.5	7.6	12-2-7	54.9	85	1	57	13	2	29	124	1	N	66.0	140	3	1	N	29	4	20	71.3	223	1	2	40	127	228	6
16	5	19.5	6.8	12-3-9	52.8	63	1	88	43	6	29	97	8		62.1	104	4	0	N	33	0	N	71.0	145	4	0	N	69	200	3
21	5	21.0	8.3	12-1-10	50.7	60	4	80	28	5	29	150	1	N	67.0	140	3	1	N	26	6	Q	74.0	275	3	6	N	80	147	1
22	7	19.5	6.9	3-3-13	51.7	70	2	81	39	6	29	140	4	N	63.5	148	4	0	N	43	6	N	66.0	150	3	6	N	117	120	1
23	5	20.5	7.6	4-6-9	54.3	77	3	85	47	6	29	130	0	N	67.3	136	4	0	N	34	3	N	75.5	206	4	3	N	10	120	6
24	6	20.5	7.1	4-5-13	53.8	65	1	87	41	6	29	125	3	20	65.3	137	4	0	N	34	1	N	68.6	151	4	1	N	120	120	6
25	6	20.0	6.9	6-2-11	55.4	71	2	67	21	5	30	143	4	20	69.6	142	4	3	20	31	0	20	73.5	183	4	0	N	147	370	4
26	7	20.0	7.3	10-7-5	54.3	68	1	81	33	6	30	128	0	08	66.6	117	3	0	08	33	4	05	73.5	208	3	3	20	90	66	6
31	6	17.0	3.6	6-6-9	47.6	62	2	91	38	6	30	135	7	N	59.9	145	4	4	N	31	6	03	69.0	186	2	2	10	80	200	1
32	6	20.5	7.9	9-2-1	52.4	64	8	84	51	1	30	115	0	N	58.9	123	3	0	N	30	6	N	66.5	160	6	1	N	50	95	1
33	5	20.0	7.7	8-3-2	53.6	85	5	86	44	3	30	135	0	20	65.3	144	3	0	Q	45	1	Q	70.0	135	4	1	Q	100	150	6
34	5	21.0	7.4	7-3-15	55.3	73	4	79	44	5	30	140	0	N	63.6	164	3	4	N	33	6	20	67.0	185	2	6	60	200	150	1
35	8	21.0	7.5	9-4-17	58.6	70	3	106	44	3	30	132	0	N	67.8	135	4	4	N	30	0	20	72.0	173	4	0	Q	66	180	4
36	1	21.0	7.0	6-1-9	52.6	61	1	81	41	6	30	124	0	N	63.9	140	3	4	N	32	0	20	68.0	150	4	4	N	68	150	1
41	7	20.0	7.9	5-3-21	54.7	70	4	80	30	5	30	135	0	10	67.1	140	4	3	Q	37	1	10	70.5	173	4	1	Q	60	113	6
42	7	21.0	8.1	6-4-15	50.6	65	4	88	46	7	30	102	4	N	58.8	115	4	0	N	38	3	06	64.0	140	4	1	Q	134	120	2
43	5	20.3	7.2	6-7-3	55.1	74	1	95	40	6	30	114	1	Q	62.3	111	4	0	Q	32	0	N	72.0	170	4	1	Q	57	140	6
44	9	20.0	6.9	7-5-13	56.1	72	1	84	44	5	30	124	2	20	67.9	132	3	4	N	33	0	10	73.0	200	4	4	Q	90	160	1
45	5	19.5	6.3	9-7-2	51.8	57	1	72	31	5	30	98	1	N	59.5	111	4	0	N	32	0	N	68.0	140	4	3	N	42	600	6
46	5	20.5	7.5	3-6-22	54.9	72	4	90	41	6	30	150	0	20	65.8	185	2	0	30	31	4	30	70.5	175	4	3	30	100	300	1
51	5	21.0	7.7	5-5-23	54.9	74	2	78	31	5	30	120	1	10	64.3	142	4	0	Q	34	0	35	71.0	195	4	0	Q	125	250	1
52	6	21.5	10.3	8-5-12	55.9	80	2	83	31	6	30	124	0	N	64.1	134	4	2	N	34	2	30	76.0	210	2	6	30	95	250	4
53	7	20.5	7.6	7-4-1	54.0	74	1	102	50	6	30	121	4	25	64.7	120	4	0	35	32	2	25	72.0	135	3	6	35	225	115	6
54	8	20.0	7.1	9-1-22	50.9	56	1	64	27	6	30	110	3	10	60.8	110	4	0	07	31	1	20	70.5	178	3	1	20	105	200	6
55	9	20.0	7.8	9-5-6	53.3	74	1	90	42	5	30	150	4	N	70.9	150	4	4	N	26	6	20	72.0	175	4	0	15	70	256	4
56	6	20.0	8.7	11-5-18	51.6	60	1	85	43	5	30	125	1	N	64.3	130	4	0	N	31	1	07	66.0	140	4	1	N	94	130	6
61	9	21.0	8.3	12-2-17	51.4	64	1	73	41	9	30	110	0	Q	64.1	126	3	3	50	37	0	07	69.0	160	4	0	20	72	195	6
62	6	19.5	5.8	8-2-20	55.9	77	1	65	20	6	30	138	1	Q	66.7	159	3	1	N	30	6	N	65.0	170	4	6	N	67	159	4
63	5	19.5	6.9	2-1-17	51.1	62	1	86	39	1	30	165	1	N	63.0	230	3	0	N	34	5	20	67.5	150	3	2	Q	101	118	6
64	9	19.0	6.4	2-5-10	50.7	53	4	72	34	5	30	135	3	10	65.2	159	4	3	30	28	5	20	73.5	155	4	0	10	88	210	6
65	6	18.5	5.8	4-4-1	48.9	54	4	87	31	6	30	104	4	15	62.8	109	4	0	Q	38	0	20	64.7	158	4	4	Q	140	160	4
66	7	19.5	6.9	12-3-6	53.3	63	5	61	21	7	30	105	0	N	64.8	122	3	1	N	35	5	20	63.0	165	0	8	Q	40	156	2

StatLab CENSUS: Girls

	CHILD-BIRTH				CHILD-TEST					MOTHER-BIRTH					MOTHER-TEST					FATH-BRTH			FATHER-TEST					FAMILY		
ID	B	LGTH	WGT	MO-D-HR	HGHT	WGT	L	PEA	RA	B	AG	WGT	O	SM	HGHT	WGT	E	O	SM	AG	O	SM	HGHT	WGT	E	O	SM	I-B	I-T	C
11	8	23.0	10.9	2-7-19	55.3	83	4	60	17	7	30	160	0	N	63.5	176	2	0	N	31	8	Q	69.0	160	2	7	Q	40	96	1
12	5	20.0	6.1	5-4-22	53.9	74	1	86	20	5	31	111	0	20	58.9	152	2	0	07	28	7	02	69.0	160	2	7	03	50	84	1
13	5	18.8	5.6	6-1-3	53.8	81	4	80	42	5	31	118	0	N	63.1	132	4	3	N	38	1	N	71.0	200	4	1	N	90	170	1
14	5	19.8	7.2	8-4-9	50.3	63	4	92	48	5	31	120	0	20	64.6	138	2	0	Q	29	3	20	69.0	180	4	3	Q	50	360	1
15	5	20.0	7.3	3-7-19	50.5	62	1	71	27	5	31	130	0	Q	65.0	135	4	0	Q	33	2	17	69.0	170	2	7	Q	50	90	1
16	6	21.0	7.9	5-5-13	56.5	95	1	92	51	4	31	121	0	N	65.6	127	4	0	N	29	0	N	74.0	230	4	0	N	72	200	6
21	5	19.5	6.6	7-5-17	52.0	63	4	75	34	2	31	91	0	09	62.1	110	4	3	12	34	6	10	68.0	145	2	3	10	66	204	1
22	5	20.0	8.0	8-2-23	55.6	74	1	76	37	6	31	125	0	Q	67.2	132	3	1	15	32	1	20	72.0	210	4	1	40	51	164	6
23	5	19.5	8.3	8-7-3	53.7	71	5	105	39	5	31	112	0	N	61.0	122	1	5	N	29	5	N	67.8	168	2	5	N	49	108	1
24	6	21.5	8.4	10-7-8	55.5	84	1	76	25	6	31	120	0	Q	66.8	143	3	0	N	32	0	Q	71.0	182	4	0	Q	67	150	4
25	6	20.0	7.1	12-3-9	50.6	61	1	82	17	5	31	102	0	01	61.7	129	4	3	N	31	0	N	67.5	185	4	0	Q	90	310	6
26	9	21.0	7.2	12-3-18	52.1	72	3	81	34	5	31	125	0	06	61.8	129	4	4	10	39	0	20	68.5	155	4	0	25	100	254	6
31	5	20.0	7.1	11-5-7	57.9	80	1	76	36	5	31	140	0	N	67.9	154	3	0	N	35	1	N	71.0	175	4	1	N	90	140	1
32	5	20.0	7.1	4-6-10	55.1	88	1	81	18	5	31	184	0	N	65.2	202	2	0	N	33	0	N	76.0	290	3	0	N	75	165	6
33	2	20.0	6.0	6-3-19	53.5	61	1	77	48	2	31	148	4	20	66.9	192	4	3	20	32	1	N	72.0	200	4	0	N	86	300	6
34	7	21.0	8.3	5-4-17	48.5	51	1	65	30	7	31	110	3	N	61.1	124	4	3	N	34	1	N	65.5	188	4	1	N	134	340	5
35	7	21.5	7.8	8-6-8	52.5	76	1	67	26	5	31	139	0	Q	61.6	129	2	1	02	31	6	10	66.0	165	2	6	20	60	110	3
36	7	20.5	6.9	6-3-23	53.5	66	1	68	18	5	31	130	5	10	66.5	142	2	5	10	32	8	N	71.5	190	2	8	N	139	224	1
41	5	21.5	7.8	12-5-9	54.1	67	4	60	16	5	31	123	1	N	61.1	135	2	1	N	31	0	N	73.0	250	2	0	N	84	180	1
42	2	20.0	7.9	10-1-10	52.3	61	1	64	34	6	31	127	1	N	62.9	131	3	1	11	32	6	20	69.5	167	2	0	30	120	174	1
43	5	19.5	5.4	1-5-12	52.2	69	3	87	29	5	31	110	1	Q	64.8	115	3	0	N	35	6	N	73.0	208	2	7	N	116	140	1
44	8	19.0	8.6	2-7-17	56.4	69	1	73	41	7	31	105	0	N	60.9	112	4	3	N	37	3	N	67.0	135	4	3	N	125	159	6
45	5	20.0	5.6	11-2-8	56.8	75	4	60	14	5	32	175	0	N	69.0	205	2	5	N	37	5	10	71.9	178	0	3	20	58	165	1
46	5	20.0	7.8	8-6-3	52.5	74	1	83	18	5	32	125	0	N	62.6	156	3	0	N	25	5	08	71.0	220	4	3	N	60	130	1
51	6	21.3	8.4	8-2-20	51.9	60	2	65	27	2	32	140	0	N	64.1	155	3	0	N	32	1	30	64.3	180	4	1	Q	80	250	6
52	5	20.0	7.4	4-5-3	55.3	66	5	80	46	5	32	130	0	N	63.0	138	4	0	N	32	0	N	71.8	175	4	0	Q	70	250	1
53	7	21.0	8.6	2-3-24	48.2	54	1	83	44	7	32	104	0	Q	58.8	115	4	0	N	40	3	04	64.0	160	4	0	Q	60	120	2
54	5	20.0	8.8	6-1-10	56.2	73	1	84	40	6	32	126	0	N	66.7	126	3	0	N	35	5	N	70.0	160	4	2	N	49	120	6
55	5	21.0	9.2	8-6-10	56.1	81	4	76	32	1	32	132	0	Q	61.4	141	4	0	N	33	0	24	71.0	180	4	0	20	60	140	1
56	6	20.0	6.6	8-7-18	53.5	67	1	94	41	6	32	127	0	N	60.9	139	4	0	N	34	0	15	70.0	170	4	0	N	120	108	1
61	5	20.0	6.6	11-7-6	53.6	66	1	75	29	5	32	115	0	N	61.3	122	1	5	N	30	5	N	68.4	169	2	5	N	56	76	1
62	7	19.5	6.1	3-6-6	50.9	53	3	75	40	7	32	120	0	N	63.8	122	4	0	N	32	0	20	72.0	145	4	2	N	81	204	1
63	5	20.0	6.6	9-1-8	52.9	64	1	77	34	5	32	117	0	20	65.4	121	2	0	40	27	2	01	70.0	230	4	0	N	90	200	6
64	5	20.5	6.9	7-4-20	57.1	73	1	77	29	5	32	158	0	15	67.7	207	3	1	Q	34	0	30	76.5	235	3	3	N	60	156	6
65	6	18.0	4.5	4-7-22	55.2	75	8	58	12	6	32	150	0	07	62.9	182	2	8	01	35	7	N	65.0	165	1	8	Q	60	97	1
66	5	20.5	7.8	12-2-13	52.9	66	4	84	48	5	32	119	0	N	66.2	130	4	0	N	39	1	Q	70.5	155	4	1	N	120	240	1

StatLab CENSUS: Girls

	CHILD-BIRTH				CHILD-TEST					MOTHER-BIRTH					MOTHER-TEST					FATH-BRTH			FATHER-TEST					FAMILY		
ID	B	LGTH	WGT	MO-D-HR	HGHT	WGT	L	PEA	RA	B	AG	WGT	O	SM	HGHT	WGT	E	O	SM	AG	O	SM	HGHT	WGT	E	O	SM	I-B	I-T	C
11	2	19.0	5.6	10-5-18	51.1	66	2	79	30	2	32	126	0	18	64.9	142	2	2	15	33	2	10	68.5	155	3	1	Q	90	134	1
12	6	20.0	6.6	5-3-10	51.4	64	4	75	41	6	32	126	1	N	61.2	132	4	0	N	35	1	30	66.0	155	4	1	15	87	120	6
13	8	20.0	7.8	4-4-17	54.1	77	4	56	10	6	32	229	5	Q	68.3	284	1	0	Q	33	7	Q	71.0	230	2	7	Q	86	60	1
14	5	20.0	7.4	7-5-15	48.9	54	1	79	34	7	32	118	0	Q	61.4	128	2	1	Q	31	5	10	68.5	162	4	0	Q	61	164	1
15	6	22.0	9.2	8-3-11	57.3	68	1	83	33	6	32	130	0	N	66.3	123	3	5	N	38	1	N	74.0	175	4	1	N	111	190	6
16	1	21.5	7.6	6-6-8	53.5	75	1	72	38	5	32	118	0	N	63.7	134	3	0	N	31	8	N	69.0	170	4	8	N	67	125	1
21	5	20.0	6.6	5-1-19	50.7	54	1	76	22	6	32	120	0	20	63.1	133	1	5	20	37	8	24	71.0	185	1	8	22	65	93	2
22	6	20.8	7.4	9-1-19	50.6	71	4	80	33	6	32	115	2	N	60.5	116	2	0	N	40	4	N	72.0	180	3	2	N	83	120	4
23	5	19.8	6.4	7-7-5	51.6	78	3	65	20	6	32	130	0	20	61.6	153	1	0	20	33	6	Q	66.0	181	2	6	N	96	96	6
24	5	21.5	8.3	10-5-23	52.4	79	1	64	17	5	32	125	7	N	63.3	145	0	5	N	36	7	Q	64.0	155	1	7	Q	46	104	1
25	9	18.5	5.4	12-2-22	49.3	52	1	61	16	6	32	114	0	20	63.4	112	1	0	30	34	6	20	65.0	145	2	0	Q	50	108	4
26	1	19.0	7.1	11-3-2	52.1	73	8	82	34	3	32	136	0	20	65.2	135	3	4	20	34	0	20	71.0	160	4	3	30	90	321	4
31	5	22.0	9.9	2-7-6	55.0	70	3	79	39	5	32	125	3	N	67.0	140	4	3	03	30	1	N	75.6	172	4	1	N	143	350	6
32	7	20.5	6.7	2-6-20	56.1	91	1	78	24	7	32	131	0	N	66.6	129	1	0	Q	37	7	20	69.5	170	2	3	N	55	200	3
33	5	19.8	6.8	11-5-2	50.1	61	1	101	39	8	32	136	4	20	64.1	161	4	4	Q	39	4	N	68.0	190	4	2	N	145	138	6
34	7	19.0	5.9	1-1-3	52.0	49	1	57	11	7	32	127	0	Q	61.9	121	1	5	08	42	6	N	66.3	201	1	7	Q	53	118	2
35	7	19.5	7.1	2-1-5	49.8	63	1	77	21	5	32	122	1	20	60.6	163	1	1	20	38	6	20	68.0	185	2	6	05	85	149	4
36	5	19.8	6.5	8-7-4	54.4	78	1	81	30	6	33	130	0	N	62.9	143	2	0	N	38	0	N	66.0	130	4	6	N	80	100	1
41	5	19.0	6.7	9-3-23	54.6	74	1	70	27	5	33	125	0	N	64.4	138	1	0	N	38	6	Q	68.0	175	1	3	Q	53	77	1
42	5	20.0	6.8	3-6-3	53.2	66	1	80	38	5	33	120	0	40	64.6	135	2	0	N	31	3	20	69.0	180	4	3	Q	80	300	1
43	6	19.5	6.9	4-7-14	58.7	97	1	87	40	5	33	129	0	25	64.3	135	2	1	40	36	6	40	74.0	205	3	6	40	90	162	4
44	9	16.0	3.3	6-3-16	51.8	64	3	87	37	5	33	105	3	N	63.2	118	4	0	N	32	1	N	68.0	185	4	1	N	93	150	4
45	6	21.0	7.5	8-4-1	53.4	66	4	91	38	8	33	125	0	15	64.6	126	3	0	12	41	8	N	72.0	170	4	8	Q	50	60	5
46	9	19.0	5.3	7-6-21	53.5	66	4	73	22	6	33	159	0	N	59.6	167	2	0	N	44	7	N	74.0	250	0	6	N	78	78	1
51	7	20.0	6.6	8-2-17	55.0	66	4	74	24	7	33	116	0	N	68.4	119	2	0	N	39	6	N	73.0	185	2	6	N	60	99	1
52	5	20.0	6.1	9-6-22	53.6	58	8	63	20	7	33	180	0	N	67.9	188	4	5	N	37	6	12	67.5	155	1	7	Q	55	105	1
53	6	20.0	7.6	10-3-1	55.1	103	4	81	21	5	33	170	0	20	62.8	181	2	1	40	36	6	20	70.5	180	1	6	40	71	184	6
54	5	21.0	7.6	9-6-16	54.6	58	4	64	43	5	33	150	0	20	66.1	158	2	1	20	39	6	10	69.0	150	1	7	Q	78	149	4
55	6	19.0	7.3	7-3-11	49.3	56	1	86	41	6	33	130	0	N	61.8	121	4	0	N	39	3	N	69.0	161	4	5	N	60	70	6
56	5	18.5	5.6	10-5-14	50.3	58	1	78	31	7	33	130	0	N	63.8	141	2	1	N	34	1	N	68.0	180	4	1	N	100	192	2
61	5	20.0	7.1	12-4-18	53.7	64	3	63	14	5	33	137	0	N	63.3	193	2	0	N	30	6	01	72.0	190	2	0	22	100	125	4
62	8	20.5	6.6	4-6-13	51.0	68	4	82	32	7	33	112	0	Q	63.2	108	4	0	Q	40	3	01	70.5	155	4	3	N	173	180	6
63	5	19.3	6.9	6-1-23	54.4	91	1	78	9	6	33	137	0	N	64.8	162	4	4	N	34	1	N	69.0	165	3	1	N	83	270	4
64	8	20.5	7.8	7-7-2	53.4	66	1	87	39	7	33	130	1	20	62.6	154	2	0	N	39	8	Q	71.0	173	8	7	N	118	90	1
65	6	21.5	8.6	11-2-9	53.7	94	1	73	21	6	33	164	0	N	63.5	181	0	0	N	36	6	20	68.0	202	3	7	20	45	27	1
66	5	21.8	8.6	2-2-17	50.6	64	1	79	34	5	33	180	0	04	63.0	192	4	0	N	27	0	N	67.0	150	4	0	N	55	110	6

	CHILD-BIRTH				CHILD-TEST					MOTHER-BIRTH					MOTHER-TEST					FATH-BRTH			FATHER-TEST					FAMILY		
ID	B	LGTH	WGT	MO-D-HR	HGHT	WGT	L	PEA	RA	B	AG	WGT	O	SM	HGHT	WGT	E	O	SM	AG	O	SM	HGHT	WGT	E	O	SM	I-B	I-T	C
11	7	20.5	6.6	3-7-17	55.3	81	1	81	40	1	33	133	0	Q	66.0	145	3	4	N	38	8	Q	72.0	210	3	8	Q	65	227	1
12	6	20.8	7.8	11-4-16	54.9	87	4	82	23	6	33	133	4	N	66.8	135	4	0	N	37	0	N	74.0	145	4	0	N	114	380	6
13	7	20.0	7.6	6-4-1	54.0	72	4	95	46	3	34	137	0	N	65.5	143	4	4	N	39	0	20	71.8	170	4	0	15	65	196	6
14	2	21.0	7.9	7-3-11	53.1	67	1	86	43	2	34	125	0	Q	65.7	128	4	5	Q	37	3	10	70.5	185	4	3	20	100	270	4
15	5	20.0	6.8	5-2-8	52.8	76	1	75	19	5	34	212	5	Q	68.8	206	1	5	Q	38	6	10	67.0	180	1	6	20	120	185	4
16	6	21.0	8.6	8-4-16	59.8	142	1	74	20	6	34	176	5	N	64.6	232	2	5	N	25	0	24	67.0	180	1	5	20	36	123	3
21	6	19.5	6.3	4-1-17	48.8	52	1	81	41	6	34	120	0	N	60.9	139	4	0	N	36	0	20	70.0	170	4	0	Q	120	101	1
22	2	19.8	7.3	9-6-23	54.9	78	1	80	32	6	34	129	0	08	63.5	126	3	0	15	37	0	20	70.5	185	4	0	Q	105	230	1
23	5	20.5	6.8	7-3-7	55.5	71	1	84	42	1	34	127	0	40	66.1	127	2	3	60	39	6	01	74.0	175	2	6	05	64	152	3
24	2	19.8	7.5	9-2-17	55.1	68	1	106	52	1	34	135	0	15	67.3	192	4	0	Q	34	1	Q	73.5	180	4	1	Q	100	250	1
25	8	20.0	7.8	9-4-2	50.5	62	1	90	46	7	34	125	0	N	63.8	167	4	0	Q	34	0	N	70.1	140	4	0	N	80	180	1
26	5	20.5	8.1	3-7-22	54.8	80	5	75	34	1	34	142	0	15	66.1	146	4	4	N	38	5	24	67.0	158	2	2	N	68	267	6
31	2	19.5	6.1	4-4-23	53.3	65	1	74	30	2	34	120	0	10	65.3	129	4	0	12	36	1	Q	70.6	177	4	1	N	70	160	5
32	7	20.5	7.1	4-1-12	49.9	72	4	73	44	6	34	132	0	25	63.8	144	3	0	30	34	3	02	66.5	160	1	3	N	75	80	1
33	5	20.0	7.1	4-2-13	50.5	55	2	58	10	7	34	125	0	06	60.7	146	2	1	10	38	8	20	66.8	161	4	0	40	60	132	1
34	2	20.0	7.2	7-6-19	50.4	72	1	76	37	7	34	120	1	Q	62.8	117	2	1	N	46	6	20	69.0	152	1	0	15	71	113	1
35	1	19.0	5.6	6-6-14	50.0	54	8	59	12	6	34	114	0	N	66.1	75	2	0	N	38	6	N	68.0	125	2	6	N	80	83	6
36	6	20.5	8.7	7-6-23	47.4	62	1	86	28	6	34	144	0	Q	65.0	144	4	0	Q	39	5	30	68.5	185	4	3	40	104	200	1
41	5	19.5	6.1	8-6-18	50.3	60	2	80	14	6	34	130	1	N	64.4	148	3	4	N	41	6	30	66.0	185	2	6	Q	115	187	4
42	2	21.0	7.5	11-5-19	52.4	53	4	86	32	6	34	155	0	Q	69.1	192	3	0	02	39	0	N	75.0	200	4	0	N	74	170	1
43	5	20.0	7.1	12-1-22	54.4	58	8	76	30	6	34	118	0	N	68.6	122	4	0	N	38	8	N	71.0	170	4	8	N	59	114	1
44	5	20.3	7.9	1-3-1	54.3	96	4	89	38	6	34	130	0	15	67.0	145	3	0	30	35	1	03	67.8	185	4	1	15	103	120	6
45	2	19.0	7.1	10-4-1	50.4	56	4	71	24	2	34	138	1	20	66.0	175	2	0	N	42	6	30	67.8	141	4	4	40	98	242	4
46	6	21.0	8.1	4-7-3	51.4	72	1	70	33	8	34	135	0	10	61.5	170	2	0	Q	35	6	20	70.0	180	1	6	15	60	144	1
51	5	22.0	8.2	1-2-13	52.1	58	4	80	43	6	35	125	0	Q	63.7	134	4	0	Q	34	0	01	73.5	175	4	4	Q	130	250	6
52	5	20.0	7.3	4-3-6	50.3	52	1	73	31	6	35	125	0	N	67.5	129	3	2	N	37	6	24	71.0	192	2	0	N	75	162	2
53	5	20.5	9.3	5-5-10	53.9	101	1	64	22	6	35	220	7	N	62.9	272	2	0	N	37	6	20	66.0	220	0	8	Q	71	76	1
54	3	21.5	7.0	5-1-9	55.8	126	1	66	49	3	35	140	0	Q	62.1	161	3	5	Q	35	7	30	63.0	155	1	8	N	37	76	1
55	9	20.0	7.3	5-2-19	60.8	102	1	91	28	2	35	119	0	20	64.1	121	2	1	20	33	4	20	72.0	180	2	4	Q	55	141	1
56	6	20.0	7.1	6-2-16	51.6	66	4	89	48	6	35	125	0	N	60.5	135	3	0	N	43	0	N	62.3	109	4	0	N	96	126	1
61	6	21.0	8.9	7-5-2	54.8	78	4	79	47	5	35	137	0	N	63.3	140	3	0	N	38	6	10	70.0	178	2	6	Q	65	110	1
62	6	21.0	6.9	9-1-3	53.8	60	1	90	36	5	35	116	0	50	65.8	107	4	5	20	40	3	50	70.0	170	4	0	Q	100	150	6
63	7	21.5	8.4	9-6-11	51.2	58	4	72	25	7	35	140	0	N	65.3	167	1	5	N	36	6	N	69.5	225	2	7	N	48	76	2
64	6	21.5	9.3	10-5-14	54.4	82	1	63	15	6	35	180	4	N	63.5	187	2	5	Q	40	2	20	68.5	180	2	2	Q	61	114	1
65	9	21.0	8.1	12-1-1	55.6	95	4	82	25	7	35	195	0	20	69.0	230	3	0	N	38	6	20	72.0	205	2	6	N	70	90	6
66	6	21.0	8.1	11-4-17	54.9	75	4	82	31	5	35	165	0	N	66.1	185	2	3	Q	36	6	N	76.0	245	2	3	N	60	153	1

	CHILD-BIRTH				CHILD-TEST					MOTHER-BIRTH					MOTHER-TEST					FATH-BRTH			FATHER-TEST					FAMILY		
ID	B	LGTH	WGT	MO-D-HR	HGHT	WGT	L	PEA	RA	B	AG	WGT	O	SM	HGHT	WGT	E	O	SM	AG	O	SM	HGHT	WGT	E	O	SM	I-B	I-T	C
11	5	21.0	8.1	3-6-21	62.0	104	8	72	38	6	35	112	0	10	65.3	135	3	0	Q	32	0	N	71.5	180	4	0	N	70	130	1
12	3	21.0	8.1	6-2-13	52.9	67	3	70	33	2	35	137	0	01	62.9	144	4	1	Q	36	1	25	70.5	158	4	0	Q	31	122	4
13	5	19.5	6.3	8-3-9	48.9	48	4	76	30	5	35	102	0	N	59.9	111	3	0	N	36	3	N	71.0	178	4	3	N	111	275	1
14	5	18.8	5.3	12-4-19	51.8	63	8	60	17	7	35	158	0	Q	62.4	170	0	0	N	34	7	N	71.0	235	3	7	N	63	84	1
15	5	21.0	7.1	1-1-13	52.9	65	1	78	37	5	35	117	0	20	65.2	115	4	0	Q	38	3	25	74.0	205	4	2	Q	120	300	1
16	6	22.5	10.8	3-1-15	52.9	89	4	68	23	6	35	155	4	N	65.6	172	2	0	N	35	6	20	66.0	200	2	6	22	55	156	1
21	7	21.5	9.1	3-4-10	57.5	108	1	88	21	7	36	119	0	05	62.5	146	2	0	01	37	6	10	72.0	175	2	6	05	60	66	1
22	6	20.0	7.0	9-4-8	53.1	67	1	81	41	5	36	150	0	N	62.5	150	3	0	N	39	8	20	71.0	175	2	8	10	60	110	3
23	5	20.0	7.9	12-3-23	54.6	80	1	72	23	5	36	200	0	N	62.0	201	1	3	N	43	6	10	71.0	155	2	7	15	54	73	1
24	5	20.0	6.7	10-3-19	48.5	48	4	65	21	5	36	155	0	N	62.8	179	0	0	N	36	6	30	68.7	216	2	6	30	60	77	1
25	5	22.3	9.3	12-3-8	56.6	84	4	74	17	5	36	150	0	N	69.4	164	2	0	N	36	5	N	71.5	175	4	0	N	70	150	3
26	6	20.0	7.6	4-7-6	51.6	54	1	68	40	6	36	94	0	N	62.3	100	4	0	N	35	0	N	66.0	155	4	0	N	56	135	5
31	5	22.5	9.3	12-3-8	53.4	106	1	75	8	7	36	167	7	Q	67.1	215	2	7	N	40	6	24	67.0	135	1	7	10	110	181	1
32	6	20.5	7.9	9-1-19	57.4	108	8	82	33	6	36	144	1	N	64.3	156	3	0	N	36	0	40	66.0	185	4	0	Q	100	216	4
33	6	21.0	7.8	9-7-17	54.6	67	1	91	52	6	36	108	0	N	66.4	113	3	0	N	35	4	Q	73.0	193	4	3	Q	250	135	1
34	9	19.5	8.0	7-6-12	52.4	73	1	87	27	6	36	106	0	20	61.9	113	3	0	20	37	0	20	68.0	156	4	0	20	60	143	3
35	5	21.5	8.4	8-3-4	58.9	84	1	96	49	6	36	125	0	N	65.1	127	4	3	N	37	7	N	74.0	200	3	6	N	65	207	1
36	5	20.0	7.0	9-6-20	57.2	92	3	72	28	5	36	124	5	N	62.3	146	1	4	N	42	6	N	65.0	150	1	6	N	50	132	1
41	6	21.0	8.1	11-2-21	55.5	73	4	84	48	6	36	125	0	08	68.4	141	4	0	Q	36	2	32	74.0	195	4	2	Q	85	215	1
42	5	19.0	6.1	6-4-1	51.8	89	1	84	45	6	36	98	0	20	60.6	102	3	0	20	40	7	30	64.0	155	3	3	Q	72	120	1
43	6	21.0	8.1	6-6-24	54.0	89	4	86	20	6	36	123	1	N	62.8	127	3	1	N	43	6	06	66.0	155	3	6	Q	149	120	1
44	1	20.0	7.6	6-1-19	54.9	76	1	84	29	2	36	128	0	40	68.0	139	4	3	10	42	0	N	71.5	165	4	0	Q	85	186	1
45	6	20.0	7.1	6-7-14	50.1	58	5	74	29	6	36	137	0	20	65.3	154	3	2	10	36	0	12	71.5	230	4	3	10	75	203	6
46	6	20.0	6.0	6-2-6	57.3	80	4	73	26	6	36	128	5	07	69.5	139	0	5	05	42	7	N	67.0	150	0	7	N	68	95	1
51	3	21.0	8.0	8-4-2	55.2	78	1	74	22	5	36	153	0	N	64.3	185	2	0	N	40	6	Q	70.0	170	2	2	Q	63	100	4
52	7	20.0	6.4	11-4-17	52.8	72	1	82	28	7	36	120	1	N	64.0	130	2	1	Q	36	5	20	68.3	179	2	7	15	90	162	5
53	5	20.0	6.9	12-6-18	50.6	57	1	73	25	5	36	153	0	N	62.5	171	2	5	N	40	8	20	73.5	190	1	7	20	60	143	3
54	6	19.5	7.0	12-1-23	48.7	56	1	88	41	5	36	120	1	30	62.8	138	3	0	20	37	1	40	71.0	130	4	1	40	131	150	6
55	5	21.0	6.9	1-5-9	53.9	75	1	62	15	5	36	140	1	N	66.6	168	2	1	N	39	6	02	72.0	203	2	6	N	90	170	2
56	5	20.5	7.6	2-6-2	52.6	68	4	89	29	5	36	135	0	N	64.3	144	3	1	N	41	1	Q	68.5	160	4	1	N	118	166	1
61	6	19.0	6.9	3-7-4	51.9	70	4	62	28	6	36	128	1	N	64.8	140	2	1	N	41	8	N	68.0	200	1	8	N	96	170	2
62	5	19.3	5.3	4-3-14	49.6	64	4	81	47	5	36	126	4	25	62.0	115	4	4	25	40	0	N	65.5	152	4	0	N	137	325	4
63	6	19.5	6.6	6-6-18	51.2	64	2	78	21	6	37	170	0	N	64.9	179	2	0	N	47	2	20	72.0	230	3	0	N	85	112	4
64	5	22.0	9.2	12-6-12	56.5	81	1	79	19	7	37	190	1	30	63.1	218	0	3	Q	38	4	N	74.0	228	1	3	N	30	250	1
65	5	20.0	7.3	10-3-17	57.4	84	1	82	33	5	37	125	0	30	64.6	156	1	0	30	36	6	30	68.3	167	2	6	N	54	120	3
66	6	21.5	8.0	10-2-17	54.3	98	4	70	24	6	37	230	5	N	66.0	255	5	4	N	39	7	N	68.0	180	2	6	N	75	96	1

StatLab CENSUS: Girls

	CHILD-BIRTH				CHILD-TEST					MOTHER-BIRTH					MOTHER-TEST					FATH-BRTH			FATHER-TEST					FAMILY		
ID	B	LGTH	WGT	MO-D-HR	HGHT	WGT	L	PEA	RA	B	AG	WGT	O	SM	HGHT	WGT	E	O	SM	AG	O	SM	HGHT	WGT	E	O	SM	I-B	I-T	C
11	7	20.0	7.2	4-5-15	54.4	73	1	96	49	8	37	144	0	N	62.6	178	4	3	N	36	5	N	72.0	210	4	5	N	70	266	6
12	9	20.5	7.9	5-2-19	52.8	99	1	86	36	8	37	157	0	N	63.0	186	3	0	N	35	1	20	70.0	160	4	1	N	50	100	6
13	6	22.5	8.9	5-2-16	57.6	98	1	84	43	2	37	126	0	20	66.6	131	2	0	20	35	2	50	71.0	205	3	0	20	49	130	6
14	6	20.5	6.8	8-1- 9	54.8	61	1	74	31	7	37	159	0	20	65.9	207	2	0	Q	38	2	50	72.0	195	1	3	Q	60	100	5
15	9	21.0	9.1	7-2-11	58.4	98	1	78	28	6	37	155	8	10	68.3	165	4	4	Q	35	5	20	73.0	200	4	0	20	110	221	5
16	5	19.0	6.4	12-3-21	58.5	82	1	78	29	5	37	185	5	03	66.0	205	2	5	N	38	6	Q	71.0	235	1	7	N	70	144	4
21	5	20.5	7.4	4-1- 9	51.1	66	8	78	14	5	37	135	1	N	64.6	142	3	4	N	46	6	N	65.0	150	0	6	N	70	144	1
22	5	20.0	7.9	7-1-11	54.1	66	4	71	42	5	37	147	0	Q	68.5	161	2	0	Q	35	2	20	74.5	184	1	0	N	123	190	6
23	5	20.0	6.6	8-3-21	51.1	59	4	89	42	6	37	165	0	N	64.4	165	3	0	Q	41	0	N	70.5	175	4	0	N	120	144	1
24	7	19.8	7.2	7-4-17	55.3	113	4	63	27	7	37	190	0	N	69.5	194	1	4	N	42	6	20	73.0	290	2	5	N	71	84	1
25	2	19.0	6.0	10-1- 6	52.2	60	2	67	32	6	37	127	1	N	63.7	164	3	0	N	39	5	20	66.0	152	0	3	15	99	280	1
26	3	20.0	6.0	3-4- 6	55.5	63	4	88	24	7	37	124	0	20	67.8	120	2	0	40	34	0	18	71.1	156	2	2	Q	101	145	4
31	9	20.0	6.7	7-6-13	57.2	88	4	83	36	6	38	155	3	Q	66.7	169	4	4	Q	37	0	Q	72.5	195	4	8	Q	62	290	4
32	5	19.8	6.5	8-3- 2	56.0	69	4	70	10	5	38	146	1	Q	67.8	171	2	0	Q	44	7	20	72.0	210	0	6	Q	86	84	4
33	5	20.0	6.4	10-2- 8	53.9	58	1	68	30	5	38	128	0	N	64.4	127	2	0	Z	41	3	Q	74.5	235	2	3	N	100	200	1
34	6	20.0	6.3	1-7- 9	51.8	64	1	80	14	1	38	123	0	N	63.1	141	4	0	N	39	6	N	68.0	168	2	6	03	50	110	1
35	5	20.0	7.4	6-5-21	56.7	78	8	81	28	5	38	133	0	15	69.8	130	3	0	20	44	8	20	71.0	200	2	8	Q	71	100	1
36	7	21.0	8.2	10-3- 2	53.2	55	4	91	44	8	38	123	0	N	66.7	126	4	0	N	37	0	N	76.6	185	4	2	N	200	350	1
41	2	19.0	6.2	4-4-19	47.0	45	1	70	29	2	38	109	0	N	59.6	107	1	0	N	44	5	N	67.0	178	2	5	N	60	91	4
42	8	21.0	7.8	7-6-16	55.2	76	8	84	36	6	38	120	0	Q	62.9	129	3	0	Q	37	0	20	71.0	175	4	2	Q	100	290	6
43	5	20.0	6.3	3-1-20	50.4	53	4	80	45	5	38	125	4	Q	62.7	127	4	0	N	39	0	N	68.5	160	4	2	N	131	257	1
44	6	20.5	6.9	10-3-17	52.3	70	1	68	19	5	39	172	0	N	63.9	167	2	0	N	44	6	N	70.5	202	1	6	N	68	108	1
45	6	20.5	7.5	7-6-12	54.8	101	1	65	39	6	39	247	5	N	64.0	185	0	0	N	50	7	22	69.0	190	0	8	22	68	42	1
46	8	20.0	7.9	1-5- 6	51.3	70	2	81	20	6	39	125	0	Q	63.8	145	4	0	03	40	0	Q	71.5	203	4	0	60	104	240	1
51	2	19.0	6.4	4-4- 4	50.7	56	4	93	24	2	39	120	0	10	60.0	131	0	5	Q	49	0	Q	67.0	155	3	0	N	58	145	2
52	6	18.0	6.8	6-1-17	54.4	72	1	87	48	6	39	145	3	20	64.7	159	4	4	Q	37	0	N	71.0	160	0	4	Q	110	500	6
53	5	20.0	6.8	6-4-22	54.5	60	1	95	32	6	39	176	4	N	64.0	156	4	0	Q	41	6	N	69.6	160	4	6	N	84	221	4
54	5	20.5	9.0	6-1-21	58.1	93	4	79	45	6	39	137	1	N	67.2	189	3	4	N	43	6	30	73.0	180	2	0	40	60	228	1
55	6	20.5	7.6	9-1- 6	54.0	64	4	76	32	5	39	105	1	N	58.9	118	2	1	N	43	6	30	70.5	210	0	6	40	100	137	1
56	7	21.0	8.3	7-4-23	57.4	96	4	64	17	6	39	138	0	06	65.9	152	0	0	03	39	7	20	66.0	220	0	7	30	50	60	1
61	7	19.0	6.7	8-7- 4	56.2	78	4	80	37	1	39	93	3	N	60.9	111	4	3	N	41	1	Q	71.5	195	4	1	Q	100	330	1
62	5	21.0	10.1	9-2-15	55.4	88	1	79	29	7	39	237	0	20	65.3	208	3	0	N	39	8	20	75.0	250	3	4	Q	70	110	2
63	5	20.5	6.7	11-2-14	57.0	89	4	79	22	1	39	156	0	30	64.9	173	3	0	40	32	4	20	72.0	140	2	3	20	75	120	1
64	2	20.0	7.2	3-3- 5	54.4	68	1	107	36	6	39	136	3	Q	64.1	143	4	3	10	34	0	N	71.0	175	4	8	Q	137	313	6
65	6	21.0	8.3	7-4-10	55.9	79	8	79	39	6	39	190	0	N	70.0	270	4	0	N	51	6	Q	67.0	145	1	6	N	72	118	1
66	7	21.0	7.6	9-1- 7	52.8	82	1	77	25	5	39	130	0	N	60.3	138	2	0	N	44	6	Q	70.5	160	2	6	N	75	84	1

	CHILD-BIRTH			CHILD-TEST					MOTHER-BIRTH					MOTHER-TEST					FATH-BRTH			FATHER-TEST					FAMILY		
ID	B	LGTH	WGT	MO-D-HR	HGHT	WGT	L	PEA	RA	B	AG	WGT	O	SM	HGHT	WGT	E	O	SM	AG	O	SM	HGHT	WGT	E	O	SM	I-B	I-T	C
11	6	19.0	6.9	1-6-2	53.9	91	1	82	29	6	39	155	1	20	65.9	162	3	4	40	40	7	N	68.0	180	2	5	N	102	245	1
12	6	20.5	6.8	12-7-10	50.4	59	1	104	37	6	39	120	0	20	63.6	114	2	2	N	42	5	15	68.0	150	3	5	20	93	124	1
13	5	21.0	8.2	12-1-9	53.9	77	1	85	34	5	39	120	5	N	62.1	134	3	1	N	45	7	20	71.5	172	1	7	20	90	170	4
14	1	20.0	8.9	9-6-4	52.0	75	1	75	24	6	40	136	3	N	63.7	155	4	3	N	39	6	20	67.0	155	3	7	30	130	250	4
15	5	18.5	4.9	8-6-7	53.9	67	1	71	14	5	40	240	0	N	66.8	261	0	0	N	40	6	N	67.0	140	1	6	N	70	72	1
16	4	20.0	7.8	5-2-12	49.6	50	4	87	31	7	40	127	3	Q	65.3	146	4	0	N	42	1	N	67.0	140	4	1	N	100	160	1
21	5	19.5	6.5	7-6-5	54.2	78	4	67	35	7	40	162	5	N	63.3	160	1	5	N	43	6	N	64.0	150	0	6	22	52	57	1
22	5	20.0	8.4	5-4-18	49.3	56	2	87	29	5	40	135	0	N	64.0	131	3	0	N	43	1	N	68.0	175	4	1	N	102	160	1
23	3	22.0	8.4	4-2-7	53.1	74	3	66	13	6	40	110	5	15	64.6	124	2	0	20	47	5	Q	71.8	175	3	2	Q	42	96	1
24	2	20.0	7.2	10-6-1	49.6	56	4	77	37	1	40	120	0	20	62.8	119	4	3	N	36	1	N	66.5	156	4	1	N	70	196	4
25	6	22.0	8.6	3-6-9	51.6	56	1	83	31	5	40	136	0	10	63.2	143	0	0	Q	45	6	10	65.0	155	0	6	N	70	96	2
26	5	20.0	6.1	1-6-22	55.1	94	4	67	15	7	40	154	0	Q	65.8	160	0	0	Q	46	6	20	68.0	190	4	2	40	115	100	3
31	8	21.0	8.2	3-4-19	55.6	130	1	65	21	8	41	147	0	N	64.0	150	1	8	N	50	7	N	74.0	250	0	7	N	54	66	1
32	7	21.5	8.1	5-5-8	56.4	74	1	90	25	8	41	116	1	N	65.1	118	2	0	Q	48	7	Q	65.0	130	2	0	Q	90	60	1
33	9	22.5	9.5	4-2-14	56.5	77	1	80	24	5	41	226	0	Q	62.5	225	2	0	N	41	2	N	71.0	180	3	1	N	76	170	1
34	5	20.5	6.7	7-3-8	55.3	101	4	76	29	5	41	184	0	N	66.6	207	1	0	Q	43	6	10	66.0	140	0	2	10	62	70	3
35	6	20.5	7.9	9-4-10	53.7	69	1	78	42	6	41	118	1	N	63.8	120	3	8	N	42	3	N	73.0	180	2	3	N	90	146	1
36	6	20.5	6.0	1-5-8	54.9	76	1	70	39	7	40	113	2	Q	65.1	130	0	0	N	43	0	30	67.0	170	0	8	55	148	130	2
41	5	21.0	7.5	4-1-10	54.4	76	4	65	35	5	41	160	0	N	64.0	175	0	0	N	37	6	N	66.0	155	2	6	N	36	55	1
42	5	20.0	7.6	4-7-19	56.7	90	4	87	42	5	42	185	7	04	64.9	184	1	0	04	49	6	N	65.0	157	3	0	N	85	100	4
43	6	22.0	9.6	5-6-8	52.6	65	1	85	31	9	42	150	0	N	65.2	173	2	4	Q	40	6	Q	70.5	180	6	6	Q	65	150	6
44	6	20.0	6.9	7-6-15	53.4	78	4	100	42	6	42	120	1	20	62.2	132	4	2	20	47	7	Q	67.0	135	0	3	Q	95	260	6
45	5	20.8	9.0	7-2-23	54.1	70	1	79	34	5	42	195	0	Q	62.5	225	2	0	N	42	0	N	71.0	180	3	1	N	84	170	1
46	6	19.5	6.3	7-3-10	53.9	94	1	70	13	5	42	129	0	05	62.6	122	0	0	20	40	5	20	66.9	178	0	5	30	87	210	1
51	9	20.8	6.8	9-4-7	52.6	72	4	84	39	6	42	123	0	N	59.7	118	3	0	N	43	6	40	67.0	170	2	6	20	69	130	3
52	5	20.0	6.6	12-3-10	50.5	58	1	84	32	5	42	110	0	10	66.4	157	2	1	10	44	8	24	71.5	185	2	4	20	100	136	1
53	6	21.0	9.3	1-5-3	56.1	82	4	62	18	8	42	150	0	N	64.7	178	2	0	N	44	7	20	73.0	210	0	6	20	32	81	1
54	5	20.5	6.8	11-5-11	54.1	116	1	92	39	5	43	141	0	Q	61.6	143	4	3	04	45	3	05	64.0	140	4	0	10	108	375	4
55	5	21.0	8.3	4-5-21	54.6	68	1	76	31	5	43	135	0	N	67.6	135	3	3	N	39	0	30	72.0	175	4	0	Q	59	133	3
56	9	18.0	5.9	9-5-16	53.6	66	1	68	36	5	43	133	0	30	63.4	128	1	0	N	38	8	20	70.5	158	1	8	40	60	86	6
61	6	22.0	10.6	3-2-10	56.3	77	4	69	21	6	43	132	0	N	65.6	141	1	0	N	39	0	N	72.0	185	2	0	N	96	150	2
62	9	20.0	7.8	9-5-14	59.2	111	5	80	14	5	44	150	5	05	61.3	180	2	5	06	42	0	24	66.0	155	1	5	20	80	100	1
63	6	21.0	7.8	12-3-15	49.9	55	8	100	16	5	44	114	0	Q	63.9	126	2	0	Q	52	6	Q	67.0	137	3	6	N	50	70	2
64	7	21.3	9.1	3-6-1	55.9	85	8	70	40	6	44	150	0	N	67.4	181	4	0	N	44	0	N	69.5	188	4	3	N	150	190	2
65	5	21.0	8.3	5-3-14	47.8	55	4	80	19	5	45	148	0	N	60.1	147	2	2	N	46	4	20	63.0	150	2	4	Q	68	138	1
66	5	18.0	6.4	9-3-24	52.0	66	4	80	27	5	46	125	2	N	62.4	144	2	3	N	46	3	20	66.0	160	0	3	20	72	300	4

StatLab CENSUS: Boys

	CHILD-BIRTH			CHILD-TEST					MOTHER-BIRTH					MOTHER-TEST					FATH-BRTH			FATHER-TEST					FAMILY		
ID	B	LGTH	WGT	MO-D-HR	HGHT	WGT	L	PEA	RA	B	AG	WGT	O	SM	HGHT	WGT	E	O	SM	AG	O	SM	HGHT	WGT	E	O	SM	I-B	I-T	C
11	5	20.0	5.7	11-7-15	54.7	94	8	64	27	5	15	125	0	Q	66.0	135	3	5	N	21	6	N	70.0	170	3	5	N	22	103	6
12	7	19.7	6.3	6-7-1	54.3	89	4	64	35	5	17	125	8	N	63.1	157	2	0	N	20	6	N	69.0	157	2	6	N	37	99	1
13	2	21.0	7.7	12-5-10	52.4	66	1	71	28	5	17	140	0	N	64.0	212	1	0	15	21	6	Q	65.5	151	2	7	20	44	120	4
14	5	20.5	7.8	4-5-17	50.9	59	1	77	43	3	17	112	0	Q	62.0	125	2	0	N	21	4	10	68.3	148	2	6	Q	48	144	6
15	5	19.5	6.3	9-4-18	51.1	56	2	74	22	7	17	130	0	Q	64.4	146	1	0	N	27	7	N	66.0	128	0	6	N	45	47	1
16	5	20.5	7.6	6-5-7	55.1	63	4	92	49	5	18	149	0	Q	66.9	133	0	0	06	25	8	N	72.0	180	3	3	N	42	96	3
21	7	20.5	6.9	6-5-20	50.8	59	4	69	21	8	18	108	0	02	61.9	119	1	0	Q	19	6	Q	68.0	155	2	6	Q	47	125	1
22	9	21.0	7.0	11-4-5	51.3	63	1	79	37	5	18	119	1	N	65.4	113	2	0	20	23	5	N	67.5	165	2	2	N	66	180	6
23	6	20.5	7.3	1-1-10	54.9	86	4	67	31	7	18	112	0	Q	63.5	112	1	0	N	23	5	N	70.0	165	2	6	20	50	161	6
24	2	20.0	8.8	5-6-16	53.0	68	4	79	30	6	19	126	0	N	63.5	135	2	0	N	22	5	N	69.0	158	2	4	N	47	125	6
25	5	20.5	8.6	6-2-12	52.3	67	4	79	13	5	19	138	1	Q	63.2	163	3	1	N	20	6	15	67.0	160	3	6	Q	104	225	2
26	6	22.5	8.8	11-2-12	55.1	61	3	81	47	6	19	132	0	20	68.0	126	2	0	20	23	8	20	73.7	212	1	8	N	56	117	6
31	6	20.5	7.8	11-4-6	55.5	66	1	89	22	6	19	135	1	Q	67.1	143	3	5	Q	23	6	N	71.0	165	3	7	N	66	104	1
32	6	19.8	6.4	12-7-2	54.4	73	1	78	22	6	19	127	5	07	65.8	147	2	0	N	20	6	10	70.0	203	2	7	06	43	120	3
33	3	22.0	7.2	10-4-10	56.1	72	1	85	38	7	19	118	0	N	65.5	150	3	1	N	22	8	N	71.0	165	3	6	N	20	165	1
34	5	22.0	11.0	11-5-18	56.9	88	1	88	19	5	19	180	1	Q	68.0	225	3	0	N	21	6	N	73.0	210	2	2	N	78	140	1
35	6	19.5	8.6	5-3-7	50.2	58	2	71	38	6	19	110	4	N	65.5	134	2	0	15	23	7	N	71.0	200	2	5	N	67	66	1
36	6	20.0	6.4	5-4-13	55.1	54	4	74	32	4	19	149	0	25	62.6	143	2	0	30	22	6	N	70.5	195	3	0	N	58	150	1
41	1	21.5	8.2	7-1-22	52.0	68	1	79	11	5	19	114	1	22	61.4	136	2	4	40	20	6	20	69.0	180	2	6	20	56	175	6
42	6	19.5	6.4	5-7-22	49.4	54	1	57	14	6	19	170	0	N	64.8	168	0	5	10	24	5	20	70.0	185	2	7	20	43	88	1
43	1	21.5	8.4	8-2-21	55.1	74	1	90	30	1	19	95	0	N	60.3	109	2	0	N	22	8	20	74.0	170	2	6	05	50	140	4
44	6	20.0	7.3	6-1-7	51.3	62	4	61	21	6	20	145	0	02	64.3	179	1	0	N	24	7	20	70.8	185	2	6	03	48	96	4
45	9	20.5	8.3	5-5-21	53.2	71	1	77	28	9	20	137	1	10	63.2	148	2	1	20	20	5	10	70.8	180	2	6	10	70	129	6
46	5	21.0	7.8	6-4-16	56.4	84	4	98	34	5	20	133	1	03	65.4	132	2	0	N	23	0	05	73.0	175	3	0	Q	67	130	1
51	6	18.5	5.8	5-4-9	55.9	78	5	85	21	5	20	132	1	30	67.9	124	1	0	20	23	6	40	75.1	195	3	6	20	78	130	4
52	7	20.5	8.6	9-5-4	55.4	72	4	84	40	5	20	157	1	N	65.1	175	2	5	N	26	2	20	73.0	180	1	5	30	54	124	6
53	6	22.0	7.7	1-7-13	49.3	55	5	69	23	5	20	120	0	12	64.4	130	2	0	02	22	6	20	67.0	170	1	6	Q	50	240	1
54	5	20.5	7.4	12-6-4	54.0	70	1	82	34	5	20	130	1	N	67.4	134	2	0	N	25	6	N	70.9	220	2	2	N	65	124	6
55	7	20.5	7.3	12-4-4	53.8	70	2	85	40	2	20	115	1	20	65.3	109	2	1	30	23	7	22	72.0	150	3	7	22	60	166	6
56	5	21.0	7.8	3-5-20	52.5	68	1	76	24	6	20	142	1	N	65.6	160	2	1	N	21	6	N	74.0	204	3	2	N	55	165	6
61	6	21.5	8.3	5-3-2	53.4	70	1	78	25	1	20	165	1	N	65.1	265	2	0	N	23	5	N	73.0	185	2	2	N	63	120	4
62	2	21.0	7.7	3-6-21	54.1	68	3	69	14	6	20	105	0	03	61.4	112	2	0	04	23	6	18	73.0	170	3	6	18	70	144	4
63	5	20.5	6.0	7-7-19	51.8	59	1	68	30	5	20	117	1	01	66.9	128	2	0	02	29	6	07	68.0	165	1	6	Q	68	60	6
64	5	20.0	7.1	8-1-21	51.1	52	4	73	24	5	20	104	1	N	62.3	117	2	0	N	25	5	20	70.0	136	2	6	Q	82	100	4
65	5	22.0	10.0	12-7-7	53.3	85	5	75	26	5	20	125	4	Q	63.1	139	2	0	N	23	2	30	70.0	165	2	2	40	160	305	4
66	6	22.0	9.8	11-3-4	53.6	62	1	68	25	6	20	155	1	03	69.3	146	2	0	Q	23	6	40	71.0	137	2	0	30	28	95	2

	CHILD-BIRTH				CHILD-TEST					MOTHER-BIRTH					MOTHER-TEST					FATH-BRTH			FATHER-TEST					FAMILY		
ID	B	LGTH	WGT	MO-D-HR	HGHT	WGT	L	PEA	RA	B	AG	WGT	O	SM	HGHT	WGT	E	O	SM	AG	O	SM	HGHT	WGT	E	O	SM	I-B	I-T	C
11	5	20.0	6.6	11-3-20	52.5	71	1	63	13	7	20	135	1	N	63.8	155	2	0	N	25	6	34	68.0	160	2	8	Q	104	100	4
12	5	21.0	8.0	12-6-9	50.1	58	2	75	16	5	20	125	0	Q	66.8	140	3	4	N	22	8	20	66.5	130	1	6	25	26	147	2
13	5	21.3	8.6	1-3-4	52.9	87	1	82	35	5	20	135	1	20	65.4	132	2	1	12	22	7	10	70.0	187	2	7	40	86	193	5
14	6	22.0	8.6	2-4-12	56.1	67	4	79	19	6	20	120	0	N	63.1	134	3	0	N	24	6	10	76.0	188	3	7	Q	64	164	1
15	5	21.0	6.9	2-3-16	53.1	56	1	61	18	5	20	131	0	10	67.2	144	3	4	10	21	8	04	73.0	250	2	7	15	42	114	1
16	5	21.0	7.6	2-7-12	50.5	60	4	86	21	6	20	126	1	05	64.8	111	2	0	05	22	6	20	70.5	175	1	6	20	79	134	1
21	6	20.0	8.5	2-4-3	53.1	71	5	77	20	6	20	113	1	04	63.1	131	2	1	N	23	7	04	74.0	195	1	7	N	85	21	3
22	9	19.0	5.6	3-3-3	52.3	60	5	103	33	5	20	130	1	N	68.0	125	2	1	N	21	5	11	70.0	145	3	6	20	75	249	6
23	6	20.0	7.0	2-5-9	51.4	58	1	61	12	6	21	128	2	Q	66.9	147	2	0	01	25	6	20	70.0	150	3	8	20	64	137	6
24	7	19.5	5.7	4-5-24	50.4	57	1	80	23	6	21	122	1	10	63.8	136	2	0	30	24	3	N	73.0	164	3	6	20	82	139	6
25	6	19.5	7.6	11-3-2	54.3	70	4	82	42	6	21	132	7	N	65.0	145	3	8	N	30	6	10	73.0	180	3	0	N	80	172	3
26	9	21.0	6.9	9-4-16	51.4	70	4	78	23	5	21	125	1	20	65.9	126	2	1	20	22	6	20	72.0	230	3	0	40	126	191	6
31	5	21.0	9.9	4-3-16	55.7	86	4	81	47	5	21	130	0	N	62.0	135	2	0	N	33	0	10	70.3	218	4	3	20	65	230	1
32	5	17.0	4.8	4-1-19	57.6	80	1	63	11	5	21	110	0	N	62.0	134	2	0	N	25	4	04	71.0	140	2	7	10	29	60	4
33	9	20.0	8.0	4-5-16	51.7	64	3	91	30	6	21	123	0	N	62.4	140	3	0	N	21	4	N	67.0	160	2	2	N	42	120	6
34	7	20.5	6.9	8-4-23	52.8	78	8	82	27	2	21	125	0	N	62.1	133	3	0	N	22	6	10	71.0	165	2	4	Q	58	114	6
35	6	20.5	8.6	5-7-4	55.9	81	1	81	21	6	21	154	0	05	64.5	205	3	5	N	25	8	20	68.0	150	2	7	10	21	150	1
36	6	20.5	6.3	8-3-10	58.4	86	1	86	36	4	21	136	0	20	66.7	135	3	0	20	24	5	05	69.4	182	3	6	10	54	120	6
41	6	21.0	7.6	6-4-8	50.9	64	8	76	33	6	21	134	0	N	64.1	149	2	0	N	25	8	04	66.0	145	1	3	05	39	104	4
42	6	20.0	6.8	9-5-22	54.9	82	1	76	44	6	21	120	0	Q	67.6	142	3	0	Q	23	2	10	71.0	165	3	2	06	42	86	3
43	6	19.5	5.8	8-6-4	56.3	90	8	80	13	6	21	118	0	03	64.7	135	2	7	07	26	5	10	73.0	158	3	2	15	45	155	3
44	6	22.0	8.2	9-2-18	54.8	67	1	83	39	5	21	126	1	N	66.4	132	3	8	N	24	4	20	72.0	160	3	0	Q	76	106	3
45	6	21.5	6.8	9-2-14	58.9	82	4	84	30	6	21	122	4	20	66.2	117	3	0	20	24	2	10	77.0	250	3	0	03	88	160	1
46	5	20.5	7.0	10-1-20	56.4	96	4	93	45	4	21	120	3	N	62.0	125	4	0	N	29	6	N	69.9	151	4	5	N	102	180	1
51	6	19.0	6.5	5-6-14	53.5	68	4	82	28	5	21	104	0	N	64.9	118	2	7	N	24	8	10	72.0	210	2	7	10	26	219	4
52	6	21.5	6.9	4-3-12	51.3	60	5	65	22	6	21	135	0	10	67.1	143	2	5	Q	25	4	N	71.0	165	3	7	N	62	116	1
53	6	20.5	6.7	11-5-15	53.2	66	6	72	28	5	21	143	0	10	67.1	180	2	0	Q	22	6	20	71.0	210	1	6	15	50	73	1
54	6	21.0	7.1	12-3-14	52.2	72	1	102	41	6	21	120	0	N	60.6	138	2	0	12	25	6	20	71.0	203	3	0	N	55	120	1
55	1	21.0	8.3	11-3-4	54.9	75	2	78	39	6	21	140	5	N	67.8	174	3	0	N	21	0	N	75.5	210	3	3	Q	20	300	1
56	6	20.0	7.3	12-2-24	55.0	70	5	123	38	5	21	141	1	12	70.0	176	3	0	20	25	6	12	74.5	210	3	2	N	70	96	6
61	8	20.5	7.4	12-4-4	56.3	96	1	69	31	7	21	155	0	Q	64.9	212	1	0	N	25	5	N	71.0	165	3	4	N	39	72	1
62	5	21.0	8.1	10-1-22	52.1	69	4	82	37	5	21	130	0	01	61.3	141	3	0	20	26	6	20	74.0	216	3	0	Q	36	120	6
63	5	20.5	8.2	4-2-13	54.9	77	4	79	28	6	21	129	0	09	61.8	166	2	0	15	26	8	10	73.0	195	2	0	40	60	120	6
64	5	21.0	7.9	11-1-3	51.4	61	4	76	26	6	21	140	1	N	63.9	197	2	0	N	35	7	Q	66.0	155	0	7	Q	72	147	6
65	6	21.5	7.8	4-3-11	51.6	60	1	82	34	5	21	115	0	N	62.2	114	2	0	N	21	6	20	67.1	159	2	8	Q	39	72	4
66	6	21.0	7.6	4-5-12	57.4	84	4	87	27	6	21	110	2	Q	65.8	126	3	3	20	25	5	10	73.0	200	3	0	Q	56	158	3

ID	|| CHILD-BIRTH			|| CHILD-TEST					|| MOTHER-BIRTH					|| MOTHER-TEST					|| FATH-BRTH			|| FATHER-TEST					|| FAMILY		
	B	LGTH	WGT	MO-D-HR	HGHT	WGT	L	PEA	RA	B	AG	WGT	O	SM	HGHT	WGT	E	O	SM	AG	O	SM	HGHT	WGT	E	O	SM	I-B	I-T	C
11	6	21.0	7.7	7-4-8	53.3	69	1	79	26	6	21	110	1	N	64.9	127	2	0	N	23	6	N	69.0	175	3	4	N	86	200	6
12	8	20.0	7.9	8-1-2	50.4	67	4	82	19	7	21	114	1	N	60.1	133	2	1	N	27	5	Q	66.0	165	3	0	N	110	190	6
13	5	20.5	7.9	8-7-24	48.6	60	1	66	40	2	21	105	0	Q	60.4	113	3	1	Q	28	5	N	69.0	167	3	8	20	43	208	6
14	9	20.0	7.6	9-3-22	53.6	73	4	76	39	6	21	120	7	20	65.1	128	2	0	25	28	2	20	70.5	180	2	2	40	120	144	6
15	7	19.5	6.6	11-6-6	56.1	72	4	77	37	1	21	128	1	15	63.6	152	2	1	17	23	6	25	71.8	165	3	3	20	99	180	4
16	6	21.5	8.9	11-6-17	53.9	64	4	81	11	5	21	127	0	30	67.9	130	2	0	40	24	6	10	69.0	135	3	2	Q	72	120	4
21	9	22.0	8.3	1-6-3	53.1	70	8	74	21	5	21	97	1	10	61.9	110	3	0	Q	24	2	N	70.0	190	3	6	N	53	96	6
22	5	21.0	8.1	2-6-18	56.1	89	1	111	40	1	21	147	8	20	66.8	153	4	4	Q	27	2	20	73.0	240	3	8	Q	56	318	6
23	6	22.0	9.3	3-6-23	55.6	70	1	83	38	5	21	135	1	N	69.3	153	3	0	N	25	6	20	74.0	197	3	0	N	71	192	1
24	6	20.0	7.1	11-2-6	51.1	56	1	58	13	5	21	110	0	Q	64.0	114	2	4	10	28	0	N	68.6	186	3	0	N	60	176	1
25	7	20.5	7.5	12-7-6	54.1	64	1	57	24	7	21	137	0	N	66.3	179	1	0	N	27	8	08	70.0	170	1	7	02	26	108	1
26	5	21.0	8.4	7-4-14	55.3	84	4	123	47	5	22	172	0	N	65.8	145	4	3	N	22	0	10	68.5	150	3	2	Q	45	176	6
31	5	20.5	7.4	10-5-22	55.0	73	8	81	29	6	22	132	0	N	66.6	134	3	0	N	30	6	20	70.0	220	2	6	25	81	100	2
32	6	22.0	7.8	3-3-22	52.4	60	4	92	34	2	22	118	0	N	58.3	129	3	1	N	25	0	05	76.0	200	4	3	N	32	192	4
33	6	20.5	7.4	11-1-7	57.4	80	4	85	33	5	22	135	0	20	67.8	145	1	0	25	25	6	10	72.0	185	1	7	20	90	144	4
34	7	21.0	8.4	9-3-2	50.9	66	1	79	19	5	22	121	1	N	60.7	133	2	0	N	23	5	N	68.0	170	3	5	N	80	48	1
35	6	20.5	8.5	10-1-24	53.1	76	1	82	35	5	22	110	8	Q	63.0	130	2	0	N	23	7	20	72.0	205	1	6	30	67	91	6
36	7	20.0	6.9	1-7-3	50.5	60	1	74	17	6	22	118	4	N	60.2	139	2	1	08	24	4	20	70.0	180	2	5	Q	37	129	6
41	4	20.0	7.3	4-5-19	53.8	63	8	90	29	3	22	114	1	N	65.4	111	3	0	N	22	0	Q	68.0	175	4	0	Q	60	150	6
42	6	20.8	6.9	8-5-1	49.0	60	1	79	9	5	22	136	0	10	64.3	120	3	7	20	23	7	20	71.5	160	3	7	28	41	87	4
43	6	19.0	6.1	4-7-14	53.8	71	1	82	41	7	22	107	0	03	61.8	103	3	3	Q	23	0	20	75.0	215	4	2	20	40	130	4
44	1	19.5	5.5	9-7-17	56.9	86	1	76	31	2	22	115	8	10	62.0	150	4	3	15	23	4	20	77.5	240	2	2	Q	75	305	6
45	5	21.0	8.3	10-1-4	50.7	63	1	99	50	5	22	119	1	N	62.9	117	3	8	N	29	6	N	69.0	235	2	6	N	40	144	1
46	5	20.0	7.1	5-5-5	53.1	64	4	62	14	5	22	104	4	N	64.9	118	2	7	N	25	5	24	72.0	210	2	7	10	42	219	4
51	6	21.0	8.2	11-2-7	53.8	68	1	89	42	8	22	124	1	10	65.3	136	3	0	N	22	0	N	71.8	170	4	4	N	60	150	1
52	6	20.8	7.6	12-1-3	54.2	64	1	80	28	1	22	133	0	N	66.0	130	1	7	N	28	4	35	70.1	162	2	5	30	83	198	2
53	7	20.0	7.4	10-5-8	53.8	72	4	69	31	2	22	160	0	N	64.9	212	1	0	N	26	5	N	71.0	165	3	4	10	36	72	1
54	5	20.5	7.9	4-4-3	52.9	60	8	70	26	5	22	165	7	20	68.0	204	2	0	20	25	6	15	68.0	160	3	6	10	67	89	6
55	7	19.5	6.8	9-3-4	51.4	76	1	72	26	5	22	112	0	15	59.0	125	3	3	Q	23	6	01	66.2	177	2	6	N	48	85	1
56	5	21.0	7.3	4-1-7	56.8	112	2	80	11	5	22	160	5	Q	64.6	158	2	5	05	30	8	N	69.5	200	2	6	N	32	131	1
61	7	21.0	7.8	5-7-12	50.4	61	4	84	40	5	22	96	4	N	58.6	105	4	1	N	24	0	Q	69.5	175	4	3	Q	24	200	6
62	5	20.5	8.2	4-6-1	54.9	64	1	68	17	6	22	135	1	Q	67.3	161	3	0	N	26	5	N	73.0	210	4	0	N	47	110	6
63	5	21.8	9.4	6-7-17	54.4	72	5	84	17	6	22	130	1	Q	67.0	136	3	0	06	23	6	24	73.0	190	4	0	15	46	150	1
64	6	21.0	7.7	7-2-18	49.4	55	1	65	13	2	22	121	0	20	51.1	125	1	0	10	23	6	06	74.0	190	2	6	Q	82	211	4
65	6	20.0	6.6	9-7-19	52.2	56	1	78	31	4	22	89	1	18	62.9	97	3	0	12	27	5	20	68.0	140	3	2	20	80	120	6
66	5	20.5	7.6	10-1-8	56.1	79	1	64	39	7	22	107	0	N	67.2	121	2	1	N	23	7	24	72.0	205	3	8	Q	45	152	4

StatLab CENSUS: Boys

	CHILD-BIRTH				CHILD-TEST					MOTHER-BIRTH					MOTHER-TEST					FATH-BRTH			FATHER-TEST					FAMILY		
ID	B	LGTH	WGT	MO-D-HR	HGHT	WGT	L	PEA	RA	B	AG	WGT	O	SM	HGHT	WGT	E	O	SM	AG	O	SM	HGHT	WGT	E	O	SM	I-B	I-T	C
11	6	20.0	6.2	11-4-10	49.3	56	1	79	21	6	22	95	0	N	57.9	133	1	0	N	27	6	Q	70.5	195	3	6	N	48	120	1
12	6	21.5	7.9	1-6-1	53.1	71	4	88	34	6	22	135	1	25	65.3	134	3	0	Q	24	6	40	70.0	180	6	19		79	120	2
13	5	19.8	7.1	1-7-2	53.1	66	1	58	29	6	22	125	1	20	64.9	144	3	2	Q	24	7	10	67.0	170	1	4	Q	75	34	1
14	5	18.5	5.8	3-5-1	48.9	66	8	82	46	5	22	107	1	20	61.9	108	3	2	25	27	0	N	66.0	150	4	2	N	81	230	1
15	7	20.5	7.5	3-1-16	53.6	77	1	87	21	7	22	172	1	20	67.0	180	4	3	20	25	5	20	74.0	220	4	2	N	81	170	1
16	6	20.5	7.5	3-1-7	53.6	66	2	89	42	6	22	97	4	20	58.5	109	3	4	18	25	5	20	74.0	190	4	0	20	77	209	6
21	2	22.0	8.2	9-5-19	52.8	66	1	78	27	5	23	122	0	Q	65.0	112	3	0	20	23	5	20	72.0	170	3	6	01	57	108	3
22	5	21.0	8.5	6-6-2	53.5	84	1	85	38	6	23	153	1	20	67.2	159	2	1	20	29	5	N	74.0	172	3	0	N	38	147	1
23	6	18.5	6.0	3-4-23	53.1	81	5	72	39	6	23	116	0	N	65.4	128	3	0	N	24	6	N	68.0	180	3	6	N	60	114	1
24	6	20.0	6.8	12-6-15	52.4	68	4	90	23	5	23	105	1	N	61.7	109	3	0	N	27	0	10	68.6	151	4	0	15	70	200	4
25	1	20.5	6.4	4-6-8	50.3	74	1	82	29	5	23	102	1	20	61.0	110	3	4	30	32	2	10	65.4	164	2	8	20	96	166	4
26	6	20.5	7.9	8-3-22	52.2	64	1	82	30	6	23	120	1	N	63.3	139	3	0	N	26	0	N	69.0	170	4	0	20	90	120	1
31	9	20.5	7.2	6-3-13	55.9	70	4	87	43	6	23	158	1	N	66.0	160	2	0	N	25	6	20	75.0	220	2	2	Q	41	144	2
32	9	21.0	7.6	8-4-15	55.4	70	1	96	29	5	23	118	0	15	65.9	119	3	0	N	26	0	20	73.0	170	3	3	18	35	800	6
33	6	18.5	6.0	7-5-9	51.9	58	1	88	41	4	23	99	3	Q	63.0	105	4	0	N	24	0	01	71.0	190	4	1	40	34	200	2
34	6	20.0	6.8	9-6-17	57.7	80	1	82	25	5	23	152	3	20	70.0	155	4	3	15	23	0	N	71.0	177	4	1	N	47	185	1
35	9	19.5	6.8	11-6-22	51.2	62	1	73	26	5	23	113	0	10	64.3	122	2	0	N	25	6	N	72.0	165	2	2	N	55	110	1
36	5	21.0	7.7	8-7-24	54.4	74	1	71	21	6	23	127	0	N	67.0	140	2	5	N	23	7	N	70.2	183	2	6	N	45	137	4
41	6	20.5	7.4	8-6-16	52.6	70	8	82	11	7	23	117	5	N	60.6	129	2	5	N	24	5	N	73.0	225	4	0	N	44	206	1
42	5	20.8	7.5	11-3-2	54.9	71	3	84	28	6	23	114	1	N	67.0	140	3	0	N	23	7	Q	73.0	225	3	0	N	67	144	6
43	5	20.0	7.3	1-2-22	51.6	66	4	79	34	6	23	128	0	08	60.5	138	2	0	12	27	6	11	71.0	203	3	0	N	62	120	1
44	5	20.0	7.8	12-2-4	58.6	112	1	75	12	6	23	145	1	N	66.2	166	3	0	N	25	0	N	74.0	215	4	2	N	124	192	6
45	5	20.5	7.2	12-4-2	51.9	54	4	86	24	6	23	106	1	N	60.6	114	3	0	N	25	2	N	71.0	154	4	2	N	89	300	4
46	5	20.0	7.2	11-3-19	54.2	72	4	92	45	6	23	215	0	17	69.1	240	3	0	40	38	8	20	69.2	172	3	6	20	10	100	6
51	5	20.0	7.5	5-6-21	52.7	70	4	83	25	7	23	128	1	15	66.0	155	3	1	Q	26	6	30	69.5	180	3	6	Q	68	114	6
52	5	21.0	7.6	4-3-8	57.0	82	4	84	20	6	23	130	0	N	62.3	146	2	1	N	28	8	N	73.5	190	3	3	N	68	172	1
53	5	20.0	7.2	4-1-22	53.6	62	1	75	21	5	23	134	0	03	67.5	157	2	1	03	27	5	N	75.0	190	2	2	N	58	165	4
54	6	20.5	7.3	3-6-18	50.8	60	2	65	35	6	23	155	0	Q	62.3	201	2	0	30	22	6	20	69.0	185	2	6	20	53	60	1
55	1	21.0	7.8	4-6-23	55.7	82	2	61	29	5	23	133	1	Q	64.1	144	3	1	N	21	6	Q	69.5	203	3	6	N	90	202	1
56	7	18.5	5.9	9-3-24	46.6	48	4	64	28	3	23	90	0	N	59.2	93	3	0	N	29	1	N	71.0	180	4	1	N	30	150	1
61	8	21.0	7.5	4-3-7	58.5	85	2	71	25	8	23	116	1	40	66.5	118	2	1	N	29	6	10	75.5	189	1	7	10	81	161	6
62	7	21.0	7.4	8-5-8	52.4	58	4	87	21	3	23	113	1	N	66.2	121	3	1	N	23	8	15	75.0	190	3	8	Q	107	170	2
63	5	21.5	7.8	6-1-21	54.6	68	4	81	27	5	23	104	1	N	64.7	118	2	3	N	22	6	20	72.0	184	1	6	20	80	157	4
64	6	21.0	7.0	1-7-8	50.7	56	1	63	8	6	23	137	0	N	64.9	126	3	0	N	26	6	03	71.0	185	3	6	N	80	115	4
65	6	20.5	6.4	1-5-7	52.7	70	1	94	39	6	23	120	1	15	63.8	125	2	0	Q	25	6	20	69.5	170	3	7	20	94	120	1
66	5	20.0	7.1	2-5-18	51.8	65	1	71	16	7	23	109	1	20	62.9	118	3	1	20	28	7	Q	72.0	185	2	5	Q	58	182	4

StatLab CENSUS: Boys

ID	CHILD-BIRTH B	LGTH	WGT	MO-D-HR	CHILD-TEST HGHT	WGT	L	PEA	RA	MOTHER-BIRTH B	AG	WGT	O	SM	MOTHER-TEST HGHT	WGT	E	O	SM	FATH-BRTH AG	O	SM	FATHER-TEST HGHT	WGT	E	O	SM	FAMILY I-B	I-T	C
11	7	20.5	7.2	3-5-18	51.4	58	1	85	37	7	23	112	4	03	63.0	112	4	4	N	27	0	N	65.4	154	4	0	N	104	85	6
12	5	20.8	7.1	6-2-15	54.4	74	1	68	22	5	24	120	0	N	64.6	142	3	0	N	24	0	N	77.5	225	4	0	N	30	150	1
13	8	19.5	6.7	7-7-12	52.5	58	1	89	21	5	24	115	0	Q	63.5	115	4	1	Q	31	0	Q	68.5	166	4	1	Q	72	130	1
14	5	19.5	7.1	12-4-18	53.5	68	4	87	33	6	24	113	1	N	67.5	130	3	0	N	25	8	10	73.7	189	3	1	15	97	135	1
15	6	19.8	7.3	11-2-5	54.6	76	4	69	28	5	24	141	0	N	61.3	129	0	0	N	27	6	06	68.0	160	0	7	N	16	68	1
16	7	22.3	9.4	2-7-19	55.8	73	5	84	37	5	24	150	0	N	66.4	187	2	0	N	34	6	20	70.0	160	2	7	20	50	100	1
21	6	19.0	6.4	12-1-5	55.2	76	1	69	33	5	24	125	0	N	63.8	172	3	0	N	27	7	N	73.0	198	0	8	N	53	36	1
22	6	20.8	8.3	5-3-13	52.5	70	1	72	19	5	24	104	0	05	60.2	118	3	0	N	27	6	15	70.0	180	3	6	Q	57	150	1
23	5	20.0	8.1	12-1-15	52.3	62	4	72	22	5	24	125	0	N	62.9	153	4	3	03	30	7	N	71.0	154	3	0	22	48	90	4
24	6	19.0	7.7	4-1-15	53.9	92	1	92	47	6	24	121	0	20	64.9	126	2	0	20	25	5	30	66.0	155	4	8	30	42	144	4
25	5	19.5	8.1	4-1-22	56.1	88	1	81	31	5	24	143	4	20	65.5	137	3	0	30	26	6	N	72.4	185	3	6	N	60	94	1
26	5	20.0	8.2	4-7-2	54.3	65	1	89	37	6	24	92	1	N	62.1	110	3	0	N	28	5	30	73.0	183	4	6	30	74	34	6
31	6	20.5	7.7	5-2-12	57.2	102	1	100	40	6	24	150	5	12	65.1	208	2	0	20	29	6	20	69.8	198	2	0	20	86	110	6
32	5	21.8	8.8	5-7-24	57.8	78	4	80	21	5	24	169	0	Q	68.9	189	2	0	N	27	6	N	74.0	215	2	0	N	60	120	6
33	6	22.5	8.9	8-7-7	52.8	65	1	69	14	5	24	125	0	N	63.3	133	0	0	N	25	8	03	68.0	190	1	7	01	42	65	2
34	6	18.0	5.6	5-2-4	51.6	57	5	84	16	6	24	114	0	N	66.0	129	3	0	N	25	6	N	65.9	145	2	6	N	39	78	3
35	6	21.0	8.7	8-6-5	58.2	86	6	80	33	5	24	141	0	N	69.8	153	2	0	N	30	6	N	72.0	180	2	0	N	60	125	4
36	5	20.5	6.8	7-5-17	51.6	65	1	90	49	5	24	122	0	N	65.6	133	4	0	N	30	0	N	68.0	145	4	0	N	58	162	1
41	5	20.5	6.1	6-1-10	53.9	62	1	85	48	5	24	113	1	N	62.2	114	2	1	N	29	6	14	72.0	180	1	6	15	73	163	2
42	1	21.5	7.4	8-4-13	61.9	99	2	84	21	5	24	160	8	20	72.0	172	2	0	N	30	4	20	71.4	211	1	4	20	77	160	3
43	6	20.5	7.8	5-6-10	52.5	68	4	94	45	6	24	122	0	15	64.3	118	4	4	N	23	4	15	75.0	201	4	0	Q	75	406	6
44	4	21.0	8.1	4-2-17	53.2	74	1	90	43	2	24	115	0	N	64.2	128	3	3	N	25	1	Q	72.0	200	4	1	N	61	142	6
45	5	21.0	9.1	11-6-19	56.8	66	1	84	38	5	24	125	1	Q	67.3	175	4	0	03	26	0	20	73.5	175	4	0	25	95	220	6
46	5	21.5	8.1	11-1-7	53.8	66	1	79	42	6	24	135	5	N	65.9	143	1	0	10	26	6	N	73.0	170	2	8	02	46	158	3
51	5	21.5	8.1	12-2-11	54.1	75	1	81	21	5	24	140	1	20	65.9	162	3	0	35	24	5	50	69.9	171	4	0	50	47	100	6
52	7	20.5	8.4	12-3-7	53.8	62	8	94	47	7	24	110	0	20	62.8	112	3	0	N	31	8	N	68.0	148	4	0	N	95	170	2
53	7	20.0	7.4	9-4-5	53.3	62	1	78	35	6	24	218	0	20	69.1	240	3	0	40	39	8	20	69.2	172	3	6	20	41	100	6
54	1	20.5	7.3	6-4-15	53.2	60	1	87	35	6	24	117	4	N	66.1	121	4	0	N	29	1	N	71.6	193	4	0	N	81	180	6
55	6	18.8	5.3	5-3-3	53.1	48	1	82	48	5	24	159	1	N	67.0	150	2	8	N	26	8	N	71.6	201	3	8	N	66	157	3
56	6	20.5	7.1	6-2-12	54.6	67	1	72	32	5	24	110	0	N	63.1	120	3	1	N	28	6	N	71.5	190	3	0	N	75	136	4
61	7	17.0	3.9	7-6-1	49.9	55	1	75	41	7	24	99	1	N	58.6	108	4	0	N	26	0	N	66.0	135	4	0	N	100	250	1
62	1	21.5	7.9	9-6-20	49.9	60	4	82	29	1	24	125	2	N	63.4	131	2	3	N	25	5	N	66.0	180	2	3	N	70	157	4
63	7	20.0	6.5	7-1-3	53.2	70	3	76	28	7	24	120	0	N	61.1	108	2	0	N	30	0	30	66.5	170	4	0	Q	81	200	5
64	6	22.0	8.3	11-2-2	54.9	80	1	69	23	5	24	140	0	20	64.7	178	2	7	N	25	8	02	70.5	185	2	7	10	71	153	4
65	5	19.5	7.8	11-5-23	52.4	59	1	86	22	6	24	130	1	15	66.7	147	3	1	N	24	6	25	71.0	190	2	3	Q	100	46	6
66	9	21.0	7.8	11-4-16	50.8	57	4	82	35	6	24	117	0	15	60.1	112	2	0	N	20	6	N	68.0	160	2	6	N	48	130	6

StatLab CENSUS: Boys

	CHILD-BIRTH				CHILD-TEST					MOTHER-BIRTH					MOTHER-TEST					FATH-BRTH			FATHER-TEST					FAMILY		
ID	B	LGTH	WGT	MO-D-HR	HGHT	WGT	L	PEA	RA	B	AG	WGT	O	SM	HGHT	WGT	E	O	SM	AG	O	SM	HGHT	WGT	E	O	SM	I-B	I-T	C
11	6	20.5	6.8	10-1-12	50.8	60	1	89	23	5	24	137	0	20	66.9	158	4	0	20	26	1	N	72.0	175	4	3	N	38	425	6
12	6	21.3	7.3	12-6-19	55.9	75	1	87	28	6	24	145	0	N	68.0	135	4	3	N	31	0	N	70.8	174	4	0	N	35	189	6
13	5	20.5	6.1	1-5-12	53.9	70	2	79	31	5	24	106	1	Q	61.6	108	2	0	Q	25	7	10	68.0	160	2	0	Q	86	96	6
14	5	19.0	8.7	10-5-11	51.4	57	4	53	27	7	24	146	0	N	63.3	150	1	0	01	28	6	Q	65.0	159	1	8	N	29	64	2
15	6	20.0	7.3	11-2-8	49.4	53	4	84	35	7	24	105	1	N	59.5	133	3	0	N	29	0	24	66.8	135	4	0	22	94	180	6
16	6	21.3	9.0	11-6-16	52.4	62	3	78	38	8	24	128	0	N	65.5	139	2	0	N	25	7	N	71.0	210	2	6	N	36	108	1
21	6	22.0	8.8	10-3-24	56.3	62	1	83	27	2	25	135	0	20	65.7	162	1	8	Q	30	8	20	75.0	190	2	8	40	75	223	1
22	8	22.0	8.2	8-2-7	57.4	90	4	79	50	6	25	142	3	Q	66.4	163	4	3	N	25	5	20	75.0	220	3	2	30	60	200	1
23	5	21.0	7.0	6-2-21	52.3	66	1	70	33	5	25	125	1	20	64.2	148	3	0	20	27	6	15	69.0	180	1	6	20	85	130	1
24	5	21.0	8.0	5-5-3	52.4	72	8	81	40	5	25	147	0	06	65.6	180	3	0	07	26	2	11	69.0	175	4	2	Q	55	220	1
25	6	20.5	7.9	1-6-4	54.3	77	2	89	36	2	25	144	0	N	69.1	174	3	0	N	25	6	15	72.0	200	3	8	N	36	116	6
26	6	21.0	7.5	11-1-1	52.6	59	1	90	39	8	24	110	0	Q	64.6	111	3	0	N	33	5	N	69.5	165	2	5	N	55	100	1
31	5	20.5	7.6	12-1-22	54.6	64	1	75	24	5	25	168	0	N	65.0	156	4	0	N	27	3	N	68.5	150	4	0	N	25	315	4
32	9	20.5	6.9	4-3-7	54.0	64	8	65	31	5	25	150	0	N	61.7	205	2	0	N	41	6	N	71.0	220	3	7	N	60	100	1
33	5	21.0	7.9	7-1-18	53.1	68	8	82	19	5	25	150	4	N	62.8	188	3	0	N	26	0	12	70.5	198	4	3	N	103	200	4
34	5	21.0	7.1	4-5-7	53.3	61	4	70	14	5	25	140	0	20	65.5	137	3	0	30	27	6	20	72.0	185	3	6	N	40	94	1
35	6	20.0	9.4	7-6-20	55.9	65	1	71	24	5	25	117	0	10	67.9	125	3	2	20	25	6	20	77.0	225	2	3	35	73	210	6
36	5	20.0	7.1	4-1-16	54.8	88	1	94	44	6	25	145	3	20	64.6	151	4	3	40	30	0	20	70.0	170	4	1	40	117	287	6
41	6	22.0	8.9	7-6-14	55.3	70	3	99	43	6	25	100	1	20	61.1	113	2	0	30	27	4	N	71.0	165	4	8	N	90	126	6
42	5	20.5	6.8	7-3-8	53.5	62	2	94	18	5	25	95	0	N	59.6	109	4	0	N	27	0	20	70.0	150	2	8	15	72	100	2
43	9	19.0	6.9	7-6-15	55.3	73	1	77	34	5	25	140	3	Q	67.6	147	4	0	Q	27	0	10	70.5	165	4	0	N	95	140	1
44	5	19.0	7.4	9-4-24	53.1	72	8	73	32	6	25	146	0	N	61.7	170	2	0	N	32	5	06	68.7	168	2	7	08	51	96	1
45	6	20.0	7.3	8-1-19	53.9	64	1	83	41	5	25	120	1	N	66.9	128	2	1	N	28	0	15	70.0	145	2	6	15	82	177	4
46	8	20.0	7.4	5-4-12	55.1	76	8	79	12	3	25	134	0	17	65.2	151	2	0	20	31	8	20	70.5	175	3	8	20	67	120	1
51	6	21.0	8.8	9-5-11	54.0	69	1	74	28	6	25	103	0	N	59.9	118	3	0	N	25	6	N	72.0	149	3	7	N	60	78	2
52	5	19.0	5.9	6-6-5	54.2	74	4	91	23	5	25	114	8	20	62.8	138	2	0	30	27	4	30	70.5	165	3	4	30	90	100	6
53	7	21.5	8.1	8-4-24	56.4	64	1	72	14	7	25	138	1	N	64.1	160	2	1	N	31	7	10	65.0	170	1	6	N	47	183	3
54	5	19.0	5.8	8-2-3	50.2	58	5	80	44	2	25	108	0	20	60.7	136	3	0	N	28	0	20	68.0	190	4	2	N	55	200	1
55	5	19.0	5.0	10-5-21	55.7	91	4	106	35	6	25	125	0	N	61.9	155	2	0	N	28	6	1	71.4	199	3	3	N	50	140	1
56	1	20.5	6.8	9-1-12	51.8	60	1	78	49	5	25	106	0	07	60.9	103	4	3	15	36	1	Q	66.0	130	4	1	Q	65	96	4
61	6	20.0	6.1	11-6-22	55.7	84	4	86	20	6	25	112	0	20	65.3	116	4	0	35	25	0	20	71.0	170	4	0	23	72	270	1
62	6	20.0	6.4	12-4-4	55.2	75	1	93	44	6	25	93	0	07	60.6	104	4	0	15	32	0	03	75.0	190	4	0	05	54	125	2
63	5	19.8	6.7	9-5-24	55.2	66	5	81	24	5	25	133	1	N	67.1	165	3	1	N	25	4	20	74.0	185	3	2	Q	96	170	4
64	7	21.0	7.4	10-7-12	54.0	69	1	81	25	6	25	105	1	07	60.7	99	2	0	20	27	6	07	68.0	148	3	0	10	75	96	6
65	6	21.0	9.6	10-4-23	52.6	70	4	82	17	6	25	147	0	N	65.6	176	2	0	N	36	5	20	66.6	159	2	6	N	75	96	6
66	5	20.5	7.3	3-1-5	54.9	70	2	83	29	5	25	169	4	N	69.3	209	4	3	N	26	0	N	76.0	195	3	0	N	61	175	4

	CHILD-BIRTH				CHILD-TEST					MOTHER-BIRTH					MOTHER-TEST					FATH-BRTH			FATHER-TEST					FAMILY		
ID	B	LGTH	WGT	MO-D-HR	HGHT	WGT	L	PEA	RA	B	AG	WGT	O	SM	HGHT	WGT	E	O	SM	AG	O	SM	HGHT	WGT	E	O	SM	I-B	I-T	C
11	5	20.0	7.2	6-2-13	50.6	60	1	61	23	5	25	155	0	20	61.9	166	2	3	10	33	6	15	69.0	160	3	6	20	49	154	4
12	6	20.5	6.2	6-6-7	50.7	53	4	61	31	6	25	128	5	15	68.6	155	2	3	20	30	6	15	70.0	150	2	0	20	85	162	1
13	6	17.5	4.5	8-4-3	54.4	78	8	79	31	6	25	200	5	20	66.2	211	1	0	40	29	6	20	72.0	172	6	3	30	76	88	2
14	9	20.0	7.0	7-2-20	51.5	74	4	91	20	6	25	103	3	N	60.0	111	4	0	Q	30	1	01	64.8	183	4	1	0	123	230	4
15	7	22.0	8.8	10-3-17	52.3	71	1	102	49	7	25	115	3	N	63.8	112	4	0	N	26	0	N	71.0	175	4	0	N	120	120	1
16	8	21.5	9.4	11-5-11	51.6	62	5	79	44	8	25	130	3	N	67.0	135	4	3	N	27	4	Q	74.9	201	4	0	N	61	276	4
21	7	20.0	5.9	10-6-19	54.1	70	8	78	30	7	25	135	0	40	65.5	165	1	0	30	34	8	40	72.0	194	3	8	02	76	156	6
22	5	21.0	8.1	9-6-24	52.2	60	1	69	21	5	25	120	0	0	62.8	132	2	8	N	27	6	15	67.0	158	3	6	10	55	110	1
23	7	21.0	8.7	11-3-1	53.8	67	1	75	38	8	25	135	0	Q	66.6	149	4	0	Q	32	0	N	68.0	155	4	0	03	80	245	6
24	7	20.0	7.4	12-6-21	51.1	60	4	86	37	7	25	112	5	Q	64.0	125	4	0	03	28	0	10	70.0	165	4	0	03	130	200	1
25	7	21.5	8.3	12-5-18	53.6	69	1	66	17	6	25	110	0	13	62.8	113	1	5	10	31	8	20	67.9	204	1	7	10	52	130	4
26	5	21.0	7.3	2-1-13	50.9	63	1	66	38	7	25	116	0	N	63.9	155	3	0	N	27	7	24	72.0	198	3	6	10	37	72	1
31	7	22.0	9.0	1-5-7	54.5	85	1	90	45	7	25	155	0	N	66.7	164	3	0	N	26	0	N	72.5	195	3	2	N	44	200	1
32	6	21.0	7.5	2-7-21	57.4	113	4	77	25	5	25	137	5	12	63.1	171	2	0	15	29	7	20	72.0	185	2	7	15	29	84	1
33	1	20.0	7.5	3-2-4	52.4	70	1	69	40	1	25	145	4	Q	65.1	154	3	0	N	29	1	Q	69.5	180	4	1	01	91	145	6
34	8	20.0	6.8	1-5-9	52.5	66	4	98	41	6	25	135	0	N	62.8	156	4	4	N	32	0	N	69.0	160	4	1	N	73	213	6
35	1	20.0	7.6	11-2-15	55.8	76	1	84	39	6	26	147	0	N	66.6	150	4	0	N	25	6	N	71.0	180	4	6	N	34	110	1
36	5	20.0	7.2	9-7-24	50.4	56	5	76	21	5	26	130	1	N	62.3	153	1	0	N	30	6	10	62.5	145	2	6	02	50	109	1
41	5	18.0	5.8	4-2-19	54.0	65	1	72	21	5	26	109	0	20	62.6	109	2	0	20	27	4	13	69.0	170	3	4	Q	67	108	1
42	5	20.8	7.4	8-5-10	52.1	69	1	76	15	6	26	133	0	N	66.6	172	2	1	Q	26	6	10	73.0	170	3	6	10	54	116	1
43	6	20.0	7.4	3-6-3	50.3	62	2	65	28	5	26	120	1	12	64.1	140	3	0	N	26	5	Q	70.5	180	4	3	N	50	175	1
44	5	20.5	7.3	10-7-6	51.2	69	1	65	16	5	26	122	1	01	65.0	130	3	1	N	30	5	20	66.9	143	2	5	18	53	170	1
45	9	19.5	7.0	4-6-23	54.3	72	3	87	46	3	26	139	0	01	66.0	149	2	0	06	27	0	03	72.1	182	4	8	Q	53	94	4
46	7	20.5	6.8	4-1-23	54.3	61	1	78	23	7	26	139	1	N	64.3	175	3	0	N	29	8	24	70.0	158	2	6	22	60	90	1
51	6	22.0	10.1	5-1-9	56.7	82	2	86	36	5	26	139	1	12	65.8	140	2	1	13	23	6	03	74.5	210	2	0	06	60	142	4
52	5	20.0	6.3	7-1-10	51.6	70	4	87	43	6	26	130	0	15	66.7	139	4	3	04	27	0	Q	70.4	202	4	0	Q	36	97	6
53	6	20.5	6.8	10-5-12	51.9	66	1	84	28	6	26	135	3	20	64.6	151	4	3	40	31	0	20	70.0	170	4	1	30	105	332	6
54	8	21.5	8.8	7-1-3	57.3	82	5	92	47	6	26	152	0	Q	67.3	154	3	0	Q	30	0	N	73.0	160	3	0	N	42	96	1
55	5	20.0	8.8	8-3-20	57.1	70	1	90	42	7	26	130	4	20	67.8	139	4	2	Q	26	0	20	72.0	155	4	0	20	74	190	4
56	2	20.0	6.8	9-4-19	54.3	68	1	89	49	5	26	105	1	N	64.1	111	3	0	N	24	4	20	72.0	195	3	2	02	96	180	1
61	6	21.0	8.3	12-2-12	51.8	55	4	84	26	6	26	166	0	N	67.1	169	3	4	N	27	8	24	75.0	220	3	5	N	80	156	3
62	7	19.0	5.8	12-7-9	52.4	64	8	66	17	7	26	99	0	0	65.6	98	2	0	N	36	6	20	70.1	166	1	6	15	47	104	3
63	6	20.5	7.8	11-1-23	53.5	67	1	67	33	5	26	122	1	Q	61.4	139	4	3	N	27	7	10	71.1	164	1	2	22	97	165	4
64	3	20.5	7.9	9-2-8	53.1	62	8	85	25	5	26	155	0	17	67.6	178	3	0	N	32	0	N	71.0	140	2	2	N	73	120	6
65	5	22.0	7.9	12-2-14	55.4	69	4	73	36	1	26	128	0	N	63.5	141	2	4	N	31	6	07	71.8	188	3	1	12	80	180	4
66	5	21.0	7.8	12-7-20	52.4	70	1	85	45	5	26	140	1	Q	59.0	172	3	1	Q	31	5	03	72.0	190	3	5	15	64	183	1

StatLab CENSUS: Boys

CHILD-BIRTH					CHILD-TEST					MOTHER-BIRTH					MOTHER-TEST					FATH-BRTH			FATHER-TEST					FAMILY			
ID	B	LGTH	WGT	MO-D-HR	HGHT	WGT	L	PEA	RA	B	AG	WGT	O	SM	HGHT	WGT	E	O	SM	AG	O	SM	HGHT	WGT	E	O	SM	I-B	I	T	C
11	6	20.5	7.4	3-2-20	53.8	66	4	87	41	6	26	108	8	N	62.2	130	3	0	N	47	6	Q	68.0	186	1	3	Q	163	100		1
12	5	20.5	6.9	3-1-12	57.5	106	8	89	21	5	26	135	4	Q	66.0	230	0	0	Q	30	6	40	72.0	230	1	7	40	110	216		1
13	1	21.0	6.9	12-3-22	57.5	74	1	81	30	6	26	181	0	20	68.9	177	2	7	10	31	7	20	73.5	180	1	2	30	50	176		3
14	5	21.0	7.1	4-6-10	56.4	81	1	81	34	5	26	140	0	Q	65.9	143	1	5	10	35	6	Q	68.0	160	2	7	Q	60	95		1
15	3	21.0	7.6	6-6-8	51.3	61	8	85	31	8	26	210	0	N	65.3	249	2	0	N	27	5	N	70.0	185	3	8	N	53	120		1
16	5	19.5	7.3	4-4-5	56.1	72	7	62	10	1	26	153	0	Q	65.6	159	1	5	N	30	6	20	72.0	170	2	7	22	21	82		4
21	6	20.5	6.5	5-7-11	51.4	60	1	77	26	6	26	115	3	N	62.6	107	4	3	N	28	6	Q	69.5	180	3	0	N	60	219		4
22	5	21.0	7.0	6-7-12	54.0	71	4	91	21	5	26	122	4	N	64.1	126	3	0	N	28	3	N	73.0	183	4	3	N	70	300		1
23	6	20.0	7.1	7-3-12	52.9	78	4	73	12	6	26	127	1	Q	62.9	151	3	1	15	27	8	N	71.0	205	2	5	N	79	188		6
24	9	21.5	7.8	6-2-22	50.8	69	1	79	47	6	26	165	0	N	64.8	189	4	3	N	25	0	10	66.5	150	4	3	Q	83	210		6
25	6	20.5	6.8	7-2-1	57.8	80	4	81	27	6	26	130	3	N	67.9	135	4	3	N	24	0	N	75.0	185	4	0	N	65	170		1
26	2	21.0	8.1	8-4-10	52.8	66	3	78	32	3	26	146	0	N	66.7	190	3	0	N	30	1	N	70.5	197	4	1	N	81	140		1
31	6	20.0	8.0	10-5-23	51.9	71	4	82	25	6	26	114	3	N	63.6	122	4	3	40	31	1	20	70.0	185	4	1	20	119	295		1
32	3	21.0	8.8	10-7-13	52.3	64	1	63	35	1	26	127	5	Q	62.0	150	1	1	Q	27	6	20	71.6	205	2	7	15	80	203		6
33	6	20.3	7.7	11-4-13	51.8	87	8	77	26	5	26	136	0	N	60.4	128	3	1	N	26	0	20	71.0	180	2	0	Q	90	211		4
34	5	20.5	9.4	10-3-12	50.0	65	1	83	49	5	26	140	0	N	66.9	159	3	0	N	26	0	N	73.5	190	4	3	N	18	135		1
35	7	19.0	6.2	10-5-13	51.0	76	1	82	20	7	26	115	3	N	62.9	121	4	3	N	28	1	N	66.0	160	4	1	N	115	207		1
36	8	23.0	10.6	12-1-19	54.3	69	4	78	36	6	26	140	0	N	66.0	144	3	2	20	31	8	Q	74.0	208	3	8		62	176		1
41	6	22.0	9.1	1-3-20	55.6	68	1	79	31	6	26	117	4	40	65.5	125	4	0	40	31	1	20	76.0	175	4	0	40	139	170		1
42	5	21.8	8.3	3-3-10	52.6	65	4	85	33	5	26	136	0	18	69.4	180	2	0	Q	29	5	N	71.8	220	3	5	N	52	120		6
43	7	19.0	5.9	12-5-11	51.4	62	4	83	23	5	26	90	5	N	60.7	101	2	2	N	32	8	N	67.0	220	3	5	N	72	65		1
44	5	20.0	6.4	3-4-20	54.6	68	1	63	20	7	26	130	8	N	61.1	152	2	5	N	29	5	N	71.0	180	3	5	N	55	96		1
45	5	20.8	7.4	2-4-20	51.4	63	1	91	53	5	26	117	4	N	54.3	118	4	0	N	27	0	N	69.5	155	4	3	N	65	450		6
46	6	20.5	7.1	2-1-11	52.6	68	4	81	24	6	26	129	0	16	66.4	136	2	0	20	33	2	20	72.0	160	4	2	20	102	162		1
51	5	21.0	8.5	4-1-22	48.9	54	3	83	16	6	27	128	0	12	63.8	148	2	0	Q	31	6	20	69.0	160	3	8	Q	60	144		6
52	2	20.5	7.4	1-2-15	53.4	68	4	80	32	1	27	135	0	40	65.6	153	8	0	40	30	6	20	72.0	195	3	7	Q	70	88		5
53	5	22.0	7.6	12-5-8	60.1	85	4	86	39	5	27	101	0	02	65.6	130	2	4	20	29	6	N	70.0	200	3	2	Q	100	220		6
54	6	21.0	8.1	8-4-23	53.3	64	1	74	19	6	27	125	0	12	66.0	128	4	4	N	27	1	N	75.3	169	4	3	N	140	540		4
55	6	20.0	6.9	4-3-13	50.4	56	8	64	12	5	27	119	0	Q	61.3	127	1	0	Q	33	7	05	66.0	179	1	6	Q	53	110		1
56	6	20.5	8.1	4-5-19	51.0	62	4	81	46	6	27	130	1	Q	65.4	138	4	0	N	32	0	20	70.0	145	3	0	Q	67	100		6
61	2	20.0	9.8	10-4-8	55.3	78	1	79	26	1	27	139	0	Q	68.5	138	2	2	Q	28	0	N	72.0	200	3	0	N	82	154		1
62	5	21.5	8.0	6-7-6	50.4	67	1	76	36	5	27	145	0	N	69.6	168	4	3	N	27	5	30	70.0	225	1	5	40	53	126		2
63	7	20.0	6.6	5-5-2	50.4	52	1	91	26	7	27	120	1	15	64.3	124	2	1	10	28	5	20	72.0	170	3	0	20	65	151		4
64	5	20.5	9.1	4-7-9	55.3	84	1	91	40	1	27	125	0	N	65.1	129	1	0	N	31	4	N	67.0	150	1	0	Q	78	250		1
65	6	20.0	6.9	5-1-4	49.6	50	1	75	23	3	27	146	0	N	64.5	191	4	0	Q	29	6	20	64.0	120	0	6	40	48	96		6
66	2	21.0	7.4	6-7-17	50.4	60	4	80	22	6	27	104	1	Q	57.3	98	3	0	Q	31	0	N	66.0	138	4	0	N	63	114		3

StatLab CENSUS: Boys

	CHILD-BIRTH				CHILD-TEST					MOTHER-BIRTH					MOTHER-TEST					FATH-BRTH			FATHER-TEST					FAMILY		
ID	B	LGTH	WGT	MO-D-HR	HGHT	WGT	L	PEA	RA	B	AG	WGT	O	SM	HGHT	WGT	E	O	SM	AG	O	SM	HGHT	WGT	E	O	SM	I-B	I-T	C
11	5	21.0	8.0	3-4-17	52.7	65	1	84	25	5	27	133	0	12	67.3	137	1	0	15	42	6	10	70.5	176	1	0	Q	72	126	4
12	5	20.5	7.6	3-6-14	50.5	56	4	88	28	6	27	99	0	17	61.1	113	2	0	30	29	4	N	71.0	165	4	8	30	81	126	6
13	6	20.5	7.6	2-1-1	50.8	62	1	62	14	3	27	140	0	25	65.2	151	2	0	30	33	8	25	70.2	175	3	2	20	70	167	1
14	8	22.0	9.3	9-3-21	59.0	93	8	78	22	8	27	189	0	N	71.0	200	3	0	N	37	6	10	74.0	198	3	6	Q	48	84	1
15	5	20.0	7.0	6-7-6	55.4	74	5	85	41	5	27	118	1	04	67.0	128	2	1	03	31	8	40	72.5	189	3	8	20	100	145	2
16	6	20.0	6.9	9-6-18	51.1	66	4	88	41	5	27	102	0	Q	61.0	112	4	4	N	29	0	N	68.5	155	4	3	N	80	290	2
21	9	18.0	5.9	10-7-6	49.8	58	1	71	28	5	27	106	1	13	60.9	111	1	1	35	30	6	30	70.1	165	2	7	35	109	184	1
22	6	20.0	7.3	10-1-6	54.7	68	4	64	31	2	27	118	1	30	66.7	138	1	0	40	31	5	22	69.5	169	3	5	40	64	100	6
23	7	21.5	8.1	11-6-1	56.4	86	4	85	46	8	27	130	1	30	68.5	130	4	1	40	27	0	20	70.5	175	3	2	20	90	174	1
24	5	19.0	6.7	10-4-20	56.6	100	1	89	30	5	27	135	8	N	63.8	136	2	0	N	27	0	N	70.0	170	3	0	N	58	144	1
25	7	21.5	7.9	11-6-13	57.4	103	1	78	23	8	27	118	0	Q	61.9	127	2	8	20	34	6	40	67.0	137	1	2	20	57	100	2
26	5	19.5	8.4	12-2-16	52.8	71	8	81	37	1	27	148	0	N	66.4	165	1	4	8	33	6		65.0	168	1	6		65	100	1
31	2	19.5	6.3	11-6-18	49.4	55	4	76	30	5	27	99	0	12	62.8	127	2	0	Q	30	2	10	68.6	158	1	6	Q	52	96	4
32	5	20.5	7.8	12-6-8	55.3	69	4	80	41	5	27	117	0	18	65.2	146	3	0	N	31	0	N	70.0	165	4	0	N	67	150	4
33	6	21.0	8.2	3-4-7	51.1	70	1	85	39	6	27	160	0	20	62.7	148	2	0	Q	32	6	Q	69.0	175	3	6	Q	60	120	1
34	6	19.5	5.7	3-5-13	54.3	86	1	94	32	6	27	118	1	N	62.3	134	3	1	N	31	6	10	69.0	185	2	6	N	120	160	6
35	6	21.3	8.6	4-3-16	52.6	65	4	81	46	5	27	111	0	Q	65.3	144	4	4	N	28	8	05	72.0	205	4	3	10	102	270	6
36	6	21.5	8.9	5-7-17	55.8	82	1	81	38	6	27	148	3	N	70.9	154	4	4	3	32	0	20	77.1	210	4	0	20	119	245	2
41	7	22.5	10.1	5-5-24	51.4	75	4	76	18	7	27	126	0	N	64.2	175	1	7	N	29	6	N	67.0	220	2	6	Q	54	103	1
42	5	21.5	7.9	6-7-11	50.6	62	1	85	35	1	27	111	4	N	64.1	158	4	4	N	26	4	N	70.5	165	4	3	N	99	254	6
43	6	20.0	6.9	5-2-16	52.3	62	2	85	43	6	27	102	3	N	64.1	112	4	0	N	31	5	03	70.0	161	4	0	01	165	500	6
44	5	21.0	9.0	6-3-18	52.3	80	1	62	22	7	27	102	0	04	59.0	130	0	0	N	27	5	N	65.1	151	0	7	06	10	97	1
45	5	22.5	9.9	7-1-21	47.7	46	1	86	41	5	27	112	1	Q	63.8	111	3	0	N	36	0	N	66.5	150	0	8	03	115	180	6
46	6	21.0	6.9	7-4-4	50.2	74	1	71	18	5	27	110	1	N	62.3	116	2	1	N	25	5	10	69.0	160	3	8	03	94	208	1
51	5	22.5	8.9	7-2-10	54.8	77	3	70	25	7	27	185	0	N	67.2	231	3	0	N	36	6	N	69.0	199	2	6	N	62	70	1
52	5	20.0	7.7	8-4-18	50.3	57	1	117	45	6	27	125	0	N	63.9	135	4	3	N	34	1	N	72.0	200	4	1	N	74	136	4
53	6	21.5	9.8	7-3-23	51.9	62	8	75	31	6	27	120	0	15	64.7	137	2	5	10	39	5	15	70.0	170	2	5	Q	55	116	1
54	7	21.0	9.5	10-3-24	54.0	70	1	94	46	8	27	128	1	N	66.1	130	0	4	01	32	1	10	76.0	245	4	1	Q	90	150	1
55	9	21.0	8.1	11-3-24	50.1	54	1	88	36	6	27	128	3	N	65.5	115	4	0	N	32	4	N	73.5	180	4	4	N	148	135	4
56	5	21.5	7.6	1-6-20	51.8	55	1	89	35	5	27	119	0	N	64.4	123	3	0	20	29	6	20	69.0	160	3	0	20	100	140	4
61	5	21.5	8.9	12-7-16	54.3	76	1	122	28	5	27	119	4	20	66.1	120	4	0	01	27	0	N	73.0	225	4	0	N	121	350	6
62	5	21.5	7.3	1-5-22	55.8	91	1	78	30	7	27	135	1	N	63.9	164	2	0	N	37	0	N	65.0	130	3	0	N	120	144	1
63	5	21.0	7.1	7-2-3	52.1	62	5	103	37	6	27	100	3	N	60.4	117	4	3	N	32	6	10	72.5	185	3	0	20	83	216	2
64	3	20.0	6.6	3-1-11	53.3	70	1	82	44	3	27	110	0	Q	65.4	125	2	0	N	27	6	Q	71.0	187	3	6	Q	64	144	4
65	7	20.0	7.7	9-4-7	52.8	66	3	78	34	7	28	129	8	N	61.4	146	2	0	N	29	6	40	66.0	150	1	6	30	75	130	4
66	5	19.0	5.9	12-7-2	51.3	58	1	64	13	5	28	122	0	N	62.8	140	2	0	N	28	6	01	66.8	147	1	7	Q	84	44	1

FAMILY			FATHER-TEST					FATH-BRTH			MOTHER-TEST					MOTHER-BIRTH					CHILD-TEST					CHILD-BIRTH				ID
I-B	I-T	C	HGHT	WGT	E	O	SM	AG	O	SM	HGHT	WGT	E	O	SM	B	AG	WGT	O	SM	HGHT	WGT	L	PEA	RA	B	LGTH	WGT	MO-D-HR	
65	172	4	72.0	200	2	2	N	29	2	N	63.6	139	2	1	N	3	28	130	0	N	50.7	66	4	80	38	3	19.5	6.5	10-4-16	11
120	300	1	66.0	148	4	0	20	31	1	Q	63.5	128	4	0	N	2	28	112	0	N	54.7	66	1	85	47	9	21.0	8.4	7-6-10	12
43	150	1	75.0	200	2	2	20	40	5	10	67.1	182	3	4	40	5	28	153	0	40	58.7	96	1	83	38	5	20.0	7.0	7-7-18	13
71	242	1	71.0	175	4	0	N	31	0	N	62.6	156	4	0	N	8	28	140	0	N	52.9	63	1	75	24	7	22.5	9.9	12-4-13	14
53	160	1	74.0	165	4	0	N	27	0	N	64.1	163	4	0	25	5	28	134	0	20	53.1	60	1	104	39	5	22.0	7.8	4-1-6	15
70	96	6	64.0	120	0	6	40	30	6	25	64.5	191	1	0	N	3	28	124	0	Q	51.2	54	1	62	17	7	20.5	7.6	6-6-14	16
108	175	4	74.0	235	2	3	20	35	6	20	66.4	150	3	0	20	5	28	141	1	20	57.9	92	5	77	15	7	20.0	7.4	4-1-20	21
76	120	1	66.0	135	3	0	N	30	6	N	58.8	101	3	0	N	5	28	97	0	N	48.9	52	1	103	42	7	19.5	6.4	6-4-3	22
34	72	1	68.5	178	4	0	N	29	0	N	64.7	110	3	0	N	7	28	103	0	N	55.3	64	4	82	41	6	21.5	8.3	6-2-13	23
38	120	1	68.0	152	4	0	N	31	0	N	65.1	123	2	0	N	6	28	135	0	N	55.9	69	4	78	29	6	22.0	8.8	7-2-8	24
84	200	1	72.0	182	4	2	Q	34	2	10	64.8	158	2	0	N	5	28	132	0	N	59.2	114	1	78	29	5	23.0	9.9	6-2-12	25
67	180	6	68.0	155	4	1	20	32	1	15	63.8	107	2	0	20	6	28	100	0	Q	53.1	71	3	76	36	2	20.0	7.9	7-4-3	26
50	99	4	67.0	168	0	6	80	36	6	20	66.1	189	2	5	N	7	28	135	0	N	51.5	66	4	62	23	1	20.5	7.3	5-5-10	31
55	120	6	68.5	167	0	2	40	29	8	30	65.1	179	1	0	Q	5	28	156	0	Q	56.1	68	1	75	19	5	21.5	8.7	8-5-13	32
70	200	1	72.0	195	3	2	02	26	4	20	64.4	114	3	0	N	5	28	104	1	N	54.0	65	1	92	42	5	20.5	7.4	1-3-21	33
42	86	1	75.0	195	2	7	20	39	7	20	65.0	195	2	5	N	5	28	210	0	N	53.8	75	5	78	28	7	22.0	8.8	10-2-10	34
60	72	1	66.0	165	1	7	N	29	6	N	60.7	126	1	0	N	5	28	135	0	N	53.8	64	4	65	16	5	20.5	7.7	12-7-10	35
70	249	1	76.0	235	3	2	N	30	6	N	66.0	170	3	4	04	5	28	175	4	01	56.3	98	4	80	39	6	20.5	7.4	11-6-17	36
70	150	4	66.1	163	4	1	N	29	0	N	59.8	108	4	0	N	6	28	102	0	N	52.8	67	4	91	38	8	21.0	8.1	10-6-2	41
51	140	1	70.5	167	4	3	N	30	6	N	64.7	137	3	0	N	6	28	125	0	N	52.5	70	4	77	33	6	21.0	8.1	1-5-13	42
69	124	4	70.5	176	3	0	10	33	0	10	66.9	159	3	5	N	6	28	134	0	10	50.7	56	1	81	26	9	21.0	7.9	4-1-18	43
43	156	3	69.1	202	2	6	N	28	5	10	64.0	138	2	1	20	5	28	118	0	06	52.8	66	4	63	23	8	21.0	8.2	12-6-15	44
90	220	1	70.0	182	3	5	Q	30	5	10	69.3	161	2	0	20	6	28	145	1	20	53.8	79	1	79	36	8	21.5	7.4	4-2-18	45
111	120	6	72.0	175	3	8	20	31	8	20	65.3	121	3	0	02	5	28	118	0	07	51.8	62	1	84	35	2	21.0	8.8	4-1-4	46
59	115	1	75.5	179	2	7	10	25	6	10	65.5	158	2	0	N	5	28	155	0	N	61.1	65	1	68	26	5	20.5	7.9	5-6-15	51
89	178	6	74.0	180	3	7	N	30	8	N	63.2	131	3	1	20	7	28	110	0	15	59.3	104	4	81	23	7	19.5	5.4	3-7-12	52
130	200	4	73.0	180	3	6	20	31	6	20	63.8	126	4	4	N	2	28	125	1	N	54.3	71	5	84	22	8	21.5	7.5	7-7-9	53
84	120	6	71.0	170	2	6	N	29	6	40	62.4	120	2	0	N	2	28	105	0	Q	49.3	54	4	64	14	1	21.0	8.8	9-4-23	54
71	134	3	69.5	160	2	6	40	27	6	20	60.7	140	2	0	20	6	28	111	0	12	50.7	70	4	79	22	6	19.3	6.3	8-3-7	55
106	300	6	71.0	205	4	3	N	37	0	N	63.6	132	2	0	04	5	28	125	0	05	52.3	74	4	83	28	5	22.0	9.0	10-5-19	56
90	72	1	71.0	165	4	2	N	41	0	N	66.8	163	2	0	Q	6	28	140	0	30	52.4	65	4	78	32	6	21.0	8.6	10-2-11	61
137	144	1	74.0	245	3	0	40	26	0	30	68.0	131	2	0	N	6	28	120	1	N	54.9	60	1	81	23	6	22.5	8.6	8-2-11	62
204	108	1	72.0	210	4	0	20	33	0	30	65.9	152	2	0	N	6	28	145	0	N	56.9	86	4	82	44	8	22.0	9.1	7-3-10	63
100	185	6	71.0	180	3	2	N	29	7	N	63.4	143	4	4	N	6	28	120	7	N	55.4	81	1	94	28	6	19.0	7.8	11-4-21	64
72	157	1	74.0	180	4	4	Q	32	4	15	58.3	131	3	1	N	7	28	128	0	Q	52.4	57	1	80	22	8	20.8	7.2	12-5-19	65
84	132	1	66.9	182	4	0	20	29	2	20	62.0	105	3	0	N	6	28	98	0	N	53.2	67	4	84	24	6	20.0	6.1	2-7-20	66

StatLab CENSUS: Boys

	CHILD-BIRTH				CHILD-TEST					MOTHER-BIRTH					MOTHER-TEST					FATH-BRTH			FATHER-TEST					FAMILY		
ID	B	LGTH	WGT	MO-D-HR	HGHT	WGT	L	PEA	RA	B	AG	WGT	O	SM	HGHT	WGT	E	O	SM	AG	O	SM	HGHT	WGT	E	O	SM	I-B	I-T	C
11	5	21.0	8.0	2-6-2	53.1	84	4	84	40	7	23	96	1	Q	58.3	114	3	0	Q	38	4	20	68.0	160	4	3	Q	135	220	2
12	6	20.5	7.1	2-7-15	52.2	62	8	92	47	6	28	115	3	10	65.7	118	4	0	Q	39	1	30	71.0	142	4	1	30	96	200	4
13	1	22.0	8.5	4-1-23	56.4	77	3	84	25	5	28	140	1	N	69.0	147	3	0	N	37	0	20	71.0	165	4	0	Q	105	170	6
14	5	21.0	7.9	4-3-18	52.0	72	1	61	13	5	28	165	7	05	63.0	180	2	5	06	38	6	N	65.0	175	0	6	N	53	126	1
15	8	22.5	10.0	4-3-13	56.8	88	1	85	16	7	28	232	1	01	67.9	264	3	0	01	27	5	N	72.0	182	3	7	N	80	77	6
16	6	19.3	5.4	3-6-20	52.9	80	3	69	30	5	28	110	1	20	63.3	126	2	1	20	30	6	20	68.0	175	2	6	15	76	174	6
21	6	19.5	7.8	10-6-6	54.6	71	4	62	10	6	28	135	0	10	65.2	179	2	0	Q	29	6	20	69.0	180	2	6	Q	70	139	1
22	6	19.5	6.6	8-1-2	53.3	64	4	67	25	5	29	137	4	N	63.0	135	3	1	Q	36	6	20	69.6	206	2	6	30	134	145	1
23	1	21.5	7.9	5-3-9	54.6	77	4	84	36	5	29	148	1	Q	68.0	133	4	0	N	34	1	N	70.8	164	4	0	N	35	170	1
24	7	21.0	8.8	11-4-6	54.7	68	1	92	39	6	29	135	0	01	67.5	133	4	0	02	35	3	N	78.0	205	4	7	Q	250	500	5
25	6	22.5	10.6	8-5-2	58.8	81	2	85	37	6	29	162	0	N	68.9	170	2	0	N	30	6	10	71.0	250	1	6	N	56	100	1
26	5	20.0	7.7	5-3-2	58.4	90	1	83	30	7	29	141	0	N	63.8	182	3	3	N	33	8	40	70.0	175	2	8	10	47	162	6
31	6	20.5	6.7	7-3-11	55.9	81	4	87	29	5	29	132	4	N	56.6	120	4	2	N	34	0	N	70.0	170	4	3	N	108	240	6
32	6	19.0	6.1	9-3-6	51.1	59	1	90	44	5	29	130	0	15	65.8	135	2	2	20	37	3	30	68.8	205	2	4	30	45	124	1
33	5	21.0	7.8	9-1-18	59.4	90	1	63	12	5	29	176	0	10	69.0	200	3	0	20	29	7	01	72.0	230	3	7	01	23	60	4
34	5	22.5	14.1	9-1-11	52.5	72	1	83	9	5	29	130	0	Q	65.2	155	1	5	08	35	7	N	74.3	246	6	6	N	30	100	1
35	9	18.0	6.5	1-7-15	50.9	56	2	86	38	1	29	88	0	N	60.0	110	4	4	N	38	0	20	66.1	166	4	0	20	87	220	1
36	2	20.5	7.5	4-7-8	52.5	62	1	75	32	6	29	143	0	09	64.0	125	2	0	20	32	2	40	70.6	168	3	3	40	76	120	1
41	7	20.5	8.3	12-1-3	54.3	73	8	93	42	8	29	148	0	N	64.6	203	3	0	N	34	0	N	70.0	175	4	0	Q	100	200	6
42	6	21.5	7.9	5-2-14	53.3	70	1	93	52	5	29	119	0	Q	63.5	128	4	2	Q	34	0	05	70.0	170	4	0	Q	81	200	4
43	6	21.5	8.7	4-4-7	57.1	77	4	86	54	2	29	130	4	N	69.2	137	2	0	N	28	1	Q	73.5	190	4	1	N	118	286	2
44	1	21.0	10.3	5-6-14	52.6	71	4	82	29	1	29	145	5	N	66.5	170	4	2	N	29	8	N	68.0	129	2	6	N	38	89	1
45	5	20.5	6.9	6-7-6	53.4	71	8	86	32	5	29	130	3	N	64.7	158	4	1	N	30	2	20	70.0	165	4	0	20	102	202	2
46	6	21.5	7.8	6-2-12	53.8	68	8	91	23	6	29	125	0	09	65.2	129	4	2	Q	31	0	N	72.0	180	4	3	40	65	173	1
51	8	19.0	6.1	7-1-3	56.0	92	1	73	32	8	29	178	0	06	68.9	200	2	0	Q	42	6	02	68.0	165	2	7	Q	63	78	1
52	5	19.8	7.4	6-3-6	50.3	63	1	65	34	5	29	130	0	20	62.3	159	2	2	20	31	6	Q	68.0	175	3	3	Q	150	200	1
53	6	21.0	6.1	5-4-16	54.3	66	4	88	28	7	29	140	3	N	68.6	163	4	4	N	37	4	N	72.0	180	4	4	N	101	260	4
54	5	19.5	6.8	7-3-4	51.6	62	8	93	44	5	29	124	1	30	67.9	136	2	1	20	25	0	05	70.0	220	2	2	Q	81	177	2
55	5	21.5	8.3	8-5-6	53.1	62	1	81	34	4	29	135	1	N	65.3	148	3	3	10	31	2	20	67.0	170	3	2	Q	94	191	1
56	6	18.5	6.4	9-3-1	48.9	55	1	73	22	5	29	117	0	15	63.1	128	2	0	20	34	0	20	66.0	142	4	4	20	120	240	1
61	5	21.5	8.9	7-1-4	58.8	84	4	87	20	5	29	132	7	N	67.0	143	3	0	N	33	8	N	78.8	280	3	8	N	76	120	1
62	5	20.5	7.0	9-2-23	54.9	68	1	77	25	5	29	124	0	Q	61.1	138	4	3	N	40	5	Q	66.8	145	4	0	Q	52	120	6
63	6	21.0	9.4	9-6-21	53.5	69	4	80	32	8	29	129	0	N	64.0	127	3	1	N	39	6	Q	71.5	180	3	6	Q	76	135	6
64	8	20.0	7.1	9-7-24	49.3	50	2	81	35	3	29	113	1	20	64.8	113	2	0	N	36	6	N	69.0	160	3	6	Q	108	180	4
65	6	20.5	7.1	11-3-13	53.1	66	1	82	35	6	29	122	0	20	61.7	161	2	2	20	30	5	20	71.6	164	6	5	30	68	115	6
66	1	20.0	7.4	12-6-23	52.0	64	4	65	11	2	29	120	0	Q	62.0	125	3	8	N	29	6	N	68.6	180	2	2	Q	59	207	1

StatLab CENSUS: Boys

ID	CHILD-BIRTH				CHILD-TEST					MOTHER-BIRTH					MOTHER-TEST					FATH-BRTH			FATHER-TEST					FAMILY		
	B	LGTH	WGT	MO-D-HR	HGHT	WGT	L	PEA	RA	B	AG	WGT	O	SM	HGHT	WGT	E	O	SM	AG	O	SM	HGHT	WGT	E	O	SM	I-B	I-T	C
11	2	21.5	8.3	3-7-2	58.8	110	3	87	33	5	29	147	1	N	64.9	147	3	8	N	29	6	20	74.0	180	3	2	Q	110	184	4
12	9	22.0	8.3	11-6-6	51.6	60	1	65	19	5	29	112	3	10	65.4	115	4	3	10	30	1	20	68.6	145	4	1	13	110	230	2
13	5	19.5	6.2	4-4-1	51.0	58	4	86	30	5	29	125	0	N	65.6	144	2	0	N	36	0	N	68.0	150	4	0	N	69	114	1
14	5	21.0	7.4	1-4-9	52.8	64	1	61	14	6	29	120	0	N	64.4	168	1	0	N	29	6	N	65.0	147	1	6	40	60	135	4
15	6	19.8	8.1	1-4-5	55.9	74	1	88	32	6	30	137	0	N	63.8	134	3	0	Q	30	2	N	71.0	170	4	2	N	100	150	1
16	6	21.8	8.4	10-1-3	56.6	76	1	77	38	5	30	152	0	Q	71.1	153	4	0	N	32	5	15	74.0	200	4	0	22	66	172	6
21	6	19.0	6.7	10-6-24	50.2	62	1	65	26	5	30	188	0	N	63.6	219	3	5	N	33	6	05	72.0	195	3	6	N	52	87	1
22	1	20.5	9.0	10-1-24	51.1	57	4	89	36	5	30	127	0	N	63.7	148	4	0	Q	36	1	Q	66.0	165	4	1	Q	110	160	1
23	5	20.5	7.2	1-2-23	54.9	76	5	71	25	5	30	128	0	N	65.5	137	2	1	02	31	6	04	72.8	202	2	2	Q	60	151	1
24	7	20.5	8.1	8-4-11	51.4	63	1	68	21	7	30	150	0	N	64.0	165	1	0	N	32	8	10	65.0	145	2	2	08	44	65	1
25	8	21.5	8.6	4-7-4	53.3	60	5	89	35	6	30	124	0	Q	65.0	128	2	0	N	31	8	20	69.0	190	2	8	Q	63	135	1
26	6	21.5	7.9	5-6-10	57.1	84	1	82	34	6	30	128	0	15	65.7	144	5	5	20	31	8	12	71.5	189	2	8	10	53	181	1
31	6	21.0	8.3	4-5-16	55.7	74	3	73	35	8	30	139	0	20	63.0	135	2	0	15	31	4	30	73.3	206	2	6	40	55	120	1
32	9	20.5	7.8	1-3-24	53.7	72	1	86	29	5	30	122	0	20	64.6	136	4	0	Q	31	0	Q	70.5	175	4	0	Q	107	200	6
33	8	20.0	6.6	4-2-8	49.1	57	4	78	54	6	30	108	0	02	61.3	111	1	1	Q	34	6	20	64.5	170	1	2	Q	63	108	4
34	6	21.0	7.1	2-2-14	50.1	55	4	75	35	6	30	150	0	N	66.0	180	2	7	Q	32	6	N	66.5	185	2	7	N	62	126	1
35	5	20.5	7.6	9-1-15	53.8	71	1	77	30	5	30	115	0	N	64.6	120	2	1	N	32	0	20	73.5	200	3	0	N	78	156	6
36	7	18.5	5.6	8-7-2	52.9	56	1	93	51	7	30	120	0	15	63.8	122	4	2	N	30	0	20	72.0	145	4	2	N	58	144	1
41	5	21.0	8.1	9-4-11	55.0	69	4	92	33	6	30	142	0	01	66.9	141	3	0	N	34	1	01	71.0	165	4	1	N	48	150	6
42	1	19.5	6.1	8-2-5	53.3	60	1	82	36	1	30	102	1	20	61.6	109	2	1	20	31	0	20	75.0	183	2	0	20	108	120	6
43	6	21.0	8.3	10-2-5	55.7	74	1	87	24	2	30	112	1	20	66.1	116	2	1	20	30	6	30	74.0	200	2	6	20	70	142	6
44	2	19.0	5.5	11-4-3	55.6	92	8	84	16	5	30	130	0	N	66.7	141	2	1	N	31	6	N	69.0	165	2	3	N	55	90	6
45	6	22.0	8.4	11-2-24	54.8	67	4	96	35	6	30	130	0	N	66.5	135	3	0	N	30	0	20	70.5	210	4	3	40	55	265	3
46	5	20.0	7.2	12-3-16	55.8	75	1	85	45	6	30	140	3	17	63.5	169	4	0	17	36	1	20	72.0	178	4	1	40	130	108	6
51	6	21.5	8.2	10-2-15	55.2	72	4	93	34	6	30	132	0	N	67.2	137	4	0	06	30	1	02	75.0	200	4	1	08	100	400	1
52	5	21.5	8.2	3-1-9	52.8	66	5	94	37	6	30	150	0	02	58.4	170	2	8	06	33	2	Q	67.0	170	2	2	Q	52	131	1
53	6	19.0	6.0	5-6-23	56.1	74	1	108	41	5	30	127	5	10	68.1	148	4	0	Q	29	0	10	74.0	145	3	8	25	98	100	1
54	6	20.0	6.7	5-6-19	52.6	77	1	82	28	8	30	140	0	15	64.4	161	3	0	10	38	2	50	74.1	225	4	0	Q	118	200	6
55	6	21.0	8.1	10-5-13	53.9	70	1	74	24	6	30	127	0	10	64.1	128	2	0	N	28	5	20	70.0	185	4	0	Q	57	160	1
56	5	20.0	5.9	5-1-22	48.6	52	4	81	23	1	30	105	0	10	62.3	126	1	0	Q	32	6	20	67.0	170	1	6	40	51	83	1
61	5	21.0	8.6	8-6-3	54.6	80	4	87	39	6	30	133	0	15	66.1	144	3	0	N	34	1	Q	70.8	160	4	1	Q	62	150	6
62	6	20.0	6.1	8-5-17	50.0	52	1	73	26	6	30	110	0	N	61.7	121	2	2	N	30	6	N	68.0	155	3	6	N	80	144	6
63	5	20.5	7.8	7-6-14	53.4	73	1	80	21	5	30	145	0	N	66.0	145	2	1	Q	34	6	N	66.9	199	2	1	Q	84	196	4
64	7	21.0	7.6	7-7-9	51.5	67	4	61	27	6	30	122	0	N	65.4	131	1	7	Q	37	6	20	68.0	163	1	6	45	63	185	4
65	5	20.5	7.5	8-5-1	50.5	58	1	77	28	6	30	115	0	N	60.7	125	2	0	N	33	6	N	64.3	143	2	6	N	70	156	1
66	1	22.5	9.9	8-4-15	54.0	72	1	83	39	2	30	140	0	N	67.0	154	3	0	N	30	1	N	67.0	160	4	1	N	55	140	1

StatLab CENSUS: Boys

	CHILD—BIRTH				CHILD—TEST					MOTHER—BIRTH					MOTHER—TEST					FATH—BRTH			FATHER—TEST					FAMILY		
ID	B	LGTH	WGT	MO—D—HR	HGHT	WGT	L	PEA	RA	B	AG	WGT	O	SM	HGHT	WGT	E	O	SM	AG	O	SM	HGHT	WGT	E	O	SM	I—B	I—T	C
11	7	22.8	10.1	10-1-3	55.4	85	3	86	34	7	30	130	1	20	64.1	175	2	0	20	35	5	20	70.5	200	4	0	20	99	220	1
12	6	22.0	8.7	10-3-18	60.0	116	1	95	49	5	30	162	0	N	69.6	167	4	4	N	31	6	N	74.6	231	4	2	N	77	225	1
13	5	21.0	8.3	9-1-20	51.9	66	1	75	34	5	30	117	0	N	61.1	137	3	1	N	36	5	N	69.0	215	4	1	01	53	111	6
14	6	20.8	7.8	10-7-19	53.2	59	1	94	33	5	30	123	0	N	66.9	124	4	0	N	34	6	20	71.0	150	3	7	30	83	110	6
15	5	21.0	8.9	12-6-21	49.6	52	1	77	19	5	30	105	0	Q	61.7	120	3	1	Q	29	7	06	67.5	157	3	7	N	80	144	1
16	6	21.0	6.8	3-2-21	56.4	99	4	78	26	6	30	156	7	30	63.5	200	2	0	30	31	6	Q	76.0	250	1	7	N	90	78	3
21	6	19.0	6.8	2-4-3	51.3	75	1	65	23	6	30	123	0	N	62.4	120	1	0	N	35	0	N	67.0	160	4	0	N	61	133	1
22	9	19.3	7.3	3-3-13	54.9	89	7	91	28	5	30	119	1	16	65.2	138	3	0	Q	41	0	35	68.5	155	3	0	Q	128	100	1
23	6	21.5	8.3	9-6-11	54.0	67	1	80	26	6	31	149	4	N	67.6	149	3	0	N	33	1	N	78.0	208	4	1	N	100	240	1
24	5	21.0	7.3	8-7-8	53.7	61	4	78	30	5	31	150	0	N	70.0	140	4	0	N	32	1	N	73.6	167	4	1	30	131	150	1
25	7	20.5	8.3	11-5-6	52.8	60	1	82	33	6	31	155	0	N	64.2	162	4	2	Q	34	6	30	72.0	173	4	3	30	121	200	1
26	8	20.0	7.6	8-5-4	56.9	78	1	68	17	7	31	155	0	20	72.1	178	2	1	20	36	5	N	71.5	230	2	5	N	67	146	1
31	5	21.0	7.8	5-3-7	53.7	67	1	83	43	5	31	194	0	20	65.4	229	2	0	Q	38	2	N	75.5	199	4	2	N	78	175	1
32	6	20.5	7.9	5-6-8	55.9	84	4	80	20	6	31	118	0	N	68.5	125	4	1	N	32	0	20	70.3	218	3	0	Q	47	114	1
33	6	22.0	7.7	6-6-11	60.2	90	5	107	27	5	31	115	3	N	62.3	126	4	0	N	38	1	N	73.0	175	4	1	N	110	120	1
34	9	20.0	7.8	6-7-16	50.8	60	4	64	27	8	31	128	1	N	65.9	138	2	0	N	59	0	N	60.8	160	4	0	N	72	83	1
35	5	20.0	6.6	5-7-4	52.2	59	4	89	25	6	31	98	1	N	61.0	95	2	1	N	35	6	15	64.9	144	2	6	Q	69	180	6
36	8	20.5	7.1	6-4-21	56.3	82	1	85	35	1	31	136	1	20	65.5	140	2	0	20	38	8	10	71.6	169	2	8	20	93	120	2
41	5	22.5	9.7	7-4-7	54.3	72	1	83	28	7	31	138	0	N	61.8	163	4	3	N	31	8	16	72.0	175	4	1	15	71	154	1
42	6	19.5	7.9	5-1-20	54.4	81	1	86	24	5	31	127	5	N	63.6	140	1	5	N	33	6	20	68.0	170	1	7	10	72	138	6
43	1	19.5	6.6	4-1-10	54.6	71	1	89	20	2	31	131	1	N	65.6	124	3	0	N	35	1	20	69.0	165	4	1	Q	84	120	1
44	5	21.0	8.4	7-3-3	58.9	94	2	99	42	6	31	126	0	N	64.9	134	4	0	N	33	1	20	68.0	167	4	1	Q	77	185	1
45	6	20.0	6.4	6-3-9	51.3	57	1	75	28	5	31	138	0	N	61.5	157	2	5	N	36	6	10	67.0	195	3	6	20	42	114	1
46	8	21.0	8.5	8-3-4	55.9	74	4	85	41	6	31	135	4	N	64.9	139	4	2	N	34	0	03	71.6	189	4	0	03	100	199	3
51	6	20.0	6.1	9-6-8	57.4	76	4	96	26	5	31	110	1	05	65.9	128	3	0	Q	36	0	30	68.0	147	3	0	10	100	175	6
52	5	21.0	8.2	10-2-7	52.4	64	1	89	44	7	31	126	0	N	61.9	116	3	0	N	31	0	10	68.0	195	2	0	N	80	168	3
53	7	22.0	7.9	11-7-12	57.3	72	1	101	38	8	31	137	4	N	67.0	146	4	0	N	27	0	20	75.0	200	3	6	Q	67	120	6
54	5	20.5	7.2	12-7-8	51.3	62	1	91	34	5	31	95	4	45	62.4	112	4	0	N	32	0	20	67.0	155	4	2	45	100	220	6
55	6	19.8	6.4	12-7-1	52.9	64	1	93	42	6	31	150	4	15	64.4	178	2	0	40	38	7	20	67.5	190	1	7	40	101	250	4
56	8	19.5	7.6	12-6-9	55.0	62	1	86	26	6	31	127	0	N	63.7	125	4	3	N	33	1	N	70.0	175	4	1	Q	100	280	3
61	9	20.5	7.4	11-1-1	53.9	63	8	91	18	5	31	123	3	N	63.8	132	4	0	N	32	2	N	72.9	214	4	0	N	152	140	4
62	7	20.5	8.1	4-1-13	53.1	62	4	73	22	7	31	155	0	N	66.5	166	2	5	N	37	2	N	72.0	198	1	8	Q	64	126	4
63	6	20.5	6.6	7-3-23	53.6	64	4	75	36	6	31	142	4	Q	70.0	145	3	0	02	38	8	30	71.4	146	3	2	20	101	140	4
64	4	20.5	8.8	11-4-12	50.7	62	1	83	32	3	31	122	4	N	64.0	130	4	0	N	33	0	N	69.5	170	4	0	N	80	250	1
65	5	20.5	6.1	9-6-11	50.9	62	2	87	33	5	31	103	1	N	60.0	95	2	0	N	30	6	N	66.6	127	2	2	N	90	180	6
66	1	21.5	7.0	9-5-9	52.3	60	1	82	28	6	31	117	4	17	61.8	122	3	4	20	31	0	17	71.0	155	4	2	10	115	95	1

	CHILD-BIRTH				CHILD-TEST					MOTHER-BIRTH					MOTHER-TEST					FATH-BRTH			FATHER-TEST					FAMILY		
ID	B	LGTH	WGT	MO-D-HR	HGHT	WGT	L	PEA	RA	B	AG	WGT	O	SM	HGHT	WGT	E	O	SM	AG	O	SM	HGHT	WGT	E	O	SM	I-B	I-T	C
11	6	22.0	9.3	12-7-24	56.3	82	1	89	40	5	31	155	0	04	68.3	151	2	0	04	41	8	Q	72.0	185	4	8	Q	65	160	1
12	2	20.5	7.8	9-3-5	54.9	75	1	86	40	6	32	125	0	N	65.3	135	4	0	N	29	0	N	68.0	170	4	6	N	48	150	1
13	8	20.5	8.3	10-5-15	52.8	83	1	68	30	7	32	230	0	N	62.6	238	4	0	N	31	6	Q	70.5	233	3	6	Q	60	78	1
14	5	21.0	7.1	10-6-10	54.8	74	1	76	23	5	32	112	0	10	62.8	122	4	0	Q	34	2	20	72.5	168	4	3	30	61	300	1
15	6	19.5	6.9	12-6-2	51.6	67	2	80	35	2	32	125	0	25	64.3	131	2	1	30	32	6	40	70.5	174	1	6	Q	70	129	4
16	7	21.0	7.8	11-7-14	49.2	58	1	90	41	7	32	109	0	30	61.9	99	3	1	20	39	6	20	70.0	175	4	6	20	80	212	1
21	6	21.5	8.3	4-7-4	53.5	68	1	86	49	5	32	194	0	N	64.8	243	2	0	N	38	5	20	66.0	150	3	0	Q	48	100	1
22	6	21.5	8.7	5-1-9	55.9	80	5	86	37	6	32	142	2	Q	59.6	172	2	0	04	46	6	Q	70.5	185	2	6	Q	75	240	6
23	6	20.0	7.8	5-4-20	53.6	65	1	99	40	1	32	158	0	N	64.6	145	4	4	N	35	1	N	71.0	160	4	1	Q	130	231	1
24	4	20.0	7.9	9-2-5	51.6	70	5	76	43	8	32	128	1	N	65.9	138	2	0	N	60	0	N	60.8	160	4	0	N	95	83	1
25	6	20.0	8.1	10-3-16	50.9	61	1	81	26	6	32	118	0	N	60.8	122	2	0	N	30	6	25	65.8	153	2	7	40	90	108	1
26	5	21.5	7.8	4-2-3	54.9	58	1	103	37	7	32	115	0	Q	66.9	135	2	0	Q	24	6	20	74.0	162	2	3	20	104	400	6
31	6	20.0	7.4	7-7-5	58.3	90	4	70	30	5	32	120	0	05	63.1	130	2	4	Q	33	2	25	73.0	210	3	2	40	99	183	1
32	5	20.5	8.1	10-3-6	54.3	88	1	79	24	5	32	170	0	20	62.8	181	2	1	40	35	8	20	70.5	180	1	6	40	64	184	6
33	6	21.0	8.2	2-4-18	54.3	70	1	80	42	6	32	145	0	17	63.5	169	4	0	17	38	1	N	72.0	178	4	1	40	81	108	6
34	6	21.0	7.9	11-5-13	54.9	82	1	93	31	6	32	125	1	N	64.7	144	2	0	N	36	0	40	74.0	197	2	0	10	85	170	6
35	6	20.8	7.5	3-3-20	56.4	88	5	78	38	6	32	130	0	N	63.5	166	2	2	N	32	7	Q	67.5	260	2	6	N	47	134	1
36	5	20.5	6.9	4-3-16	52.5	65	8	84	24	6	32	122	0	N	59.6	151	2	0	N	33	6	N	68.6	200	2	3	N	70	200	1
41	6	21.8	10.0	6-7-5	56.4	86	8	78	46	8	32	215	0	N	67.1	220	4	0	N	34	0	N	72.0	160	4	0	N	102	180	1
42	2	20.8	6.4	5-1-22	52.2	60	1	71	31	1	32	97	0	20	61.9	118	2	0	20	33	7	20	74.0	180	2	2	20	36	96	1
43	6	21.0	8.1	6-2-19	53.4	60	7	86	33	5	32	110	0	05	63.8	120	3	0	05	36	6	Q	71.0	165	2	6	Q	72	114	1
44	5	22.0	10.1	11-4-23	54.6	70	1	83	52	5	32	155	0	N	65.8	179	3	3	N	33	4	15	73.0	176	3	2	Q	60	137	4
45	5	21.0	8.1	9-2-7	50.7	54	4	63	22	6	32	130	1	N	59.0	143	2	1	N	34	6	20	69.0	160	1	7	30	138	200	1
46	5	22.8	9.5	11-6-11	53.4	65	1	82	43	6	32	132	0	Q	66.1	141	3	1	04	32	2	23	70.0	190	3	4	20	88	145	1
51	5	18.5	5.1	2-5-12	49.9	54	4	79	23	5	32	130	0	N	59.1	146	0	8	N	42	6	20	72.0	250	2	7	Q	70	137	2
52	2	22.5	8.8	4-4-12	50.1	62	1	91	24	2	32	138	7	10	61.0	140	0	4	N	37	7	N	66.1	209	1	7	N	50	220	1
53	5	21.5	8.4	9-6-13	53.8	57	4	79	36	5	33	113	0	N	65.2	128	2	2	25	32	0	20	69.0	150	3	5	30	107	190	6
54	2	19.0	6.4	8-6-21	49.8	52	8	61	15	2	33	102	4	30	62.8	103	4	4	30	33	3	15	68.0	165	2	6	Q	70	210	1
55	5	20.0	8.4	6-7-19	57.2	82	2	84	35	5	33	152	0	07	66.9	184	4	8	08	35	0	10	68.0	165	3	0	10	63	146	1
56	5	20.5	9.1	5-6-18	56.4	86	8	72	28	5	33	140	0	Q	65.2	184	3	0	03	43	7	10	71.0	185	1	7	20	60	92	2
61	8	21.5	7.1	11-7-10	56.3	73	1	92	34	7	33	148	0	10	67.8	157	4	3	Q	31	2	20	70.0	170	4	0	20	81	198	6
62	5	21.8	9.7	4-7-3	60.3	85	8	76	29	5	33	143	0	N	65.9	176	4	3	N	38	6	N	69.0	185	2	6	N	60	172	1
63	6	21.0	9.6	6-5-21	59.1	100	3	62	25	6	33	166	1	N	64.5	190	2	1	30	34	8	07	72.0	225	2	8	N	78	120	1
64	5	22.5	9.1	9-4-14	56.4	102	4	76	29	2	33	201	0	N	64.2	191	2	1	N	37	3	N	72.0	254	4	3	N	42	143	2
65	5	19.5	6.6	8-7-23	49.6	56	4	69	13	5	33	98	0	0	58.7	106	4	3	N	35	2	24	67.0	165	3	4	20	90	240	1
66	6	19.0	6.5	9-7-12	56.1	82	1	82	18	6	33	145	0	N	63.9	157	2	0	N	36	3	N	69.0	180	3	2	N	120	200	6

STATLAB CENSUS: Boys

ID	CHILD-BIRTH B	LGTH	WGT	MO-D-HR	CHILD-TEST HGHT	WGT	L	PEA	RA	MOTHER-BIRTH B	AG	WGT	O	SM	MOTHER-TEST HGHT	WGT	E	O	SM	FATH-BRTH AG	O	SM	FATHER-TEST HGHT	WGT	E	O	SM	FAMILY I-B	I-T	C
11	6	22.5	9.7	5-7-10	52.9	90	1	76	24	5	33	125	0	N	60.8	146	2	0	N	34	0	N	69.0	175	3	0	N	47	100	1
12	6	21.0	7.8	6-1-17	55.1	80	4	90	51	8	33	156	1	30	67.3	163	4	0	15	35	6	N	70.0	185	4	1	15	125	200	4
13	1	23.0	10.6	7-4-2	56.3	78	3	63	17	6	33	185	0	N	66.9	175	1	0	N	33	6	N	73.0	187	2	6	N	80	65	6
14	1	21.0	9.8	8-5-13	55.7	74	4	55	12	6	33	121	0	Q	66.1	155	2	0	N	35	6	20	71.0	190	3	7	20	37	84	1
15	2	20.0	7.3	11-6-3	55.1	65	1	76	22	2	33	120	0	N	64.6	123	2	8	N	38	2	N	72.0	185	1	2	Q	70	102	1
16	6	21.0	7.3	8-5-19	52.9	60	6	71	29	6	33	130	0	20	66.2	151	3	0	Q	34	1	Q	71.0	205	4	1	Q	90	210	1
21	5	20.0	5.5	2-5-11	52.5	79	1	78	13	6	33	107	3	N	59.5	129	4	3	N	46	7	05	71.0	180	2	3	20	108	162	1
22	5	22.0	9.7	3-2-4	53.9	77	2	88	44	6	33	145	0	20	64.1	142	4	3	20	38	8	N	74.8	208	2	2	N	90	208	1
23	9	20.5	7.3	3-5-6	52.9	68	3	81	24	8	33	105	0	N	62.1	109	3	0	N	32	0	N	71.0	160	4	0	N	125	150	2
24	6	20.5	6.7	2-2-10	50.6	60	1	97	47	7	34	122	0	Q	64.9	126	3	0	Q	35	0	Q	72.0	179	4	0	40	81	130	6
25	8	21.5	6.8	8-2-9	54.6	74	4	85	35	6	34	137	0	20	63.3	162	4	0	N	34	0	N	73.0	170	4	0	N	98	168	4
26	6	21.0	8.4	7-4-19	56.1	69	4	77	15	6	34	167	0	N	64.3	176	3	0	N	38	8	18	74.0	210	3	6	N	70	156	1
31	9	21.0	8.2	4-7-15	51.1	66	1	73	41	5	34	123	0	30	63.5	119	2	0	40	36	0	30	71.0	170	4	3	10	100	100	4
32	5	20.5	6.6	6-3-8	54.0	74	1	63	10	5	34	130	1	N	62.0	140	2	1	N	36	6	N	72.0	215	1	6	N	68	174	1
33	6	20.5	8.3	5-6-23	53.3	67	1	88	30	6	34	127	1	N	66.6	130	3	1	N	36	6	Q	70.0	155	3	0	N	66	141	2
34	6	19.5	6.2	4-4-22	57.0	126	4	83	44	6	34	118	1	N	62.6	126	3	1	N	30	0	N	69.5	180	4	0	N	65	200	4
35	6	20.5	6.2	5-6-14	52.2	74	4	78	42	8	34	117	0	20	61.3	118	4	1	N	42	2	N	65.5	155	4	1	N	65	113	3
36	5	20.5	7.4	6-1-6	57.0	87	1	83	39	5	34	173	0	N	64.5	189	3	0	N	32	6	18	69.0	200	3	2	15	59	100	1
41	9	20.5	7.4	6-5-17	59.3	83	1	84	28	7	34	141	0	35	63.7	165	2	1	40	37	6	20	73.0	180	2	6	Q	58	158	1
42	6	22.0	8.3	8-5-18	61.4	94	1	96	25	6	34	140	4	N	68.3	141	4	0	N	38	0	N	68.5	160	3	5	N	100	108	6
43	2	22.0	8.3	7-7-13	61.1	118	1	90	31	2	34	145	0	Q	67.4	164	4	3	N	37	0	N	69.0	235	4	1	N	76	172	4
44	7	20.0	8.2	8-1-16	53.3	67	4	107	34	7	34	150	8	N	63.3	195	3	0	N	45	3	N	70.0	225	3	3	N	100	100	6
45	6	21.0	8.4	6-4-7	58.3	75	4	89	38	6	34	118	0	N	66.4	149	4	2	N	36	6	N	71.0	165	3	8	N	78	140	3
46	6	20.0	7.4	10-2-2	50.1	62	3	78	33	1	34	105	0	Q	60.0	85	2	0	10	50	6	20	67.0	157	3	0	N	60	110	6
51	6	20.5	7.5	9-1-12	54.8	79	4	84	24	6	34	155	3	N	64.1	172	4	3	N	39	5	Q	74.0	250	3	5	N	110	240	2
52	8	20.0	7.6	3-4-19	55.2	87	4	86	32	7	34	125	5	N	61.3	144	4	0	N	38	6	20	67.0	175	1	6	N	128	78	1
53	5	19.0	5.3	3-7-13	52.8	62	4	80	35	5	34	190	0	N	64.0	180	2	0	N	35	8	12	71.0	230	1	3	N	47	84	1
54	7	18.8	5.8	4-7-21	51.9	62	4	81	25	7	34	100	0	06	62.0	106	4	0	05	39	0	20	70.0	145	4	2	15	73	170	3
55	7	21.0	8.6	6-2-4	48.9	54	4	86	44	7	34	107	0	N	60.7	114	4	0	N	34	0	N	69.3	146	4	0	N	80	150	1
56	5	21.8	7.6	6-7-21	53.3	63	1	84	36	1	34	133	0	N	64.6	130	4	0	N	36	3	N	65.4	142	4	1	N	60	140	1
61	5	20.5	7.6	8-4-4	53.4	63	4	84	23	5	34	110	0	N	64.3	112	4	3	N	36	0	N	70.0	135	4	0	N	100	184	4
62	7	21.0	7.2	8-7-21	53.3	69	8	82	27	5	34	128	0	20	65.8	144	4	4	25	38	0	N	71.0	175	4	0	Q	156	318	6
63	5	22.5	10.0	5-1-20	58.6	115	1	88	21	5	34	110	4	40	64.6	129	1	5	40	29	8	N	71.0	270	4	8	N	90	221	6
64	6	20.0	7.9	2-7-23	50.8	53	1	81	35	6	34	128	0	N	64.0	135	4	0	N	36	0	N	70.5	170	4	0	N	72	200	1
65	8	20.0	6.4	7-7-6	54.6	59	4	90	35	6	35	111	0	N	65.0	105	3	0	N	34	8	N	66.4	134	4	2	N	69	180	1
66	9	22.0	9.3	7-7-19	51.4	54	1	81	41	6	35	116	0	Q	65.6	130	4	0	N	42	3	N	69.1	168	3	3	N	84	150	6

StatLab CENSUS: Boys

	CHILD-BIRTH				CHILD-TEST					MOTHER-BIRTH					MOTHER-TEST					FATH-BRTH			FATHER-TEST					FAMILY		
ID	B	LGTH	WGT	MO-D-HR	HGHT	WGT	L	PEA	RA	B	AG	WGT	O	SM	HGHT	WGT	E	O	SM	AG	O	SM	HGHT	WGT	E	O	SM	I-B	I-T	C
11	6	22.0	9.8	5-4-6	57.0	83	1	81	29	6	35	154	0	N	66.4	171	4	4	N	38	6	N	70.5	170	3	6	N	84	240	1
12	3	22.5	9.1	1-5-10	55.0	66	8	72	40	8	35	156	0	N	61.9	156	2	0	N	34	6	N	69.5	165	2	7	N	70	107	5
13	5	21.5	9.5	10-7-15	54.4	72	4	91	31	6	35	130	0	N	67.5	125	4	3	N	33	1	20	69.3	184	4	5	Q	78	247	3
14	2	21.0	8.9	3-3-9	55.5	86	1	82	34	6	35	136	4	20	65.8	145	4	3	25	33	1	20	72.0	180	4	5	Q	55	139	6
15	5	21.5	8.0	3-7-6	53.4	70	8	79	24	5	35	136	8	N	59.9	158	3	0	N	33	6	N	72.0	175	3	5	N	95	96	1
16	5	20.0	6.4	8-2-14	53.0	65	4	79	27	5	35	110	8	N	62.6	120	2	0	N	38	1	N	66.0	147	4	1	N	107	165	3
21	5	22.5	9.9	4-7-14	52.8	69	1	91	37	1	35	128	0	N	68.4	146	4	0	N	41	0	N	68.5	160	4	0	N	56	150	4
22	6	21.0	7.8	4-7-11	51.2	69	8	89	41	6	35	124	0	N	63.9	138	2	0	N	35	6	N	70.0	200	2	6	N	100	180	4
23	9	22.5	11.4	5-1-15	55.1	104	4	86	26	5	35	149	0	Q	62.6	158	2	1	05	38	1	20	70.5	172	4	1	20	62	214	6
24	9	22.0	9.1	6-5-14	57.9	93	8	98	33	6	35	129	0	N	67.6	141	4	0	N	31	1	20	73.0	215	4	1	20	45	120	6
25	1	20.5	6.5	11-2-18	55.9	76	4	84	21	6	35	110	0	N	63.1	111	3	0	20	37	2	Q	65.7	143	3	2	N	100	130	1
26	7	21.5	7.6	2-2-8	50.3	51	3	87	27	5	35	102	8	N	61.4	104	0	3	N	34	8	N	70.0	175	3	0	N	60	120	4
31	6	20.5	7.8	6-6-21	52.1	63	1	87	30	7	35	181	4	N	69.0	184	4	0	N	35	2	20	70.0	190	4	2	08	158	200	6
32	9	21.0	7.1	7-4-19	49.9	71	8	74	19	7	35	112	0	N	61.0	120	2	1	N	35	8	40	67.4	163	3	8	40	88	231	1
33	6	19.5	6.9	6-4-11	50.3	60	4	85	34	5	35	127	0	Q	62.3	135	3	1	10	37	6	Q	65.0	145	3	3	Q	87	153	1
34	5	20.0	7.6	12-1-2	57.8	129	7	63	15	5	35	152	8	Q	65.1	167	2	1	20	44	6	20	71.0	210	1	3	N	100	148	1
35	5	21.0	8.1	12-4-14	49.9	54	4	67	19	5	35	130	0	10	61.8	159	2	1	20	41	5	30	73.0	160	2	5	30	60	174	6
36	7	21.0	9.1	1-5-19	49.4	55	1	76	21	6	35	113	0	N	60.6	118	1	0	N	37	6	N	65.6	157	2	6	N	60	113	1
41	2	20.5	7.3	1-7-15	49.3	66	1	80	34	2	35	106	0	N	63.7	116	2	2	N	36	5	07	70.0	160	2	2	N	52	120	4
42	5	21.5	7.8	3-5-7	49.4	59	4	55	36	5	35	128	1	06	65.2	146	4	0	02	38	0	N	68.0	158	4	0	N	150	200	6
43	5	21.0	7.4	2-4-15	52.2	60	3	70	19	6	36	140	0	N	61.1	176	3	4	10	41	1	N	70.0	180	4	2	N	95	200	1
44	6	20.0	6.5	7-3-24	52.0	84	2	81	17	6	36	115	0	10	65.2	136	4	1	Q	41	6	Q	68.0	202	2	3	N	100	216	1
45	9	21.0	9.3	7-1-22	54.9	82	1	89	39	1	36	142	0	N	63.2	142	4	0	Q	42	0	Q	70.0	183	4	0	N	60	120	4
46	2	20.5	7.9	5-5-4	53.6	61	2	83	27	6	36	115	0	N	63.9	113	3	4	N	38	1	N	70.5	175	4	1	N	72	220	1
51	9	20.0	7.7	5-4-22	55.9	73	8	81	36	6	36	115	0	N	64.6	122	2	0	N	36	0	N	71.0	175	3	0	N	60	90	1
52	5	19.8	6.8	6-6-4	56.3	76	1	81	16	6	36	159	1	20	59.6	172	2	0	40	39	8	0	71.1	187	3	8	0	144	120	6
53	6	21.8	8.7	4-1-5	52.3	73	1	83	31	5	36	144	0	02	61.1	144	2	0	20	36	6	20	69.6	143	0	8	40	51	45	1
54	1	21.0	8.3	7-5-16	58.1	96	1	92	26	5	36	265	1	N	67.0	200	3	0	N	37	5	N	72.0	177	1	5	Q	58	100	4
55	3	21.0	7.9	6-6-4	57.2	87	1	87	26	3	36	126	0	12	66.1	133	2	0	20	44	0	30	74.0	185	3	0	20	65	230	4
56	5	21.0	8.9	7-1-12	51.4	94	2	105	42	6	36	140	0	20	62.5	139	2	0	Q	41	3	N	66.5	158	4	3	N	28	120	6
61	6	20.0	7.0	8-3-3	55.1	76	1	83	29	6	36	126	0	N	63.9	140	2	0	N	36	8	Q	75.8	198	3	8	Q	66	144	4
62	6	20.0	7.1	9-4-22	53.1	76	1	91	28	5	36	140	1	Q	59.4	135	2	0	N	34	2	20	67.0	145	2	0	30	120	160	5
63	5	20.0	7.3	8-7-23	53.0	69	1	64	20	5	36	128	0	N	62.2	132	1	5	N	37	5	40	71.5	160	2	6	40	46	132	4
64	7	21.5	8.3	12-3-8	59.1	82	1	77	23	5	36	137	1	20	65.2	149	1	1	Q	44	7	20	71.8	205	0	7	Q	90	193	4
65	9	19.5	7.1	5-1-3	50.9	52	1	90	21	5	36	118	0	N	63.1	124	1	0	N	40	6	N	68.0	147	2	6	N	95	120	6
66	9	21.0	8.3	5-5-22	54.6	98	2	80	21	5	36	125	5	N	62.2	132	1	5	N	38	6	07	69.0	145	2	6	N	82	174	1

	CHILD-BIRTH				CHILD-TEST					MOTHER-BIRTH					MOTHER-TEST					FATH-BRTH			FATHER-TEST					FAMILY		
ID	B	LGTH	WGT	MO-D-HR	HGHT	WGT	L	PEA	RA	B	AG	WGT	O	SM	HGHT	WGT	E	O	SM	AG	O	SM	HGHT	WGT	E	O	SM	I-B	I-T	C
11	5	20.5	7.4	7-4-3	60.1	105	4	83	23	5	36	140	7	15	66.3	149	1	0	18	42	5	Q	78.0	205	3	2	Q	78	90	1
12	6	17.5	4.7	6-6-17	52.8	67	1	71	25	5	36	120	0	20	63.3	121	1	0	20	29	6	20	70.0	160	1	6	N	65	65	1
13	6	21.5	8.2	4-2-23	48.8	56	1	77	37	5	37	143	0	N	63.4	161	4	3	N	35	1	N	70.5	180	4	1	N	200	380	2
14	6	20.8	7.8	8-3-16	50.9	56	4	90	36	6	37	162	0	N	62.3	119	3	0	N	41	6	N	73.0	170	2	6	Z	82	128	1
15	5	20.8	8.3	5-1-24	52.3	61	3	69	21	6	37	140	0	Q	65.9	128	2	0	10	40	0	10	67.0	140	4	2	Q	95	200	4
16	6	19.5	6.3	5-7-21	50.7	54	1	85	44	6	37	125	0	20	62.1	111	2	0	15	36	6	25	66.0	165	2	0	Q	55	130	1
21	2	20.0	6.6	6-3-4	55.3	72	2	79	34	6	37	128	0	N	64.8	129	1	0	N	49	6	N	70.0	194	0	6	N	51	78	1
22	1	21.5	8.4	1-6-22	55.5	76	1	99	41	1	37	170	3	10	64.3	198	4	3	Q	35	0	N	74.0	172	4	0	N	103	266	4
23	5	21.0	6.9	10-7-6	52.3	62	1	83	21	5	37	124	0	N	61.7	121	2	0	Q	43	1	Q	67.0	140	4	1	Q	90	108	1
24	8	18.5	5.1	6-2-6	53.2	66	4	82	30	6	37	128	0	06	62.2	114	3	0	N	36	0	20	68.0	155	4	4	20	69	52	6
25	5	20.0	5.8	9-7-11	59.9	95	5	90	18	6	37	130	0	Q	65.1	119	2	0	N	41	5	N	72.0	185	3	2	N	80	160	1
26	7	20.0	7.0	12-6-23	59.3	85	1	68	21	7	37	162	0	20	64.0	180	3	0	15	43	6	N	74.0	195	2	7	1	118	77	1
31	5	21.0	8.7	5-7-24	54.4	74	1	74	19	5	37	121	0	N	61.9	129	2	0	N	37	8	04	65.0	160	1	3	01	110	100	2
32	5	21.0	7.9	4-6-6	54.0	70	4	66	24	5	37	159	0	N	65.3	154	2	0	N	43	7	N	69.0	245	1	6	N	59	118	3
33	5	20.3	8.0	8-7-9	52.8	60	1	70	14	6	37	142	0	N	61.7	168	3	1	Q	42	6	20	69.0	169	3	6	Q	104	158	1
34	5	20.5	8.0	8-4-10	49.9	52	3	95	44	5	37	135	4	40	66.3	150	4	4	30	31	0	N	67.5	155	4	0	Q	75	240	6
35	1	22.0	10.2	8-1-16	52.6	72	1	70	18	6	37	98	7	N	61.1	109	2	0	Q	35	6	15	65.4	151	1	7	N	112	72	4
36	9	20.8	8.4	11-5-14	56.8	81	2	77	29	5	37	135	0	20	67.1	140	2	0	10	37	6	Q	70.1	182	4	1	Q	120	200	4
41	5	22.5	9.4	2-2-11	54.6	67	1	77	38	5	37	112	0	N	64.3	112	4	1	N	41	0	N	72.0	220	4	0	N	90	150	6
42	5	22.0	7.2	4-3-3	54.4	100	1	89	21	5	37	139	0	N	63.9	157	2	0	Q	41	8	06	68.0	195	3	6	N	62	100	1
43	7	21.0	7.3	8-5-13	52.8	86	1	73	29	5	38	109	0	N	67.0	100	2	1	20	40	8	20	69.3	182	3	3	40	76	188	1
44	5	22.0	7.9	6-2-8	56.6	86	1	55	9	5	38	250	0	N	64.0	232	0	0	N	44	8	N	68.3	244	0	8	N	40	35	1
45	5	23.5	9.2	4-7-13	58.6	82	4	82	33	6	38	124	1	N	63.9	117	3	0	N	43	0	N	71.0	150	1	3	N	82	68	6
46	5	18.0	6.6	4-3-20	53.9	73	1	93	35	6	38	113	0	20	63.8	126	2	2	10	34	0	30	67.5	140	4	0	20	66	135	6
51	6	21.5	9.3	5-4-19	53.4	68	1	71	37	5	38	160	0	Q	68.8	161	4	2	Q	34	2	10	67.5	125	4	0	20	62	145	3
52	6	22.0	8.8	5-1-16	52.6	70	1	83	45	6	38	162	0	N	66.4	175	3	0	N	50	5	N	71.8	175	2	5	N	57	90	4
53	6	19.0	4.8	5-5-23	51.8	61	1	82	45	6	38	135	0	28	67.6	150	2	0	30	39	3	N	70.0	175	4	3	N	200	200	1
54	6	20.5	7.4	10-1-12	53.2	70	1	89	36	7	38	120	0	N	62.4	121	4	4	N	44	0	N	68.0	165	4	0	N	121	184	1
55	1	21.0	8.8	4-3-8	53.2	86	4	103	42	2	38	130	1	Q	62.8	143	2	1	Q	50	2	N	70.0	162	4	2	N	130	312	6
56	5	20.0	8.1	4-5-16	50.1	62	1	113	43	6	38	125	0	10	65.2	127	4	0	20	36	0	N	70.5	145	4	0	20	112	240	3
61	5	21.0	8.6	4-5-18	56.3	79	4	87	29	5	39	138	0	Q	63.9	143	3	0	Q	39	0	N	71.0	156	4	0	N	99	230	1
62	6	20.5	8.3	12-2-22	51.8	60	4	95	41	6	39	121	0	Q	63.9	134	3	0	04	41	3	20	72.0	168	4	3	15	52	90	6
63	6	20.0	6.6	5-4-16	51.6	66	1	86	43	5	39	100	0	N	61.4	98	4	0	N	45	0	N	67.0	160	4	0	N	100	200	4
64	5	19.8	7.7	4-4-14	55.3	68	1	87	34	7	39	144	0	40	66.8	144	2	0	30	37	8	40	72.0	203	3	6	40	40	96	1
65	5	20.5	9.3	6-2-16	55.8	72	8	85	48	5	39	134	0	Q	64.3	148	3	0	Q	40	4	30	74.0	190	3	4	30	90	144	1
66	6	19.0	6.7	12-1-2	51.6	76	4	84	24	6	39	122	1	Q	62.8	135	2	0	2	37	5	14	71.0	165	3	2	14	104	118	1

StatLab CENSUS: Boys

	CHILD-BIRTH				CHILD-TEST					MOTHER-BIRTH					MOTHER-TEST					FATH-BRTH			FATHER-TEST					FAMILY		
ID	B	LGTH	WGT	MO-D-HR	HGHT	WGT	L	PEA	RA	B	AG	WGT	O	SM	HGHT	WGT	E	O	SM	AG	O	SM	HGHT	WGT	E	O	SM	I-B	I-T	C
11	6	21.0	8.0	7-7-16	51.2	61	4	91	40	6	39	125	2	N	63.5	129	2	0	N	42	6	15	69.0	170	2	6	Q	89	120	1
12	5	19.0	7.3	9-1-18	52.9	93	3	77	26	7	39	120	5	15	60.1	159	1	7	N	40	8	10	64.3	169	1	0	22	39	74	1
13	9	20.5	7.9	10-5-22	53.9	66	1	98	29	6	39	122	0	N	63.2	134	4	0	N	36	4	N	73.0	175	4	4	N	80	130	1
14	1	19.5	6.8	4-4-4	49.3	56	4	94	36	1	39	135	0	N	65.2	130	3	0	N	43	1	0	67.0	185	4	1	N	100	190	1
15	6	19.5	6.8	4-7-12	49.2	53	1	85	22	5	39	107	0	N	63.0	109	2	0	N	41	6	Q	66.0	143	3	0	Q	85	115	6
16	6	18.0	4.5	6-4-3	52.1	58	1	88	22	5	39	136	0	N	61.0	166	3	0	N	57	6	10	65.5	140	2	6	Q	70	100	1
21	6	21.0	9.5	9-3-1	52.9	64	5	75	17	5	39	140	0	N	61.9	147	2	0	N	36	5	10	68.0	200	2	5	02	50	100	6
22	8	21.0	7.8	8-3-22	55.4	88	4	66	14	7	39	228	0	N	65.0	230	2	0	N	38	7	20	70.0	220	2	0	02	75	110	1
23	6	20.0	6.2	8-7-22	56.9	78	1	67	21	6	39	151	0	Q	69.4	151	1	0	Q	39	6	05	72.0	195	0	6	04	47	64	1
24	7	21.0	7.9	12-1-13	50.4	54	1	80	32	7	39	134	0	N	61.5	127	0	0	N	49	7	N	68.0	189	0	7	N	55	54	1
25	5	22.5	8.3	2-2-23	54.4	68	1	80	34	6	39	119	0	N	63.8	130	2	0	N	38	0	N	69.0	160	3	2	N	75	156	1
26	8	18.5	5.3	1-4-10	51.7	64	3	90	21	5	39	125	0	10	62.0	120	1	0	12	39	7	04	66.5	180	1	7	04	48	99	4
31	6	21.0	7.1	6-6-20	58.6	86	1	99	37	5	40	139	1	20	66.0	145	2	0	15	45	0	Q	74.0	190	2	2	Q	89	150	1
32	5	21.0	8.1	7-3-20	55.6	96	1	102	42	5	40	141	4	N	63.3	139	1	0	N	35	0	N	70.0	180	2	0	Q	73	135	4
33	6	22.5	8.6	9-6-9	55.4	81	5	84	39	6	40	125	0	Q	63.2	132	2	0	N	43	6	30	69.5	165	2	6	N	65	120	2
34	7	21.0	8.2	9-2-11	56.8	81	2	94	44	6	40	160	4	Q	64.8	151	4	8	20	44	6	N	72.0	190	2	6	N	100	132	6
35	6	21.5	9.4	5-7-16	56.4	86	4	67	15	6	40	130	1	N	67.6	154	1	0	N	40	5	20	70.0	180	2	5	30	105	110	4
36	7	21.0	8.4	7-3-10	53.6	80	4	94	33	5	40	118	7	N	63.8	126	2	1	N	41	0	40	69.0	180	4	0	Q	95	150	1
41	7	21.0	6.5	3-5-19	51.4	55	1	77	33	5	40	107	0	N	63.9	133	2	1	N	31	0	20	68.0	160	3	2	20	80	322	2
42	5	20.0	7.0	6-1-3	53.6	68	1	95	42	6	41	117	3	N	62.6	117	4	3	N	41	1	N	62.0	150	4	1	N	110	292	1
43	1	21.0	8.2	10-4-24	50.8	51	5	83	32	7	41	139	3	N	65.3	146	4	0	N	43	1	30	67.6	135	4	1	40	99	160	1
44	6	21.5	8.4	6-6-4	56.0	78	1	89	34	5	41	132	1	04	64.0	125	3	1	Q	45	8	N	67.6	189	2	0	40	73	180	1
45	6	20.0	7.9	2-6-3	51.8	62	8	65	15	5	41	110	0	N	59.5	118	2	1	N	45	0	34	70.5	210	0	7	N	101	190	1
46	7	18.0	5.3	11-3-24	57.5	92	4	99	41	7	41	137	1	N	67.5	145	2	0	N	36	6	N	70.5	163	2	6	N	100	120	1
51	2	22.5	8.5	12-1-13	55.4	71	8	60	13	2	41	191	0	N	56.5	175	2	0	N	43	6	Q	66.5	150	3	6	Q	70	100	4
52	9	20.0	8.1	7-4-10	55.1	73	1	97	48	7	41	130	0	N	67.9	135	3	0	N	48	8	30	73.0	190	2	0	30	80	146	1
53	6	19.5	5.9	9-4-6	54.8	95	1	95	35	6	41	122	0	Q	70.0	125	3	0	N	46	5	N	70.0	180	2	5	N	71	120	6
54	5	22.0	9.9	5-1-12	56.2	74	4	130	56	5	42	151	0	N	67.7	155	3	0	N	39	0	N	75.0	195	4	0	N	100	170	1
55	5	21.0	7.9	8-4-19	55.3	123	1	68	20	1	42	154	5	06	63.3	148	0	5	N	48	8	N	71.0	200	0	7	N	53	127	1
56	7	22.0	9.2	1-3-19	55.0	87	4	63	20	7	42	122	0	20	51.6	126	0	0	15	57	8	N	69.4	169	0	8	Q	60	87	3
61	5	20.0	6.8	4-7-22	48.8	47	1	72	43	2	43	107	0	Q	60.6	111	3	4	N	46	0	N	67.0	150	4	1	N	102	228	4
62	1	19.0	6.6	12-1-19	54.6	71	1	128	45	7	43	124	0	N	64.9	127	4	3	N	51	0	N	73.0	180	4	1	N	142	66	2
63	2	21.0	8.4	6-1-17	53.8	65	4	83	35	5	43	135	0	N	65.6	166	2	0	Q	45	8	N	72.0	220	3	8	N	89	132	5
64	6	20.0	6.1	3-4-7	52.6	65	4	65	4	6	43	116	0	20	64.0	131	0	0	20	52	3	03	68.0	139	4	3	20	175	175	6
65	6	19.8	7.3	5-5-15	54.4	66	4	91	40	6	44	157	0	N	63.1	175	4	0	N	45	6	27	71.8	210	2	6	N	47	105	1
66	5	22.5	8.9	1-5-10	58.2	110	1	64	16	7	45	172	0	20	68.2	193	3	5	N	51	8	N	68.4	217	7	7	N	96	121	2

APPENDIX A

GLOSSARY OF StatLab SYMBOLS AND FORMULAS

Symbol	Denotation	Page Reference
x	The value of an observation.	25
n	Sample size.	13
$\bar{x}$	Arithmetic mean of a sample (read: "x bar").	55
$\tilde{x}$	Median of a sample (read: "x tilde").	46
$\tilde{\mu}$	Median of a population (read: "mu tilde").	51
μ	Arithmetic mean of a population (read: "mu").	62
s	Standard deviation of a sample.	73
σ	Standard deviation of a population (read: "sigma").	78
Σ	Add up the following terms.	55
$|d|$	The numerical value of d, taken as positive regardless of whether d itself is positive or negative. A set of vertical lines enclosing a value denotes the absolute value of the enclosed.	72
df	Degrees of freedom of a test, indicating the number of terms in a sample (or samples) that are free to vary independently.	191
f	The frequency of occurrence of observations in a specified class interval, a category, or the segment of a distribution.	25
F	The cumulative frequency of occurrence of observations to a specified point on a continuum from the lowest to highest value.	27
p, q	Proportions of a sample such that $p + q = 1$.	124
P, Q	Proportions of a population such that $P + Q = 1$.	122
P value	The probability, in case the hypothesis under test is correct, of getting so extreme a sample value as the one actually observed. (In the case of a z-score test, obtained by reference to the normal table.)	165
z (z score)	The standardized score of a value.	103
t	A variant of the z score, applicable for normal samples even when n is small.	191
H	The hypothesis under statistical test.	164
χ^2	An overall measure of the discrepancies between expected and observed frequencies within the C categories into which a distribution may be divided (read: "chi square").	222
r	Pearson's product-moment correlation coefficient for a sample.	244
ρ	Pearson's product-moment correlation coefficient for a population —a measure of the degree to which two quantitative variables are associated (read: "rho").	251

Formula	Descriptive Note	Page Reference
1. $\bar{x} = \dfrac{\Sigma fx}{n}$	Arithmetic mean of a sample for ungrouped data; direct computation.	55
2. $\bar{x} = C + w\bar{x}'$	Arithmetic mean of a sample for data grouped into intervals of width w; coded computation with guessed mean C.	59
3. Mean deviation $= \dfrac{\Sigma f|d|}{n}$	Mean deviation of a sample, where $d = x - \bar{x}$.	72
4. $\mathrm{Var} = \dfrac{\Sigma fd^2}{n}$	Variance of a sample.	73
5. $s = \sqrt{\dfrac{\Sigma fd^2}{n}}$	Standard deviation of a sample; computed in terms of deviations from sample mean ($d = x - \bar{x}$).	73
6. $s = \sqrt{\dfrac{\Sigma fx^2}{n} - \bar{x}^2}$	Standard deviation of a sample; computed directly from x values (shortcut method).	75
7. $z = \dfrac{x - \mu}{\sigma}$	Standard score or z score for x when μ and σ are available.	103
8. $z = \dfrac{x - \bar{x}}{s}$	Standard score or z score for x, where $\bar{x}$ and s estimate μ and σ.	103
9. $\mathrm{SD}_{\bar{x}} = \dfrac{\sigma}{\sqrt{n}}$	Standard deviation of $\bar{x}$ distribution when σ is available.	116
10. $\mathrm{SE}_{\bar{x}} = \dfrac{s}{\sqrt{n}}$	Standard error of $\bar{x}$ distribution, where s estimates σ.	116
11. $\sigma = \sqrt{PQ}$	Population standard deviation of an indicator, where P and Q are population proportions such that $P + Q = 1.00$.	124
12. $\mathrm{SD}_p = \sqrt{\dfrac{PQ}{n}}$	Standard deviation of sample proportion when population P and Q parameters are known.	124
13. $\mathrm{SE}_p = \sqrt{\dfrac{pq}{n}}$	Standard error of sample proportion, where p and q estimate population P and Q.	124
14. $\mathrm{SD}_s = \dfrac{\sigma}{\sqrt{2n}}$	Standard deviation of s distribution when sampling from a normal population and σ is available.	131
15. $\mathrm{SE}_s = \dfrac{s}{\sqrt{2n}}$	Standard error of s distribution when sampling from a normal population and where s estimates σ.	131
16. $\mathrm{SD}_{\tilde{x}} = \dfrac{1.25\sigma}{\sqrt{n}}$	Standard deviation of sample median from a normal population when σ is known.	133
17. $\mathrm{SD}_{\tilde{x}} = \dfrac{1.25\sigma_h}{\sqrt{n}}$	Standard deviation of sample median (even for a heavy-tailed population) where σ_h is the standard deviation of the hypothetical normal fitted to the central portion.	134
18. $\mathrm{SD}_{\tilde{x}} = \dfrac{1.25s_h}{\sqrt{n}}$	Standard error of sample median from a heavy-tailed population, where s_h estimates σ_h.	134

	Formula	Descriptive Note	Page Reference

19. $\mathrm{SD}_{\bar{x}_1 - \bar{x}_2} = \sqrt{\dfrac{\sigma_1{}^2}{n_1} + \dfrac{\sigma_2{}^2}{n_2}}$

Standard deviation of the difference between two independently drawn sample means when σ_1 and σ_2 are available.

144

20. $\mathrm{SE}_{\bar{x}_1 - \bar{x}_2} = \sqrt{\dfrac{s_1{}^2}{n_1} + \dfrac{s_2{}^2}{n_2}}$

Standard error of the difference between two independently drawn sample means, where s_1 and s_2 estimate σ_1 and σ_2.

144

21. $\mathrm{SD}_{p_1 - p_2} = \sqrt{\dfrac{P_1 Q_1}{n_1} + \dfrac{P_2 Q_2}{n_2}}$

Standard deviation of the difference between two independently drawn sample proportions when the P's and Q's are available.

147

22. $\mathrm{SE}_{p_1 - p_2} = \sqrt{\dfrac{p_1 q_1}{n_1} + \dfrac{p_2 q_2}{n_2}}$

Standard error of the difference between two independently drawn sample proportions, where the p's and q's estimate the P's and Q's.

147

23. $\mathrm{SE}_{\tilde{x}_1 - \tilde{x}_2} = \sqrt{\dfrac{(1.25 s_{h_1})^2}{n_1} + \dfrac{(1.25 s_{h_2})^2}{n_2}}$

Standard error of the difference between two independently drawn sample medians.

147

24. $\mathrm{SE}_{s_1 - s_2} = \sqrt{\dfrac{s_1{}^2}{2n_1} + \dfrac{s_2{}^2}{2n_2}}$

Standard error of the difference between two independently drawn sample standard deviations, where both samples are drawn from normal populations.

148

25. $z = \dfrac{\bar{x} - \mu_H}{\sigma/\sqrt{n}}$

z score of the departure of $\bar{x}$ from the hypothetical value of μ when σ is available. Used to test the hypothesis $\mu = \mu_H$.

164

26. $z = \dfrac{\bar{x} - \mu_H}{s/\sqrt{n}}$

z score of the departure of $\bar{x}$ from the hypothetical value of μ, where s estimates σ. Used to test the hypothesis $\mu = \mu_H$.

167

27. $z = \dfrac{\bar{x}_1 - \bar{x}_2}{\sqrt{\sigma_1{}^2/n_1 + \sigma_2{}^2/n_2}}$

z score of the difference between two independently drawn sample means when σ_1 and σ_2 are available. Used to test the null hypothesis $\mu_1 - \mu_2 = 0$.

169

28. $z = \dfrac{\bar{x}_1 - \bar{x}_2}{\sqrt{s_1{}^2/n_1 + s_2{}^2/n_2}}$

z score of the difference between two independently drawn sample means, where s_1 and s_2 estimate σ_1 and σ_2. Used to test the hypothesis $\mu_1 - \mu_2 = 0$.

170

29. $z = \dfrac{p - P_n}{\sqrt{P_H Q_H/n}}$

z score of the departure of sample p from the hypothetical value of P. Used to test the hypothesis $P = P_H$.

175

30. $z = \dfrac{p_1 - p_2}{\sqrt{pq(1/n_1 + 1/n_2)}}$

z score of the difference between two independently drawn sample proportions with combined values of p and q. Used to test the null hypothesis $P_1 - P_2 = 0$.

186

31. $z = \dfrac{\tilde{x} - \tilde{\mu}_H}{1.25 s_h/\sqrt{n}}$

z score of the departure of a sample median from the hypothetical value of $\tilde{\mu}$. Used to test the hypothesis $\tilde{\mu} = \tilde{\mu}_H$.

187

	Formula	Descriptive Note	Page Reference

32. $z = \dfrac{\bar{x}_1 - \bar{x}_2}{\sqrt{\dfrac{(1.25s_{h_1})^2}{n_1} + \dfrac{(1.25s_{h_2})^2}{n_2}}}$

z score of the difference between two independently drawn sample medians. Used to test the null hypothesis $\tilde{\mu}_1 - \tilde{\mu}_2 = 0$. — 187

33. $t = \dfrac{\bar{x} - \mu_H}{s/\sqrt{n-1}}$

Student's t for the departure of a sample mean of a normal population from the hypothetical value of μ. Used in Student's test (sometimes referred to as the t test) to test the hypothesis $\mu = \mu_H$. — 191

34. Mean of V distribution = $\frac{1}{4}n(n+1)$

Mean value of V (sum of ranks) in the one-sample Wilcoxon test. Valid if $\mu = \mu_H$. — 200

35. $\text{SD}_V = \sqrt{\frac{1}{24}n(n+1)(2n+1)}$

Standard deviation of V (sum of ranks) in the one-sample Wilcoxon test. Valid if $\mu = \mu_H$. — 200

36. $z = \dfrac{V - \frac{1}{4}n(n+1)}{\text{SD}_V}$

z score for the departure of a sample V from the hypothetical mean of V (sum of ranks) in the one-sample Wilcoxon test. Used to test the hypothesis $\mu = \mu_H$, assuming that the population is symmetric. — 200

37. $s = \sqrt{\dfrac{n_1 s_1^2 + n_2 s_2^2}{n_1 + n_2 - 2}}$

Estimate for the assumed common σ of two independently drawn samples. Used in the two-sample t test. — 205

38. $t = \dfrac{\bar{x}_1 - \bar{x}_2}{s\sqrt{1/n_1 + 1/n_2}}$

Student's t for the departure from zero of the difference between two independently drawn sample means. Used in Student's test to test the null hypothesis, $\mu_1 - \mu_2 = 0$, for two normal populations when $\sigma_1 = \sigma_2$. — 205

39. Mean of W distribution = $n_2 \dfrac{n_1 + n_2 + 1}{2}$

Mean of the W distribution, where W is the sum of ranks of the second sample (as determined in the Wilcoxon two-sample test) and where $\mu = \mu_H$. — 214

40. $\text{SD}_W = \sqrt{n_1 n_2 \dfrac{n_1 + n_2 + 1}{12}}$

Standard deviation of the W distribution, where W is the sum of ranks of the second sample (as determined in the Wilcoxon two-sample test). — 214

41. $z = \dfrac{W - n_2[(n_1 + n_2 + 1)/2]}{\text{SD}_W}$

z score for the departure of W from its hypothetical mean in the Wilcoxon two-sample test. Used to test the null hypothesis $\mu_1 - \mu_2 = 0$. Valid if the two populations have the same shape and dispersion. — 214

42. $\chi^2 = \dfrac{(np_1 - nP_1)^2}{nP_1} + \cdots + \dfrac{(np_C - nP_C)^2}{nP_C}$

χ^2 for the difference between expected and observed values within categories in which a distribution has been divided. By referring χ^2 to its appropriate distribution, its P value may be determined. Used to test for independence among categories. — 222

43. $\text{Cov} = \dfrac{\Sigma(x - \bar{x})(y - \bar{y})}{n}$

Covariance of a sample: a measure of the extent to which x and y vary together in a bivariate distribution. — 243

Formula	Descriptive Note	Page Reference
44. $r = \dfrac{\Sigma(x - \bar{x})(y - \bar{y})}{n s_x s_y}$	Pearson's product-moment correlation, a measure of the degree to which two quantities in a bivariate sample are associated; expressed in terms of Cov and standard deviation of x and y.	244
45. $r = \dfrac{\Sigma z_x z_y}{n}$	Pearson's product-moment correlation expressed in terms of x and y in z scores.	245
46. $r = \dfrac{n(\Sigma xy) - (\Sigma x)(\Sigma y)}{\sqrt{(n\Sigma x^2) - (\Sigma x)^2}\ \sqrt{(n\Sigma y^2) - (\Sigma y)^2}}$	Pearson's product-moment correlation—the "machine formula" for use with calculator and expressed in original x and y values.	250
47. $z = r\sqrt{n - 1}$	z score for r appropriate for testing the hypothesis of independence between x and y if n is large.	257
48. $r' = 1 - \dfrac{6\Sigma d^2}{n(n^2 - 1)}$	Spearman's rank-order correlation coefficient, a measure of the degree to which the ranks of two quantities in a bivariate sample are associated, where Σd^2 is the sum of the square of the rank difference between the two quantities.	261
49. $z = \dfrac{6\Sigma d^2 - n(n^2 - 1)}{n(n + 1)\sqrt{n - 1}}$	z score in terms of the sum of the squares of the rank differences between two quantities in a bivariate distribution—gives an approximate P value for Spearman's nonparametric test of independence, in case n is large.	264
50. $y^* = \bar{y} + b(x - \bar{x})$	Equation for the regression of y on x, where y^* is the predicted value for y for a given x and where b is the slope of the regression line.	277
51. $b = \dfrac{r s_y}{s_x}$	Equation for the slope of a regression line when the regression line is fitted by least squares.	277
52. $y^* = \bar{y} + \dfrac{r s_y}{s_x}(x - \bar{x})$	Equation for the regression line including the least-squares equation for slope.	278
53. $x^* = \bar{x} + \dfrac{r s_x}{s_y}(y - \bar{y})$	Equation for the regression of x on y, where x^* is the predicted value for x for a given y.	280
54. $\mathrm{SE}_{\text{prediction}} = s_y\sqrt{1 - r^2}$	Standard error of prediction of y from x. Applicable to regression lines for precisely normal populations; in other cases, this formula is to be understood as an average value taken over all x classes in the bivariate sample.	281

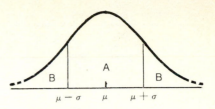

THE NORMAL TABLE

A is the area under the normal curve between $\mu - z\sigma$ and $\mu + z\sigma$. B is the area to the left of $\mu - z\sigma$; it is also the area to the right of $\mu + z\sigma$. The total area under the curve is $1 = B + A + B$.

1. Table of A for Given z

z	.00	.01	.02	.03	.04	.05	.06	.07	.08	.09
.0	.0000	.0080	.0160	.0239	.0319	.0399	.0478	.0558	.0638	.0717
.1	.0797	.0876	.0955	.1034	.1113	.1192	.1271	.1350	.1428	.1507
.2	.1585	.1663	.1741	.1819	.1897	.1974	.2051	.2128	.2205	.2282
.3	.2358	.2434	.2510	.2586	.2661	.2737	.2812	.2886	.2961	.3035
.4	.3108	.3182	.3255	.3328	.3401	.3473	.3545	.3616	.3688	.3759
.5	.3829	.3899	.3969	.4039	.4108	.4177	.4245	.4313	.4381	.4448
.6	.4515	.4581	.4647	.4713	.4778	.4843	.4907	.4971	.5035	.5098
.7	.5161	.5223	.5285	.5346	.5407	.5467	.5527	.5587	.5646	.5705
.8	.5763	.5821	.5878	.5935	.5991	.6047	.6102	.6157	.6211	.6265
.9	.6319	.6372	.6424	.6476	.6528	.6579	.6629	.6680	.6729	.6778
1.0	.6827	.6875	.6923	.6970	.7017	.7063	.7109	.7154	.7199	.7243
1.1	.7287	.7330	.7373	.7415	.7457	.7499	.7540	.7580	.7620	.7660
1.2	.7699	.7737	.7775	.7813	.7850	.7887	.7923	.7959	.7995	.8029
1.3	.8064	.8098	.8132	.8165	.8198	.8230	.8262	.8293	.8324	.8355
1.4	.8385	.8415	.8444	.8473	.8501	.8529	.8557	.8584	.8611	.8638
1.5	.8664	.8690	.8715	.8740	.8764	.8789	.8812	.8836	.8859	.8882
1.6	.8904	.8926	.8948	.8969	.8990	.9011	.9031	.9051	.9070	.9090
1.7	.9109	.9127	.9146	.9164	.9181	.9199	.9216	.9233	.9249	.9265
1.8	.9281	.9297	.9312	.9328	.9342	.9357	.9371	.9385	.9399	.9412
1.9	.9426	.9439	.9451	.9464	.9476	.9488	.9500	.9512	.9523	.9534
2.0	.9545	.9556	.9566	.9576	.9586	.9596	.9606	.9615	.9625	.9634
2.1	.9643	.9651	.9660	.9668	.9676	.9684	.9692	.9700	.9707	.9715
2.2	.9722	.9729	.9736	.9743	.9749	.9756	.9762	.9768	.9774	.9780
2.3	.9786	.9791	.9797	.9802	.9807	.9812	.9817	.9822	.9827	.9832
2.4	.9836	.9840	.9845	.9849	.9853	.9857	.9861	.9865	.9869	.9872
2.5	.9876	.9879	.9883	.9886	.9889	.9892	.9895	.9898	.9901	.9904
2.6	.9907	.9909	.9912	.9915	.9917	.9920	.9922	.9924	.9926	.9929
2.7	.9931	.9933	.9935	.9937	.9939	.9940	.9942	.9944	.9946	.9947
2.8	.9949	.9950	.9952	.9953	.9955	.9956	.9958	.9959	.9960	.9961
2.9	.9963	.9964	.9965	.9966	.9967	.9968	.9969	.9970	.9971	.9972
3.0	.9973	.9974	.9975	.9976	.9976	.9977	.9978	.9979	.9979	.9980
3.1	.9981	.9981	.9982	.9983	.9983	.9984	.9984	.9985	.9985	.9986
3.2	.9986	.9987	.9987	.9988	.9988	.9988	.9989	.9989	.9990	.9990
3.3	.9990	.9991	.9991	.9991	.9992	.9992	.9992	.9992	.9993	.9993
3.4	.9993	.9994	.9994	.9994	.9994	.9994	.9995	.9995	.9995	.9995
3.5	.9995	.9996	.9996	.9996	.9996	.9996	.9996	.9996	.9997	.9997

A remains at .9997 for z from 3.58 through 3.66.
A remains at .9998 for z from 3.67 through 3.79.
A remains at .9999 for z from 3.80 through 4.05.
A remains at 1.0000 for z = 4.06 or larger.

APPENDIX B

2. Table of B for Given z

z	.00	.01	.02	.03	.04	.05	.06	.07	.08	.09
.0	.5000	.4960	.4920	.4880	.4840	.4801	.4761	.4721	.4681	.4641
.1	.4602	.4562	.4522	.4483	.4443	.4404	.4364	.4325	.4286	.4247
.2	.4207	.4168	.4129	.4090	.4052	.4013	.3974	.3936	.3897	.3859
.3	.3821	.3783	.3745	.3707	.3669	.3632	.3594	.3557	.3520	.3483
.4	.3446	.3409	.3372	.3336	.3300	.3264	.3228	.3192	.3156	.3121
.5	.3085	.3050	.3015	.2981	.2946	.2912	.2877	.2843	.2810	.2776
.6	.2743	.2709	.2676	.2643	.2611	.2578	.2546	.2514	.2483	.2451
.7	.2420	.2389	.2358	.2327	.2296	.2266	.2236	.2206	.2177	.2148
.8	.2119	.2090	.2061	.2033	.2005	.1977	.1949	.1922	.1894	.1867
.9	.1841	.1814	.1788	.1762	.1736	.1711	.1685	.1660	.1635	.1611
1.0	.1587	.1562	.1539	.1515	.1492	.1469	.1446	.1423	.1401	.1379
1.1	.1357	.1335	.1314	.1292	.1271	.1251	.1230	.1210	.1190	.1170
1.2	.1151	.1131	.1112	.1093	.1075	.1056	.1038	.1020	.1003	.0985
1.3	.0968	.0951	.0934	.0918	.0901	.0885	.0869	.0853	.0838	.0823
1.4	.0808	.0793	.0778	.0764	.0749	.0735	.0721	.0708	.0694	.0681
1.5	.0668	.0655	.0643	.0630	.0618	.0606	.0594	.0582	.0571	.0559
1.6	.0548	.0537	.0526	.0516	.0505	.0495	.0485	.0475	.0465	.0455
1.7	.0446	.0436	.0427	.0418	.0409	.0401	.0392	.0384	.0375	.0367
1.8	.0359	.0351	.0344	.0336	.0329	.0322	.0314	.0307	.0301	.0294
1.9	.0287	.0281	.0274	.0268	.0262	.0256	.0250	.0244	.0239	.0233
2.0	.0228	.0222	.0217	.0212	.0207	.0202	.0197	.0192	.0188	.0183
2.1	.0179	.0174	.0170	.0166	.0162	.0158	.0154	.0150	.0146	.0143
2.2	.0139	.0136	.0132	.0129	.0125	.0122	.0119	.0116	.0113	.0110
2.3	.0107	.0104	.0102	.0099	.0096	.0094	.0091	.0089	.0087	.0084
2.4	.0082	.0080	.0078	.0075	.0073	.0071	.0069	.0068	.0066	.0064
2.5	.0062	.0060	.0059	.0057	.0055	.0054	.0052	.0051	.0049	.0048
2.6	.0047	.0045	.0044	.0043	.0041	.0040	.0039	.0038	.0037	.0036
2.7	.0035	.0034	.0033	.0032	.0031	.0030	.0029	.0028	.0027	.0026
2.8	.0026	.0025	.0024	.0023	.0023	.0022	.0021	.0021	.0020	.0019
2.9	.0019	.0018	.0018	.0017	.0016	.0016	.0015	.0015	.0014	.0014
3.0	.0013	.0013	.0013	.0012	.0012	.0011	.0011	.0011	.0010	.0010
3.1	.0010	.0009	.0009	.0009	.0008	.0008	.0008	.0008	.0007	.0007
3.2	.0007	.0007	.0006	.0006	.0006	.0006	.0006	.0005	.0005	.0005
3.3	.0005	.0005	.0005	.0004	.0004	.0004	.0004	.0004	.0004	.0003

B remains at .0003 for z from 3.39 through 3.48.
B remains at .0002 for z from 3.49 through 3.61.
B remains at .0001 for z from 3.62 through 3.89.
B remains at .0000 for $z = 3.90$ or larger.

3. Table of z for Selected Values of B and A

B	A	z	B	A	z	B	A	z
.00010	.9998	3.719	.0025	.995	2.807	.10	.80	1.282
.00025	.9995	3.481	.005	.99	2.576	.15	.70	1.036
.0005	.999	3.291	.01	.98	2.326	.20	.60	.842
.0010	.998	3.090	.05	.90	1.645	.25	.50	.674

THE SQUARE-ROOT TABLE

N	$\sqrt{N}$	$\sqrt{10N}$	N	$\sqrt{N}$	$\sqrt{10N}$	N	$\sqrt{N}$	$\sqrt{10N}$
1.00	1.000	3.162	1.50	1.225	3.873	2.00	1.414	4.472
1.01	1.005	3.178	1.51	1.229	3.886	2.01	1.418	4.483
1.02	1.010	3.194	1.52	1.233	3.899	2.02	1.421	4.494
1.03	1.015	3.209	1.53	1.237	3.912	2.03	1.425	4.506
1.04	1.020	3.225	1.54	1.241	3.924	2.04	1.428	4.517
1.05	1.025	3.240	1.55	1.245	3.937	2.05	1.432	4.528
1.06	1.030	3.256	1.56	1.249	3.950	2.06	1.435	4.539
1.07	1.034	3.271	1.57	1.253	3.962	2.07	1.439	4.550
1.08	1.039	3.286	1.58	1.257	3.975	2.08	1.442	4.561
1.09	1.044	3.302	1.59	1.261	3.987	2.09	1.446	4.572
1.10	1.049	3.317	1.60	1.265	4.000	2.10	1.449	4.583
1.11	1.054	3.332	1.61	1.269	4.012	2.11	1.453	4.593
1.12	1.058	3.347	1.62	1.273	4.025	2.12	1.456	4.604
1.13	1.063	3.362	1.63	1.277	4.037	2.13	1.459	4.615
1.14	1.068	3.376	1.64	1.281	4.050	2.14	1.463	4.626
1.15	1.072	3.391	1.65	1.285	4.062	2.15	1.466	4.637
1.16	1.077	3.406	1.66	1.288	4.074	2.16	1.470	4.648
1.17	1.082	3.421	1.67	1.292	4.087	2.17	1.473	4.658
1.18	1.086	3.435	1.68	1.296	4.099	2.18	1.476	4.669
1.19	1.091	3.450	1.69	1.300	4.111	2.19	1.480	4.680
1.20	1.095	3.464	1.70	1.304	4.123	2.20	1.483	4.690
1.21	1.100	3.479	1.71	1.308	4.135	2.21	1.487	4.701
1.22	1.105	3.493	1.72	1.311	4.147	2.22	1.490	4.712
1.23	1.109	3.507	1.73	1.315	4.159	2.23	1.493	4.722
1.24	1.114	3.521	1.74	1.319	4.171	2.24	1.497	4.733
1.25	1.118	3.536	1.75	1.323	4.183	2.25	1.500	4.743
1.26	1.122	3.550	1.76	1.327	4.195	2.26	1.503	4.754
1.27	1.127	3.564	1.77	1.330	4.207	2.27	1.507	4.764
1.28	1.131	3.578	1.78	1.334	4.219	2.28	1.510	4.775
1.29	1.136	3.592	1.79	1.338	4.231	2.29	1.513	4.785
1.30	1.140	3.606	1.80	1.342	4.243	2.30	1.517	4.796
1.31	1.145	3.619	1.81	1.345	4.254	2.31	1.520	4.806
1.32	1.149	3.633	1.82	1.349	4.266	2.32	1.523	4.817
1.33	1.153	3.647	1.83	1.353	4.278	2.33	1.526	4.827
1.34	1.158	3.661	1.84	1.356	4.290	2.34	1.530	4.837
1.35	1.162	3.674	1.85	1.360	4.301	2.35	1.533	4.848
1.36	1.166	3.688	1.86	1.364	4.313	2.36	1.536	4.858
1.37	1.170	3.701	1.87	1.367	4.324	2.37	1.539	4.868
1.38	1.175	3.715	1.88	1.371	4.336	2.38	1.543	4.879
1.39	1.179	3.728	1.89	1.375	4.347	2.39	1.546	4.889
1.40	1.183	3.742	1.90	1.378	4.359	2.40	1.549	4.899
1.41	1.187	3.755	1.91	1.382	4.370	2.41	1.552	4.909
1.42	1.192	3.768	1.92	1.386	4.382	2.42	1.556	4.919
1.43	1.196	3.782	1.93	1.389	4.393	2.43	1.559	4.930
1.44	1.200	3.795	1.94	1.393	4.405	2.44	1.562	4.940
1.45	1.204	3.808	1.95	1.396	4.416	2.45	1.565	4.950
1.46	1.208	3.821	1.96	1.400	4.427	2.46	1.568	4.960
1.47	1.212	3.834	1.97	1.404	4.438	2.47	1.572	4.970
1.48	1.217	3.847	1.98	1.407	4.450	2.48	1.575	4.980
1.49	1.221	3.860	1.99	1.411	4.461	2.49	1.578	4.990

THE SQUARE-ROOT TABLE (*Continued*)

N	$\sqrt{N}$	$\sqrt{10N}$	N	$\sqrt{N}$	$\sqrt{10N}$	N	$\sqrt{N}$	$\sqrt{10N}$
2.50	1.581	5.000	3.00	1.732	5.477	3.50	1.871	5.916
2.51	1.584	5.010	3.01	1.735	5.486	3.51	1.873	5.925
2.52	1.587	5.020	3.02	1.738	5.495	3.52	1.876	5.933
2.53	1.591	5.030	3.03	1.741	5.505	3.53	1.879	5.941
2.54	1.594	5.040	3.04	1.744	5.514	3.54	1.881	5.950
2.55	1.597	5.050	3.05	1.746	5.523	3.55	1.884	5.958
2.56	1.600	5.060	3.06	1.749	5.532	3.56	1.887	5.967
2.57	1.603	5.070	3.07	1.752	5.541	3.57	1.889	5.975
2.58	1.606	5.079	3.08	1.755	5.550	3.58	1.892	5.983
2.59	1.609	5.089	3.09	1.758	5.559	3.59	1.895	5.992
2.60	1.612	5.099	3.10	1.761	5.568	3.60	1.897	6.000
2.61	1.616	5.109	3.11	1.764	5.577	3.61	1.900	6.008
2.62	1.619	5.119	3.12	1.766	5.586	3.62	1.903	6.017
2.63	1.622	5.128	3.13	1.769	5.595	3.63	1.905	6.025
2.64	1.625	5.138	3.14	1.772	5.604	3.64	1.908	6.033
2.65	1.628	5.148	3.15	1.775	5.612	3.65	1.910	6.042
2.66	1.631	5.158	3.16	1.778	5.621	3.66	1.913	6.050
2.67	1.634	5.167	3.17	1.780	5.630	3.67	1.916	6.058
2.68	1.637	5.177	3.18	1.783	5.639	3.68	1.918	6.066
2.69	1.640	5.187	3.19	1.786	5.648	3.69	1.921	6.075
2.70	1.643	5.196	3.20	1.789	5.657	3.70	1.924	6.083
2.71	1.646	5.206	3.21	1.792	5.666	3.71	1.926	6.091
2.72	1.649	5.215	3.22	1.794	5.675	3.72	1.929	6.099
2.73	1.652	5.225	3.23	1.797	5.683	3.73	1.931	6.107
2.74	1.655	5.234	3.24	1.800	5.692	3.74	1.934	6.116
2.75	1.658	5.244	3.25	1.803	5.701	3.75	1.936	6.124
2.76	1.661	5.254	3.26	1.806	5.710	3.76	1.939	6.132
2.77	1.664	5.263	3.27	1.808	5.718	3.77	1.942	6.140
2.78	1.667	5.273	3.28	1.811	5.727	3.78	1.944	6.148
2.79	1.670	5.282	3.29	1.814	5.736	3.79	1.947	6.156
2.80	1.673	5.292	3.30	1.817	5.745	3.80	1.949	6.164
2.81	1.676	5.301	3.31	1.819	5.753	3.81	1.952	6.173
2.82	1.679	5.310	3.32	1.822	5.762	3.82	1.954	6.181
2.83	1.682	5.320	3.33	1.825	5.771	3.83	1.957	6.189
2.84	1.685	5.329	3.34	1.828	5.779	3.84	1.960	6.197
2.85	1.688	5.339	3.35	1.830	5.788	3.85	1.962	6.205
2.86	1.691	5.348	3.36	1.833	5.797	3.86	1.965	6.213
2.87	1.694	5.357	3.37	1.836	5.805	3.87	1.967	6.221
2.88	1.697	5.367	3.38	1.838	5.814	3.88	1.970	6.229
2.89	1.700	5.376	3.39	1.841	5.822	3.89	1.972	6.237
2.90	1.703	5.385	3.40	1.844	5.831	3.90	1.975	6.245
2.91	1.706	5.394	3.41	1.847	5.840	3.91	1.977	6.253
2.92	1.709	5.404	3.42	1.849	5.848	3.92	1.980	6.261
2.93	1.712	5.413	3.43	1.852	5.857	3.93	1.982	6.269
2.94	1.715	5.422	3.44	1.855	5.865	3.94	1.985	6.277
2.95	1.718	5.431	3.45	1.857	5.874	3.95	1.987	6.285
2.96	1.720	5.441	3.46	1.860	5.882	3.96	1.990	6.293
2.97	1.723	5.450	3.47	1.863	5.891	3.97	1.992	6.301
2.98	1.726	5.459	3.48	1.865	5.899	3.98	1.995	6.309
2.99	1.729	5.468	3.49	1.868	5.908	3.99	1.997	6.317

THE SQUARE-ROOT TABLE (*Continued*)

N	√N	√10N	N	√N	√10N	N	√N	√10N
4.00	2.000	6.325	4.50	2.121	6.708	5.00	2.236	7.071
4.01	2.002	6.332	4.51	2.124	6.716	5.01	2.238	7.078
4.02	2.005	6.340	4.52	2.126	6.723	5.02	2.241	7.085
4.03	2.007	6.348	4.53	2.128	6.731	5.03	2.243	7.092
4.04	2.010	6.356	4.54	2.131	6.738	5.04	2.245	7.099
4.05	2.012	6.364	4.55	2.133	6.745	5.05	2.247	7.106
4.06	2.015	6.372	4.56	2.135	6.753	5.06	2.249	7.113
4.07	2.017	6.380	4.57	2.138	6.760	5.07	2.252	7.120
4.08	2.020	6.387	4.58	2.140	6.768	5.08	2.254	7.127
4.09	2.022	6.395	4.59	2.142	6.775	5.09	2.256	7.134
4.10	2.025	6.403	4.60	2.145	6.782	5.10	2.258	7.141
4.11	2.027	6.411	4.61	2.147	6.790	5.11	2.261	7.148
4.12	2.030	6.419	4.62	2.149	6.797	5.12	2.263	7.155
4.13	2.032	6.427	4.63	2.152	6.804	5.13	2.265	7.162
4.14	2.035	6.434	4.64	2.154	6.812	5.14	2.267	7.169
4.15	2.037	6.442	4.65	2.156	6.819	5.15	2.269	7.176
4.16	2.040	6.450	4.66	2.159	6.826	5.16	2.272	7.183
4.17	2.042	6.458	4.67	2.161	6.834	5.17	2.274	7.190
4.18	2.045	6.465	4.68	2.163	6.841	5.18	2.276	7.197
4.19	2.047	6.473	4.69	2.166	6.848	5.19	2.278	7.204
4.20	2.049	6.481	4.70	2.168	6.856	5.20	2.280	7.211
4.21	2.052	6.488	4.71	2.170	6.863	5.21	2.283	7.218
4.22	2.054	6.496	4.72	2.173	6.870	5.22	2.285	7.225
4.23	2.057	6.504	4.73	2.175	6.877	5.23	2.287	7.232
4.24	2.059	6.512	4.74	2.177	6.885	5.24	2.289	7.239
4.25	2.062	6.519	4.75	2.179	6.892	5.25	2.291	7.246
4.26	2.064	6.527	4.76	2.182	6.899	5.26	2.293	7.253
4.27	2.066	6.535	4.77	2.184	6.907	5.27	2.296	7.259
4.28	2.069	6.542	4.78	2.186	6.914	5.28	2.298	7.266
4.29	2.071	6.550	4.79	2.189	6.921	5.29	2.300	7.273
4.30	2.074	6.557	4.80	2.191	6.928	5.30	2.302	7.280
4.31	2.076	6.565	4.81	2.193	6.935	5.31	2.304	7.287
4.32	2.078	6.573	4.82	2.195	6.943	5.32	2.307	7.294
4.33	2.081	6.580	4.83	2.198	6.950	5.33	2.309	7.301
4.34	2.083	6.588	4.84	2.200	6.957	5.34	2.311	7.308
4.35	2.086	6.595	4.85	2.202	6.964	5.35	2.313	7.314
4.36	2.088	6.603	4.86	2.205	6.971	5.36	2.315	7.321
4.37	2.090	6.611	4.87	2.207	6.979	5.37	2.317	7.328
4.38	2.093	6.618	4.88	2.209	6.986	5.38	2.319	7.335
4.39	2.095	6.626	4.89	2.211	6.993	5.39	2.322	7.342
4.40	2.098	6.633	4.90	2.214	7.000	5.40	2.324	7.348
4.41	2.100	6.641	4.91	2.216	7.007	5.41	2.326	7.355
4.42	2.102	6.648	4.92	2.218	7.014	5.42	2.328	7.362
4.43	2.105	6.656	4.93	2.220	7.021	5.43	2.330	7.369
4.44	2.107	6.663	4.94	2.223	7.029	5.44	2.332	7.376
4.45	2.110	6.671	4.95	2.225	7.036	5.45	2.335	7.382
4.46	2.112	6.678	4.96	2.227	7.043	5.46	2.337	7.389
4.47	2.114	6.686	4.97	2.229	7.050	5.47	2.339	7.396
4.48	2.117	6.693	4.98	2.232	7.057	5.48	2.341	7.403
4.49	2.119	6.701	4.99	2.234	7.064	5.49	2.343	7.409

THE SQUARE ROOT TABLE (*Continued*)

N	$\sqrt{N}$	$\sqrt{10N}$	N	$\sqrt{N}$	$\sqrt{10N}$	N	$\sqrt{N}$	$\sqrt{10N}$
5.50	2.345	7.416	6.00	2.449	7.746	6.50	2.550	8.062
5.51	2.347	7.423	6.01	2.452	7.752	6.51	2.551	8.068
5.52	2.349	7.430	6.02	2.454	7.759	6.52	2.553	8.075
5.53	2.352	7.436	6.03	2.456	7.765	6.53	2.555	8.081
5.54	2.354	7.443	6.04	2.458	7.772	6.54	2.557	8.087
5.55	2.356	7.450	6.05	2.460	7.778	6.55	2.559	8.093
5.56	2.358	7.457	6.06	2.462	7.785	6.56	2.561	8.099
5.57	2.360	7.463	6.07	2.464	7.791	6.57	2.563	8.106
5.58	2.362	7.470	6.08	2.466	7.797	6.58	2.565	8.112
5.59	2.364	7.477	6.09	2.468	7.804	6.59	2.567	8.118
5.60	2.366	7.483	6.10	2.470	7.810	6.60	2.569	8.124
5.61	2.369	7.490	6.11	2.472	7.817	6.61	2.571	8.130
5.62	2.371	7.497	6.12	2.474	7.823	6.62	2.573	8.136
5.63	2.373	7.503	6.13	2.476	7.829	6.63	2.575	8.142
5.64	2.375	7.510	6.14	2.478	7.836	6.64	2.577	8.149
5.65	2.377	7.517	6.15	2.480	7.842	6.65	2.579	8.155
5.66	2.379	7.523	6.16	2.482	7.849	6.66	2.581	8.161
5.67	2.381	7.530	6.17	2.484	7.855	6.67	2.583	8.167
5.68	2.383	7.537	6.18	2.486	7.861	6.68	2.585	8.173
5.69	2.385	7.543	6.19	2.488	7.868	6.69	2.587	8.179
5.70	2.387	7.550	6.20	2.490	7.874	6.70	2.588	8.185
5.71	2.390	7.556	6.21	2.492	7.880	6.71	2.590	8.191
5.72	2.392	7.563	6.22	2.494	7.887	6.72	2.592	8.198
5.73	2.394	7.570	6.23	2.496	7.893	6.73	2.594	8.204
5.74	2.396	7.576	6.24	2.498	7.899	6.74	2.596	8.210
5.75	2.398	7.583	6.25	2.500	7.906	6.75	2.598	8.216
5.76	2.400	7.589	6.26	2.502	7.912	6.76	2.600	8.222
5.77	2.402	7.596	6.27	2.504	7.918	6.77	2.602	8.228
5.78	2.404	7.603	6.28	2.506	7.925	6.78	2.604	8.234
5.79	2.406	7.609	6.29	2.508	7.931	6.79	2.606	8.240
5.80	2.408	7.616	6.30	2.510	7.937	6.80	2.608	8.246
5.81	2.410	7.622	6.31	2.512	7.944	6.81	2.610	8.252
5.82	2.412	7.629	6.32	2.514	7.950	6.82	2.612	8.258
5.83	2.415	7.635	6.33	2.516	7.956	6.83	2.613	8.264
5.84	2.417	7.642	6.34	2.518	7.962	6.84	2.615	8.270
5.85	2.419	7.649	6.35	2.520	7.969	6.85	2.617	8.276
5.86	2.421	7.655	6.36	2.522	7.975	6.86	2.619	8.283
5.87	2.423	7.662	6.37	2.524	7.981	6.87	2.621	8.289
5.88	2.425	7.668	6.38	2.526	7.987	6.88	2.623	8.295
5.89	2.427	7.675	6.39	2.528	7.994	6.89	2.625	8.301
5.90	2.429	7.681	6.40	2.530	8.000	6.90	2.627	8.307
5.91	2.431	7.688	6.41	2.532	8.006	6.91	2.629	8.313
5.92	2.433	7.694	6.42	2.534	8.012	6.92	2.631	8.319
5.93	2.435	7.701	6.43	2.536	8.019	6.93	2.632	8.325
5.94	2.437	7.707	6.44	2.538	8.025	6.94	2.634	8.331
5.95	2.439	7.714	6.45	2.540	8.031	6.95	2.636	8.337
5.96	2.441	7.720	6.46	2.542	8.037	6.96	2.638	8.343
5.97	2.443	7.727	6.47	2.544	8.044	6.97	2.640	8.349
5.98	2.445	7.733	6.48	2.546	8.050	6.98	2.642	8.355
5.99	2.447	7.740	6.49	2.548	8.056	6.99	2.644	8.361

THE SQUARE-ROOT TABLE (*Continued*)

N	$\sqrt{N}$	$\sqrt{10N}$	N	$\sqrt{N}$	$\sqrt{10N}$	N	$\sqrt{N}$	$\sqrt{10N}$
7.00	2.646	8.367	7.50	2.739	8.660	8.00	2.828	8.944
7.01	2.648	8.373	7.51	2.740	8.666	8.01	2.830	8.950
7.02	2.650	8.379	7.52	2.742	8.672	8.02	2.832	8.955
7.03	2.651	8.385	7.53	2.744	8.678	8.03	2.834	8.961
7.04	2.653	8.390	7.54	2.746	8.683	8.04	2.835	8.967
7.05	2.655	8.396	7.55	2.748	8.689	8.05	2.837	8.972
7.06	2.657	8.402	7.56	2.750	8.695	8.06	2.839	8.978
7.07	2.659	8.408	7.57	2.751	8.701	8.07	2.841	8.983
7.08	2.661	8.414	7.58	2.753	8.706	8.08	2.843	8.989
7.09	2.663	8.420	7.59	2.755	8.712	8.09	2.844	8.994
7.10	2.665	8.426	7.60	2.757	8.718	8.10	2.846	9.000
7.11	2.666	8.432	7.61	2.759	8.724	8.11	2.848	9.006
7.12	2.668	8.438	7.62	2.760	8.729	8.12	2.850	9.011
7.13	2.670	8.444	7.63	2.762	8.735	8.13	2.851	9.017
7.14	2.672	8.450	7.64	2.764	8.741	8.14	2.853	9.022
7.15	2.674	8.456	7.65	2.766	8.746	8.15	2.855	9.028
7.16	2.676	8.462	7.66	2.768	8.752	8.16	2.857	9.033
7.17	2.678	8.468	7.67	2.769	8.758	8.17	2.858	9.039
7.18	2.680	8.473	7.68	2.771	8.764	8.18	2.860	9.044
7.19	2.681	8.479	7.69	2.773	8.769	8.19	2.862	9.050
7.20	2.683	8.485	7.70	2.775	8.775	8.20	2.864	9.055
7.21	2.685	8.491	7.71	2.777	8.781	8.21	2.865	9.061
7.22	2.687	8.497	7.72	2.778	8.786	8.22	2.867	9.066
7.23	2.689	8.503	7.73	2.780	8.792	8.23	2.869	9.072
7.24	2.691	8.509	7.74	2.782	8.798	8.24	2.871	9.077
7.25	2.693	8.515	7.75	2.784	8.803	8.25	2.872	9.083
7.26	2.694	8.521	7.76	2.786	8.809	8.26	2.874	9.088
7.27	2.696	8.526	7.77	2.787	8.815	8.27	2.876	9.094
7.28	2.698	8.532	7.78	2.789	8.820	8.28	2.877	9.099
7.29	2.700	8.538	7.79	2.791	8.826	8.29	2.879	9.105
7.30	2.702	8.544	7.80	2.793	8.832	8.30	2.881	9.110
7.31	2.704	8.550	7.81	2.795	8.837	8.31	2.883	9.116
7.32	2.706	8.556	7.82	2.796	8.843	8.32	2.884	9.121
7.33	2.707	8.562	7.83	2.798	8.849	8.33	2.886	9.127
7.34	2.709	8.567	7.84	2.800	8.854	8.34	2.888	9.132
7.35	2.711	8.573	7.85	2.802	8.860	8.35	2.890	9.138
7.36	2.713	8.579	7.86	2.804	8.866	8.36	2.891	9.143
7.37	2.715	8.585	7.87	2.805	8.871	8.37	2.893	9.149
7.38	2.717	8.591	7.88	2.807	8.877	8.38	2.895	9.154
7.39	2.718	8.597	7.89	2.809	8.883	8.39	2.897	9.160
7.40	2.720	8.602	7.90	2.811	8.888	8.40	2.898	9.165
7.41	2.722	8.608	7.91	2.812	8.894	8.41	2.900	9.171
7.42	2.724	8.614	7.92	2.814	8.899	8.42	2.902	9.176
7.43	2.726	8.620	7.93	2.816	8.905	8.43	2.903	9.182
7.44	2.728	8.626	7.94	2.818	8.911	8.44	2.905	9.187
7.45	2.729	8.631	7.95	2.820	8.916	8.45	2.907	9.192
7.46	2.731	8.637	7.96	2.821	8.922	8.46	2.909	9.198
7.47	2.733	8.643	7.97	2.823	8.927	8.47	2.910	9.203
7.48	2.735	8.649	7.98	2.825	8.933	8.48	2.912	9.209
7.49	2.737	8.654	7.99	2.827	8.939	8.49	2.914	9.214

THE SQUARE-ROOT TABLE (*Continued*)

N	$\sqrt{N}$	$\sqrt{10N}$	N	$\sqrt{N}$	$\sqrt{10N}$	N	$\sqrt{N}$	$\sqrt{10N}$
8.50	2.915	9.220	9.00	3.000	9.487	9.50	3.082	9.747
8.51	2.917	9.225	9.01	3.002	9.492	9.51	3.084	9.752
8.52	2.919	9.230	9.02	3.003	9.497	9.52	3.085	9.757
8.53	2.921	9.236	9.03	3.005	9.503	9.53	3.087	9.762
8.54	2.922	9.241	9.04	3.007	9.508	9.54	3.089	9.767
8.55	2.924	9.247	9.05	3.008	9.513	9.55	3.090	9.772
8.56	2.926	9.252	9.06	3.010	9.518	9.56	3.092	9.778
8.57	2.927	9.257	9.07	3.012	9.524	9.57	3.094	9.783
8.58	2.929	9.263	9.08	3.013	9.529	9.58	3.095	9.788
8.59	2.931	9.268	9.09	3.015	9.534	9.59	3.097	9.793
8.60	2.933	9.274	9.10	3.017	9.539	9.60	3.098	9.798
8.61	2.934	9.279	9.11	3.018	9.545	9.61	3.100	9.803
8.62	2.936	9.284	9.12	3.020	9.550	9.62	3.102	9.808
8.63	2.938	9.290	9.13	3.022	9.555	9.63	3.103	9.813
8.64	2.939	9.295	9.14	3.023	9.560	9.64	3.105	9.818
8.65	2.941	9.301	9.15	3.025	9.566	9.65	3.106	9.823
8.66	2.943	9.306	9.16	3.027	9.571	9.66	3.108	9.829
8.67	2.944	9.311	9.17	3.028	9.576	9.67	3.110	9.834
8.68	2.946	9.317	9.18	3.030	9.581	9.68	3.111	9.839
8.69	2.948	9.322	9.19	3.031	9.586	9.69	3.113	9.844
8.70	2.950	9.327	9.20	3.033	9.592	9.70	3.114	9.849
8.71	2.951	9.333	9.21	3.035	9.597	9.71	3.116	9.854
8.72	2.953	9.338	9.22	3.036	9.602	9.72	3.118	9.859
8.73	2.955	9.343	9.23	3.038	9.607	9.73	3.119	9.864
8.74	2.956	9.349	9.24	3.040	9.612	9.74	3.121	9.869
8.75	2.958	9.354	9.25	3.041	9.618	9.75	3.122	9.874
8.76	2.960	9.359	9.26	3.043	9.623	9.76	3.124	9.879
8.77	2.961	9.365	9.27	3.045	9.628	9.77	3.126	9.884
8.78	2.963	9.370	9.28	3.046	9.633	9.78	3.127	9.889
8.79	2.965	9.375	9.29	3.048	9.638	9.79	3.129	9.894
8.80	2.966	9.381	9.30	3.050	9.644	9.80	3.130	9.899
8.81	2.968	9.386	9.31	3.051	9.649	9.81	3.132	9.905
8.82	2.970	9.391	9.32	3.053	9.654	9.82	3.134	9.910
8.83	2.972	9.397	9.33	3.055	9.659	9.83	3.135	9.915
8.84	2.973	9.402	9.34	3.056	9.664	9.84	3.137	9.920
8.85	2.975	9.407	9.35	3.058	9.670	9.85	3.138	9.925
8.86	2.977	9.413	9.36	3.059	9.675	9.86	3.140	9.930
8.87	2.978	9.418	9.37	3.061	9.680	9.87	3.142	9.935
8.88	2.980	9.423	9.38	3.063	9.685	9.88	3.143	9.940
8.89	2.982	9.429	9.39	3.064	9.690	9.89	3.145	9.945
8.90	2.983	9.434	9.40	3.066	9.695	9.90	3.146	9.950
8.91	2.985	9.439	9.41	3.068	9.701	9.91	3.148	9.955
8.92	2.987	9.445	9.42	3.069	9.706	9.92	3.150	9.960
8.93	2.988	9.450	9.43	3.071	9.711	9.93	3.151	9.965
8.94	2.990	9.455	9.44	3.072	9.716	9.94	3.153	9.970
8.95	2.992	9.460	9.45	3.074	9.721	9.95	3.154	9.975
8.96	2.993	9.466	9.46	3.076	9.726	9.96	3.156	9.980
8.97	2.995	9.471	9.47	3.077	9.731	9.97	3.158	9.985
8.98	2.997	9.476	9.48	3.079	9.737	9.98	3.159	9.990
8.99	2.998	9.482	9.49	3.081	9.742	9.99	3.161	9.995

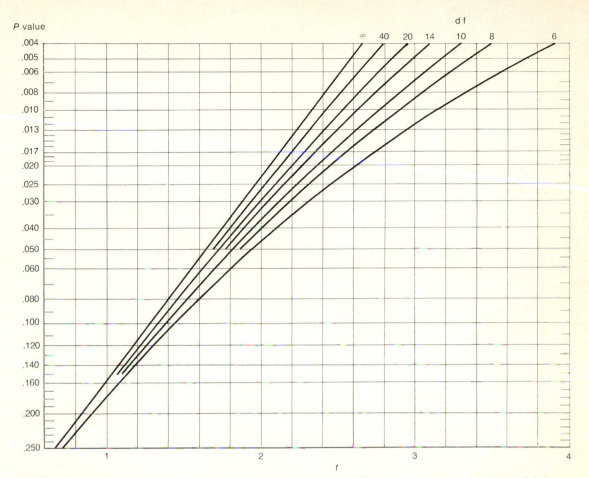

FIGURE 30 Relation between t and one-sided P values for the t test. The seven curves correspond to 7 degrees of freedom (df), as labeled: Interpolate for intermediate values. The vertical axis shows P values of the one-sided t test, plotted on the normal probability scale: The P value must be doubled if a two-sided test is used. The horizontal scale shows the corresponding value of t—positive or negative, whichever is relevant for the test. As explained in Unit 21, this chart may also be used for a two-sample t test.

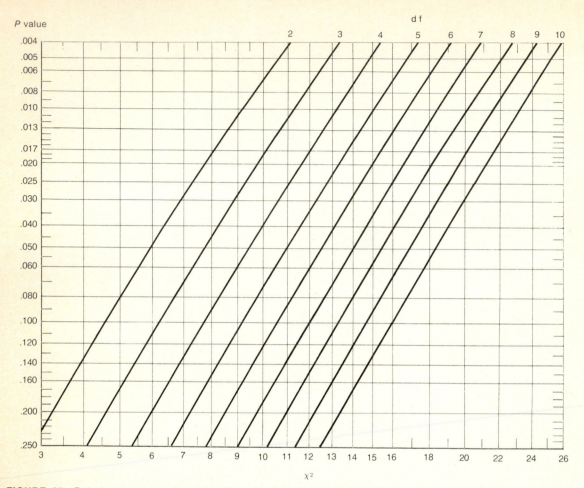

FIGURE 35 Relation between χ^2 and P values for the chi-square test. The nine curves correspond to 9 degrees of freedom (df), as labeled. The vertical axis shows P values for the chi-square test, plotted on the normal probability scale. The horizontal axis shows values of χ^2.

INDEX